THEORY
WING SECT

Including a Summary of Airfoil Data

By IRA H. ABBOTT

DIRECTOR OF AERONAUTICAL AND SPACE RESEARCH
NATIONAL AERONAUTICS AND SPACE ADMINISTRATION

and ALBERT E. VON DOENHOFF

RESEARCH ENGINEER, NASA

DOVER PUBLICATIONS, INC.
NEW YORK

Published in Canada by General Publishing Company, Ltd., 30 Lesmill Road, Don Mills, Toronto,
Ontario.
Published in the United Kingdom by Constable
and Company, Ltd.

This Dover edition, first published in 1959, is an
unabridged and corrected republication of the first
edition first published in 1949 by the McGraw-Hill
Book Company, Inc. This Dover edition includes a
new Preface by the authors.

Standard Book Number: 486-60586-8

Library of Congress Catalog Card Number: 60-1601

Manufactured in the United States of America
Dover Publications, Inc.
180 Varick Street
New York, N.Y. 10014

PREFACE TO DOVER EDITION

The new edition of this book originally published in 1949 results from the continuing demand for a concise compilation of the subsonic aerodynamic characteristics of modern NACA wing sections together with a description of their geometry and associated theory. These wing sections, or their derivatives, continue to be the most commonly used ones for airplanes designed for both subsonic and supersonic speeds, and for application to helicopter rotor blades, propeller blades, and high performance fans.

A number of errors in the original version have been corrected in the present publication. The authors are pleased to acknowledge their debt to the many readers who called attention to these errors.

Since original publication many new contributions have been made to the understanding of the boundary layer, the methods of boundary-layer control, and the effects of compressibility at supercritical speeds. Proper treatment of each of these subjects would require a book in itself. Inasmuch as these subjects are only peripherally involved with the main material of this book, and could not, in any case, be treated adequately in this volume, it was considered best to expedite republication by foregoing extensive revision.

IRA H. ABBOTT

CHEVY CHASE, MD.   ALBERT E. VON DOENHOFF

June, 1958

PREFACE

In preparing this book an attempt has been made to present concisely the most important and useful results of research on the aerodynamics of wing sections at subcritical speeds. The theoretical and experimental results included are those found by the authors to be the most useful. Alternative theoretical approaches to the problem and many experimental data have been rigorously excluded to keep the book at a reasonable length. This exclusion of many interesting approaches to the problem prevents any claim to complete coverage of the subject but should permit easier use of the remaining material.

The book is intended to serve as a reference for engineers, but it should also be useful to students as a supplementary text. To a large extent, these two uses are not compatible in that they require different arrangements and developments of the material. Consideration has been given to the needs of students and engineers with a limited background in theoretical aerodynamics and mathematics. A knowledge of differential and integral calculus and of elementary mechanics is presupposed. Care has been taken in the theoretical developments to state the assumptions and to review briefly the elementary principles involved. An attempt has been made to keep the mathematics as simple as is consistent with the difficulties of the problems treated.

The material presented is largely the result of research conducted by the National Advisory Committee for Aeronautics over the last several years. Although the authors have been privileged to participate in this research, their contributions have been no greater than those of other members of the research team. The authors wish to acknowledge especially the contributions of Eastman N. Jacobs, who inspired and directed much of the research. The authors are pleased to acknowledge the important contributions of Theodore Theodorsen, I. E. Garrick, H. Julian Allen, Robert M. Pinkerton, John Stack, Robert T. Jones, and the many others whose names appear in the list of references. The authors also wish to acknowledge the contributions to the attainment of low-turbulence air streams made by Dr. Hugh L. Dryden and his former coworkers at the National Bureau of Standards, and to express their appreciation for the inspiration and support of the late Dr. George W. Lewis.

IRA H. ABBOTT
ALBERT E. VON DOENHOFF

CHEVY CHASE, MD.
July, 1949

vii

CONTENTS

ix

CHAPTER 1

THE SIGNIFICANCE OF WING-SECTION CHARACTERISTICS

1.1. Symbols.

A aspect ratio

A_n coefficients of the Fourier series for the span-load distribution

C_D drag coefficient

C_{D_i} induced drag coefficient

C_L lift coefficient

$C_{L_{max}}$ maximum lift coefficient

C_M pitching-moment coefficient

$C_{M_{ac}}$ pitching-moment coefficient about the aerodynamic center

D drag

E Jones[50] edge-velocity factor, equals ratio of the semiperimeter of the plan form of the wing under consideration to the span of the wing

E a factor (see Fig. 13)

G a factor (see Fig. 14)

H a factor (see Fig. 15)

J a factor (see Fig. 9)

L lift

L_a "additional" loading coefficient

L_b "basic" loading coefficient

M pitching moment

S wing area

V speed

X_{ac} longitudinal distance between the aerodynamic center of the root section and the aerodynamic center of the wing, positive to the rear

a wing lift-curve slope

a_e effective section lift-curve slope, a_0/E

a_0 section lift-curve slope

ac aerodynamic center

b wing span

c wing chord

$\bar{c}$ mean geometric chord, S/b

c' mean aerodynamic chord

c_d section drag coefficient

c_{d_i} section induced-drag coefficient

c_l section lift coefficient

$c_{l_{al}}$ local "additional" section lift coefficient for a wing lift coefficient equal to unity

c_{l_b} local "basic" section lift coefficient

$c_{l_{max}}$ section maximum lift coefficient

c_m section-moment coefficient

$c_{m_{ac}}$ section-moment coefficient about the aerodynamic center

c_s root chord

c_t tip chord

d section drag
f a factor (see Fig. 8)
k a spanwise station
l section lift
l_a "additional" section lift
l_b "basic" section lift
m section moment
r an even number of stations used in the Fourier analysis of the span-load
     distribution
u a factor (see Fig. 10)
v a factor (see Fig. 11)
w a factor (see Fig. 12)
x projected distance in the plane of symmetry from the wing reference point
     to the aerodynamic center of the wing section, measured parallel to the chord
     of the root section, positive to the rear
y distance along the span
z projected distance in the plane of symmetry from the wing reference point
     to the aerodynamic center of the wing section measured perpendicular to the
     root chord, positive upward
α angle of attack
α_0 section angle of attack
α_e effective angle of attack
α_i angle of downwash
α_{l_0} section angle of attack for zero lift
$\alpha_{l_{0s}}$ angle of zero lift of the root section
α_s wing angle of attack measured from the chord of the root section
$\alpha_{s(L=0)}$ angle of attack of the root section for zero lift of the wing
β angle of sweepback
ϵ aerodynamic twist from root to tip
η_{m1} multiplier for obtaining the wing characteristics
η_{mn} multiplier for obtaining the span-load distribution
θ $\cos^{-1}\left(-\dfrac{2y}{b}\right)$
λ_{mk} multiplier for obtaining the induced-angle distribution
π ratio of the circumference of a circle to its diameter
ρ mass density of air

1.2. The Forces on Wings.   The surfaces that support the aircraft by
means of dynamic reaction on the air are called wings.   An aircraft may
have several wings which may either be fixed with respect to the fuselage
or have any of several motions as in the case of helicopters or ornithopters.
Regardless of the type of lifting surface, its aerodynamic characteristics
will be strongly affected by the shape of the wing section. The wing characteristics may be predicted from the known aerodynamic characteristics
of the wing section if the span is large with respect to the chord, if the
Mach numbers are subcritical, and if the chordwise component of velocity
is large compared with the spanwise component.   Thus the wing-section
characteristics considered in this volume have a large field of applicability.
A complete discussion of the application of section characteristics to the

prediction of wing characteristics is beyond the scope of this work, but some indication of the methods is given for the case of the monoplane wing in steady straight flight without roll or sideslip.

The monoplane wing supports the airplane by means of a lift force generated by the motion through the air. This lift is defined as the component of force acting in the plane of symmetry in a direction perpendicular

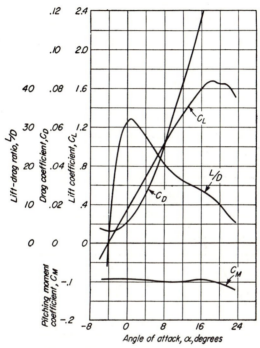

Fig. 1. Typical wing characteristics.

to the line of flight. In addition to the lift, a force directly opposing the motion of the wing through the air is always present and is called the "drag." For a given attitude of geometrically similar wings, the forces tend to vary directly with the density of the air, the wing area, and the square of the speed. It is accordingly convenient to express these forces in terms of nondimensional coefficients that are functions primarily of the attitude of the wing. The lift and drag are given by the following expressions:

$$L = \tfrac{1}{2}\rho V^2 S C_L \qquad (1.1)$$
$$D = \tfrac{1}{2}\rho V^2 S C_D \qquad (1.2)$$

The lift and drag forces may be considered to act at a fixed point with respect to the wing. A complete specification of the forces acting on the wing requires a knowledge of the moment about this fixed point. For a symmetrical wing moving with translation only in the plane of symmetry, the side force perpendicular to the lift and drag is equal to zero, and the moment acts in the plane of symmetry. This moment tends to change

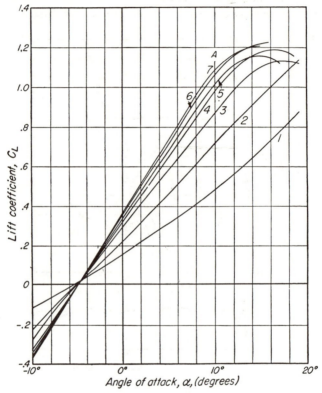

Fig. 2. Lift coefficients plotted as function of angle of attack for aspect ratios of 7 to 1.

the angle of attack of the wing. It is accordingly called the "pitching moment" and may be expressed as follows:

$$M = \tfrac{1}{2}\rho V^2 S c C_M \tag{1.3}$$

A convenient way of describing the aerodynamic characteristics of a wing is to plot the values of the coefficients against the angle of attack, which is the angle between the plane of the wing and the direction of motion. Such a plot is shown in Fig. 1. The lift coefficient increases almost linearly with angle of attack until a maximum value is reached, whereupon the wing is said to "stall." The drag coefficient has a minimum value at a low lift coefficient, and the shape of the curve is approximately parabolic at angles of attack below the stall. If the point about which the moment is taken is properly chosen (the aerodynamic center), the moment coefficient is essentially constant up to maximum lift. A measure of the efficiency of the wing as a lifting surface is given by the lift-drag

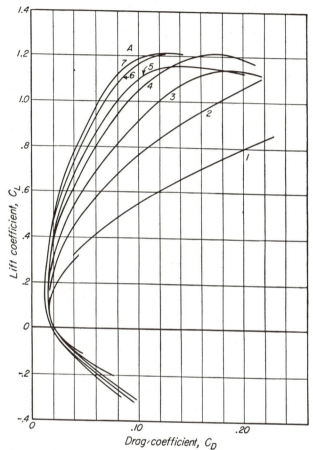

FIG. 3. Polar diagrams for seven wings with aspect ratios of 7 to 1.

ratio, which is also plotted in Fig. 1. This ratio increases from zero at zero lift to a maximum value at a moderate lift coefficient, after which it decreases relatively slowly as the angle of attack is further increased.

It is desirable for the wing to have the smallest possible drag. Inasmuch as the high-speed lift coefficient is usually substantially less than that corresponding to the best lift-drag ratio, one of the best ways of reducing the wing drag is to reduce the wing area. This reduction of area is usually limited by considerations of stalling speed or maneuverability. These considerations are directly influenced by the maximum lift coefficient obtainable. The wing should therefore have a high maximum lift coefficient combined with low drag coefficients for high-speed and cruising flight. This combination of desirable qualities can be obtained only to a limited extent by a single wing configuration. It is therefore customary to use some retractable device such as flaps to improve the maximum lift characteristics of the wing.

1.3. Effect of Aspect Ratio. Aspect ratio is defined as the ratio of the span squared to the wing area (b^2/S), which reduces to the ratio of the

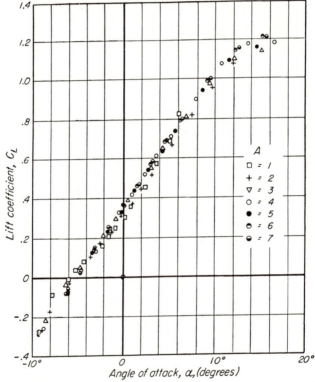

Fig. 4.   Lift coefficients as function of angle of attack, reduced to aspect ratio of 5.

span to the chord in the case of a rectangular wing. Early wind-tunnel investigations of wing characteristics showed that the rates of change of the lift and drag coefficients with angle of attack were strongly affected by the aspect ratio of the model. Wings of high aspect ratio were observed to have higher lift-curve slopes and lower drag coefficients at high lift coefficients than wings of low aspect ratio. The effect of aspect ratio on the lift curve is shown in Fig. 2.[88] The wings of various aspect ratios are shown to have about the same angle of attack at zero lift, but the slope of the lift curve increases progressively with increase of aspect ratio. The effect of aspect ratio on the drag coefficient is shown in Fig. 3.[88] Although the drag coefficients for all the models of various aspect ratios are substantially equal at zero lift, marked reductions in the drag

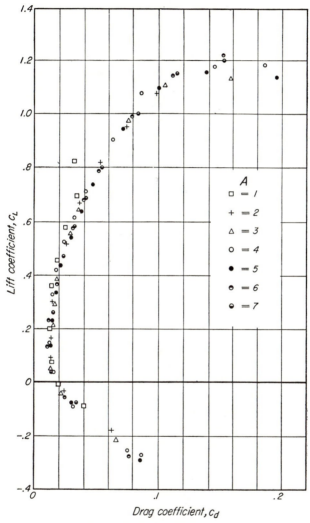

FIG. 5. Polar diagrams reduced to aspect ratio of 5.

coefficient occur at the higher lift coefficients as the aspect ratio is increased.

As a result of such observations, the Lanchester-Prandtl wing theory was developed. This theory shows that, for wings having elliptical spanwise distributions of lift, the following simple expressions relate the drag coefficients and angles of attack as functions of aspect ratio at constant lift coefficients

$$C_D' = C_D + \frac{C_L^2}{\pi}\left(\frac{1}{A'} - \frac{1}{A}\right) \tag{1.4}$$

$$\alpha' = \alpha + \frac{C_L}{\pi}\left(\frac{1}{A'} - \frac{1}{A}\right) \tag{1.5}$$

where C_D' and α' correspond, respectively, to the drag coefficient and angle of attack (radians) of a wing of aspect ratio A'.

Application of Eqs. (1.4) and (1.5) to reduce the data of Figs. 2 and 3 to an aspect ratio of five results in the data of Figs. 4 and 5.[88] These figures show that the characteristics of a wing of one aspect ratio may be predicted with considerable accuracy from data obtained from tests of a wing of widely different aspect ratio.

Equations (1.4) and (1.5) may be simplified by the concept of infinite aspect ratio. If c_d and α_0 indicate the drag coefficient and angle of attack of a wing of infinite aspect ratio, the characteristics of an elliptical wing of aspect ratio A may be expressed as

$$C_D = c_d + \frac{C_L^2}{\pi A} \tag{1.6}$$

$$\alpha = \alpha_0 + \frac{C_L}{\pi A} \tag{1.7}$$

A wing of infinite aspect ratio would have the same flow pattern in all planes perpendicular to the span. In other words, there would be no components of flow along the span, and the flow over the wing section would be two-dimensional. Infinite aspect ratio characteristics are accordingly commonly called "section characteristics." The section characteristics are intrinsically associated with the shape of the wing sections as contrasted with wing characteristics, which are strongly affected by the wing plan form. The detailed study of wings is greatly simplified by the concept of wing-section characteristics because wing theory offers a method for obtaining the properties of wings of arbitrary plan form from a summation of the characteristics of the component sections.

1.4. Application of Section Data to Monoplane Wings. *a. Basic Concepts of Lifting-line Theory.* The simplest three-dimensional wing theory is that based on the concept of the lifting line.[88, 37] In this theory the wing is replaced by a straight line. The circulation about the wing associated with the lift is replaced by a vortex filament. This vortex filament lies along the straight line; and, at each spanwise station, the strength of the vortex is proportional to the local intensity of the lift. According to Helmholtz's theorem, a vortex filament cannot terminate in the fluid. The variation of vortex strength along the straight line is therefore assumed to result from superposition of a number of horseshoe-shaped vortices, as shown in Fig. 6. The portions of the vortices lying along the span are called the "bound vortices." The portions of the vortices extending downstream indefinitely are called the "trailing vortices."

The effect of trailing vortices corresponding to a positive lift is to induce a downward component of velocity at and behind the wing. This downward component is called the "downwash." The magnitude of the downwash at any section along the span is equal to the sum of the effects of all the trailing vortices along the entire span. The effect of the downwash

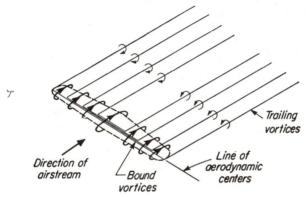

FIG. 6. Vortex pattern representing a lifting wing.

is to change the relative direction of the air stream over the section. The section is assumed to have the same aerodynamic characteristics with respect to the rotated air stream as it has in normal two-dimensional flow. The rotation of the flow effectively reduces the angle of attack. Inasmuch as the downwash is proportional to the lift coefficient, the effect of the trailing vortices is to reduce the slope of the lift curve. The rotation of the flow also causes a corresponding rotation of the lift vector to produce a drag component in the direction of motion. This induced-drag coefficient varies as the square of the lift coefficient because the amount of rotation and the magnitude of the lift vector increase simultaneously.

The problem of evaluating the downwash at each point is difficult because of the interrelation of the downwash, lift distribution, and plan form. A comparatively simple solution was obtained by Prandtl[88] for an elliptical lift distribution. In this case the downwash is constant between the wing tips and the induced drag is less than that for any other type of lift distribution. Equations (1.6) and (1.7) give the relation between the section and wing characteristics for an elliptical lift distribution.

b. Solutions for Linear Lift Curves. Glauert[37] applied a Fourier series analysis to the problem and developed methods for obtaining solutions for wings of any plan form and twist. Anderson[10] applied Glauert's methods to the determination of the characteristics of wings with a wide range of aspect ratio and straight taper and with a linear spanwise variation of twist. Anderson considered the spanwise lift distribution for any typical wing to consist of two parts. One part, called the "basic distribution," is the distribution that depends principally on the twist of the wing and occurs when the total lift of the wing is zero; it does not change with the angle of attack of the wing. The second part of the lift distribution, called the "additional distribution," is the lift due to change of the wing angle

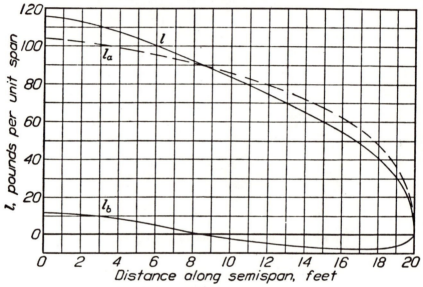

FIG. 7. Typical semispan lift distribution. $C_L = 1.2$

of attack; it is independent of the twist of the wing and maintains the same form throughout the reasonably straight part of the lift curve. A typical distribution of lift over the semispan of a twisted, tapered wing at a moderately high lift coefficient is shown in Fig. 7.

Anderson presented his solutions in the form of tables and charts that are easy to apply. In order to apply these solutions the wing geometry must be known in the following terms:

S wing area
b span
A aspect ratio
c_t/c_s taper ratio where c_t is the tip chord and c_s is the root chord
ϵ aerodynamic twist in degrees from root to tip, measured between the zero lift directions of the center and tip sections, positive for wash in
β angle of sweepback, measured between the lateral axis and a line through the aerodynamic centers of the wing sections
a_0 section lift-curve slope

a_e effective lift-curve slope, a_0/E

E Jones[50] edge-velocity factor, equals ratio of the semiperimeter of the wing under consideration to the span of the wing

$c_{m_{ac}}$ section-moment coefficient about the aerodynamic center

c chord at any spanwise station

Anderson considered only the case of a linear distribution of twist between the center section and the tip. The types of fairing ordinarily used for wings result in nonlinear distributions of twist if the wing is tapered. The departure from the assumed linear distribution may not be negligible if appreciable twist is combined with a large taper ratio.

The data required to find the spanwise lift distribution are given in Tables 1 and 2. The basic and additional loading coefficients L_b and L_a are presented at various spanwise stations for a wide range of aspect ratio and taper ratio. The local basic section lift coefficient c_{l_b} for a wing lift coefficient equal to zero is given by the expression

$$c_{l_b} = \frac{\epsilon a_e S}{cb} L_b$$

The local additional section lift coefficient $c_{l_{a1}}$ for a wing lift coefficient equal to unity is given by the expression

$$c_{l_{a1}} = \frac{S}{cb} L_a$$

The actual or total local section lift coefficient c_l for a wing lift coefficient equal to C_L is obtained from

$$c_l = c_{l_b} + C_L c_{l_{a1}}$$

The lift-curve slope per degree a of the wing is obtained from

$$a = f \frac{a_e}{1 + (57.3 a_e / \pi A)}$$

where the effective section lift-curve slope a_e is taken as the average for the sections composing the wing and the factor f is given in Fig. 8 on page 16.

The Jones edge-velocity correction,[50] which had not been derived when Anderson[10] obtained these results, has been applied to Anderson's formulas by the use of an effective section lift-curve slope. This effective slope a_e equals a_0/E where E is the ratio of the semiperimeter to the span of the wing under consideration. It will be noted that this effective lift-curve slope is not solely a property of the section but that it varies with the plan form of the wing.

The angle of attack of the wing corresponding to any value of the wing lift coefficient C_L is given by

$$\alpha_s = \frac{C_L}{a} + \alpha_{l_{0_s}} + J\epsilon$$

TABLE 1.—BASIC SPAN LIFT-DISTRIBUTION DATA

Values of L_b for tapered wings with rounded tips $c_{l_b} = \dfrac{\epsilon a_0 S}{cb} L_b$

c_t/c_s A	0	0.1	0.2	0.3	0.4	0.5	0.6	0.7	0.8	0.9	1.0
					Spanwise station $y/(b/2) = 0$						
2	−0.118	−0.121	−0.122	−0.122	−0.122	−0.121	−0.121	−0.121	−0.120	−0.120	−0.120
3	−0.153	−0.160	−0.162	−0.163	−0.165	−0.164	−0.164	−0.163	−0.162	−0.161	−0.160
4	−0.183	−0.192	−0.197	−0.199	−0.199	−0.199	−0.198	−0.197	−0.196	−0.194	−0.192
5	−0.211	−0.221	−0.224	−0.226	−0.225	−0.225	−0.224	−0.224	−0.221	−0.219	−0.218
6	−0.235	−0.248	−0.253	−0.253	−0.252	−0.252	−0.250	−0.247	−0.244	−0.243	−0.242
7	−0.256	−0.269	−0.275	−0.276	−0.274	−0.272	−0.270	−0.268	−0.264	−0.261	−0.258
8	−0.274	−0.288	−0.293	−0.293	−0.291	−0.290	−0.288	−0.285	−0.282	−0.279	−0.276
10	−0.304	−0.318	−0.322	−0.323	−0.321	−0.320	−0.318	−0.315	−0.311	−0.305	−0.299
12	−0.329	−0.342	−0.350	−0.349	−0.348	−0.345	−0.341	−0.337	−0.331	−0.323	−0.317
14	−0.350	−0.364	−0.370	−0.370	−0.368	−0.365	−0.360	−0.355	−0.350	−0.342	−0.334
16	−0.367	−0.380	−0.386	−0.385	−0.382	−0.379	−0.375	−0.370	−0.362	−0.358	−0.348
18	−0.384	−0.399	−0.405	−0.403	−0.400	−0.393	−0.387	−0.380	−0.376	−0.368	−0.360
20	−0.398	−0.411	−0.417	−0.415	−0.410	−0.404	−0.399	−0.392	−0.386	−0.378	−0.369
					Spanwise station $y/(b/2) = 0.2$						
2	−0.076	−0.080	−0.082	−0.085	−0.086	−0.086	−0.086	−0.085	−0.085	−0.084	−0.083
3	−0.098	−0.108	−0.111	−0.112	−0.113	−0.113	−0.113	−0.113	−0.112	−0.110	−0.108
4	−0.117	−0.130	−0.135	−0.138	−0.137	−0.137	−0.137	−0.137	−0.137	−0.135	−0.132
5	−0.131	−0.148	−0.156	−0.159	−0.159	−0.158	−0.158	−0.158	−0.158	−0.156	−0.152
6	−0.145	−0.162	−0.173	−0.176	−0.176	−0.176	−0.176	−0.176	−0.175	−0.172	−0.170
7	−0.156	−0.178	−0.189	−0.192	−0.192	−0.192	−0.191	−0.191	−0.190	−0.190	−0.189
8	−0.168	−0.189	−0.200	−0.204	−0.204	−0.205	−0.205	−0.206	−0.205	−0.204	−0.204
10	−0.182	−0.207	−0.220	−0.224	−0.225	−0.225	−0.226	−0.226	−0.225	−0.225	−0.225
12	−0.197	−0.226	−0.239	−0.240	−0.239	−0.238	−0.238	−0.238	−0.237	−0.237	−0.237
14	−0.206	−0.234	−0.248	−0.249	−0.248	−0.248	−0.248	−0.248	−0.248	−0.248	−0.248
16	−0.212	−0.242	−0.256	−0.258	−0.257	−0.256	−0.256	−0.256	−0.256	−0.256	−0.255
18	−0.219	−0.247	−0.260	−0.264	−0.265	−0.265	−0.265	−0.265	−0.265	−0.264	−0.262
20	−0.222	−0.255	−0.269	−0.271	−0.271	−0.271	−0.272	−0.272	−0.272	−0.272	−0.270
					Spanwise station $y/(b/2) = \overset{.}{0}.4$						
2	−0.006	−0.011	−0.013	−0.015	−0.016	−0.016	−0.016	−0.016	−0.016	−0.016	−0.015
3	−0.002	−0.010	−0.012	−0.015	−0.016	−0.016	−0.016	−0.016	−0.017	−0.018	−0.018
4	0	−0.006	−0.011	−0.012	−0.016	−0.016	−0.018	−0.019	−0.020	−0.020	−0.021
5	0.004	−0.004	−0.008	−0.010	−0.012	−0.016	−0.018	−0.020	−0.021	−0.022	−0.023
6	0.009	−0.002	−0.008	−0.012	−0.016	−0.018	−0.020	−0.021	−0.022	−0.024	−0.026
7	0.012	−0.001	−0.010	−0.013	−0.017	−0.018	−0.020	−0.022	−0.025	−0.027	−0.029
8	0.014	0	−0.008	−0.012	−0.017	−0.019	−0.021	−0.025	−0.029	−0.030	−0.030
10	0.021	0.007	−0.002	−0.010	−0.017	−0.020	−0.022	−0.027	−0.030	−0.032	−0.032
12	0.028	0.009	−0.001	−0.010	−0.017	−0.021	−0.025	−0.029	−0.032	−0.036	−0.038
14	0.036	0.013	0	−0.010	−0.017	−0.021	−0.028	−0.031	−0.035	−0.040	−0.042
16	0.043	0.019	0.002	−0.008	−0.016	−0.022	−0.029	−0.034	−0.038	−0.041	−0.045
18	0.049	0.022	0.004	−0.008	−0.015	−0.022	−0.031	−0.038	−0.041	−0.043	−0.046
20	0.050	0.023	0.006	−0.006	−0.014	−0.022	−0.031	−0.038	−0.041	−0.046	−0.049
					Spanwise station $y/(b/2) = 0.6$						
2	0.052	0.052	0.051	0.050	0.050	0.050	0.050	0.050	0.049	0.049	0.048
3	0.070	0.069	0.068	0.068	0.068	0.068	0.068	0.068	0.068	0.068	0.068
4	0.085	0.082	0.081	0.080	0.080	0.080	0.080	0.080	0.080	0.080	0.080
5	0.099	0.095	0.092	0.091	0.091	0.091	0.091	0.091	0.090	0.090	0.090
6	0.109	0.107	0.104	0.102	0.101	0.101	0.100	0.100	0.100	0.100	0.100
7	0.119	0.117	0.114	0.112	0.111	0.110	0.110	0.110	0.110	0.109	0.108
8	0.128	0.122	0.121	0.120	0.120	0.119	0.119	0.118	0.118	0.117	0.116
10	0.139	0.138	0.135	0.132	0.131	0.130	0.130	0.129	0.128	0.126	0.124
12	0.148	0.145	0.141	0.140	0.140	0.139	0.137	0.135	0.134	0.132	0.130
14	0.155	0.152	0.150	0.148	0.145	0.142	0.141	0.140	0.139	0.138	0.135
16	0.160	0.158	0.154	0.151	0.149	0.146	0.143	0.141	0.140	0.139	0.136
18	0.165	0.162	0.160	0.158	0.152	0.148	0.145	0.142	0.140	0.139	0.138
20	0.170	0.169	0.165	0.159	0.152	0.148	0.147	0.143	0.141	0.140	0.140

TABLE 1.—BASIC SPAN LIFT-DISTRIBUTION DATA.—(*Continued.*)

Spanwise station $y/(b/2) = 0.8$

A \ c_l/c_s	0	0.1	0.2	0.3	0.4	0.5	0.6	0.7	0.8	0.9	1.0
2	0.072	0.079	0.080	0.082	0.083	0.085	0.085	0.086	0.086	0.084	0.081
3	0.088	0.098	0.101	0.102	0.104	0.108	0.109	0.110	0.110	0.108	0.106
4	0.100	0.113	0.120	0.123	0.125	0.128	0.128	0.130	0.130	0.130	0.129
5	0.109	0.125	0.135	0.138	0.140	0.143	0.147	0.148	0.14ʝ	0.148	0.149
6	0.115	0.135	0.148	0.152	0.156	0.160	0.160	0.162	0.163	0.164	0.165
7	0.121	0.142	0.158	0.163	0.169	0.172	0.173	0.173	0.174	0.174	0.175
8	0.126	0.149	0.164	0.174	0.180	0.182	0.182	0.183	0.183	0.184	0.184
10	0.136	0.160	0.178	0.188	0.195	0.200	0.201	0.202	0.203	0.201	0.198
12	0.145	0.170	0.188	0.200	0.208	0.212	0.214	0.216	0.216	0.214	0.210
14	0.152	0.182	0.200	0.210	0.216	0.221	0.223	0.227	0.228	0.225	0.220
16	0.159	0.186	0.205	0.216	0.222	0.229	0.232	0.233	0.236	0.232	0.220
18	0.161	0.197	0.215	0.224	0.230	0.235	0.239	0.242	0.243	0.242	0.229
20	0.166	0.201	0.220	0.232	0.237	0.241	0.245	0.248	0.248	0.248	0.247

Spanwise station $y/(b/2) = 0.9$

A	0	0.1	0.2	0.3	0.4	0.5	0.6	0.7	0.8	0.9	1.0
2	0.059	0.068	0.072	0.073	0.075	0.076	0.075	0.075	0.075	0.075	0.075
3	0.068	0.083	0.092	0.098	0.099	0.100	0.100	0.100	0.100	0.100	0.100
4	0.074	0.098	0.111	0.118	0.121	0.122	0.123	0.123	0.123	0.123	0.123
5	0.081	0.107	0.122	0.131	0.138	0.140	0.141	0.141	0.142	0.142	0.142
6	0.087	0.117	0.136	0.148	0.154	0.159	0.160	0.160	0.160	0.160	0.160
7	0.090	0.123	0.146	0.160	0.167	0.171	0.171	0.172	0.172	0.172	0.172
8	0.092	0.131	0.153	0.170	0.179	0.182	0.183	0.184	0.185	0.186	0.187
10	0.098	0.139	0.166	0.184	0.197	0.201	0.203	0.205	0.207	0.209	0.210
12	0.100	0.147	0.178	0.198	0.210	0.218	0.221	0.225	0.228	0.229	0.230
14	0.102	0.156	0.188	0.208	0.220	0.231	0.238	0.241	0.243	0.245	0.246
16	0.103	0.161	0.197	0.219	0.231	0.241	0.249	0.253	0.258	0.259	0.260
18	0.105	0.166	0.202	0.228	0.243	0.252	0.260	0.263	0.269	0.271	0.275
20	0.107	0.172	0.211	0.233	0.248	0.260	0.268	0.273	0.279	0.282	0.285

Spanwise station $y/(b/2) = 0.95$

A	0	0.1	0.2	0.3	0.4	0.5	0.6	0.7	0.8	0.9	1.0
2	0.038	0.051	0.058	0.059	0.060	0.060	0.060	0.060	0.059	0.059	0.058
3	0.044	0.063	0.073	0.078	0.079	0.080	0.080	0.080	0.080	0.079	0.078
4	0.050	0.072	0.076	0.092	0.095	0.097	0.099	0.100	0.100	0.100	0.099
5	0.052	0.083	0.100	0.107	0.110	0.112	0.113	0.114	0.116	0.117	0.116
6	0.054	0.088	0.109	0.119	0.122	0.128	0.130	0.132	0.132	0.131	0.130
7	0.056	0.093	0.116	0.130	0.135	0.140	0.144	0.148	0.150	0.149	0.145
8	0.057	0.100	0.125	0.140	0.146	0.152	0.158	0.160	0.161	0.160	0.159
10	0.058	0.107	0.138	0.152	0.162	0.171	0.178	0.182	0.186	0.187	0.183
12	0.059	0.112	0.143	0.165	0.179	0.189	0.198	0.200	0.202	0.205	0.204
14	0.060	0.116	0.151	0.174	0.190	0.202	0.211	0.215	0.218	0.221	0.222
16	0.061	0.121	0.159	0.184	0.203	0.218	0.222	0.229	0.233	0.236	0.238
18	0.061	0.126	0.166	0.194	0.213	0.229	0.236	0.241	0.248	0.251	0.255
20	0.061	0.128	0.173	0.203	0.225	0.239	0.245	0.251	0.259	0.265	0.271

Spanwise station $y/(b/2) = 0.975$

A	0	0.1	0.2	0.3	0.4	0.5	0.6	0.7	0.8	0.9	1.0
2	0.019	0.030	0.035	0.037	0.037	0.037	0.037	0.036	0.036	0.035	0.034
3	0.022	0.039	0.045	0.049	0.050	0.051	0.052	0.054	0.053	0.052	0.051
4	0.026	0.043	0.054	0.060	0.062	0.064	0.068	0.069	0.069	0.068	0.067
5	0.029	0.051	0.065	0.070	0.071	0.075	0.078	0.081	0.082	0.083	0.083
6	0.030	0.055	0.071	0.079	0.082	0.088	0.091	0.094	0.097	0.097	0.097
7	0.030	0.060	0.078	0.087	0.091	0.098	0.101	0.107	0.110	0.110	0.110
8	0.030	0.062	0.081	0.091	0.100	0.107	0.112	0.120	0.121	0.121	0.121
10	0.031	0.067	0.090	0.105	0.115	0.124	0.132	0.138	0.141	0.142	0.143
12	0.031	0.069	0.095	0.115	0.131	0.141	0.149	0.153	0.160	0.161	0.162
14	0.031	0.071	0.102	0.127	0.143	0.155	0.163	0.171	0.175	0.177	0.178
16	0.031	0.077	0.111	0.138	0.156	0.169	0.178	0.182	0.188	0.190	0.191
18	0.032	0.083	0.121	0.150	0.169	0.182	0.191	0.197	0.200	0.201	0.202
20	0.032	0.086	0.128	0.158	0.178	0.193	0.202	0.208	0.210	0.212	0.213

TABLE 2.—ADDITIONAL SPAN LIFT-DISTRIBUTION DATA

Values of L_a for tapered wings with rounded tips, $c_{l_{a1}} = \dfrac{S}{cb} L_a$

c_t/c_s \ A	0	0.1	0.2	0.3	0.4	0.5	0.6	0.7	0.8	0.9	1.0

Spanwise station $y/(b/2) = 0$

A	0	0.1	0.2	0.3	0.4	0.5	0.6	0.7	0.8	0.9	1.0
2	1.439	1.400	1.367	1.339	1.316	1.301	1.298	1.292	1.290	1.287	1.282
3	1.489	1.430	1.385	1.350	1.322	1.302	1.288	1.275	1.263	1.253	1.246
4	1.527	1.452	1.400	1.360	1.329	1.302	1.279	1.2C0	1.242	1.226	1.211
5	1.559	1.473	1.414	1.369	1.333	1.301	1.272	1.248	1.225	1.204	1.186
6	1.585	1.492	1.428	1.378	1.338	1.300	1.267	1.237	1.211	1.187	1.163
7	1.609	1.510	1.440	1.386	1.340	1.300	1.264	1.232	1.203	1.176	1.149
8	1.629	1.534	1.456	1.392	1.344	1.300	1.264	1.229	1.198	1.165	1.135
10	1.661	1.558	1.473	1.409	1.355	1.306	1.264	1.222	1.187	1.152	1.120
12	1.686	1.578	1.490	1.420	1.361	1.308	1.261	1.219	1.180	1.143	1.109
14	1.708	1.592	1.502	1.429	1.366	1.309	1.260	1.214	1.172	1.136	1.100
16	1.726	1.610	1.513	1.433	1.368	1.309	1.255	1.208	1.165	1.127	1.090
18	1.741	1.623	1.525	1.441	1.370	1.308	1.252	1.203	1.160	1.118	1.080
20	1.755	1.632	1.531	1.446	1.372	1.307	1.250	1.199	1.152	1.109	1.070

Spanwise station $y/(b/2) = 0.2$

A	0	0.1	0.2	0.3	0.4	0.5	0.6	0.7	0.8	0.9	1.0
2	1.369	1.329	1.300	1.279	1.267	1.260	1.258	1.256	1.253	1.250	1.248
3	1.405	1.346	1.308	1.279	1.260	1.248	1.241	1.234	1.228	1.221	1.214
4	1.434	1.363	1.318	1.284	1.260	1.243	1.232	1.220	1.209	1.198	1.186
5	1.459	1.377	1.324	1.288	1.260	1.240	1.223	1.208	1.194	1.181	1.168
6	1.477	1.388	1.329	1.290	1.259	1.236	1.218	1.200	1.184	1.169	1.151
7	1.491	1.393	1.332	1.291	1.259	1.236	1.214	1.193	1.174	1.157	1.138
8	1.502	1.401	1.338	1.294	1.261	1.236	1.212	1.189	1.168	1.148	1.129
10	1.513	1.411	1.347	1.299	1.265	1.236	1.209	1.182	1.158	1.137	1.114
12	1.520	1.417	1.349	1.302	1.265	1.233	1.202	1.172	1.148	1.126	1.102
14	1.527	1.423	1.354	1.307	1.268	1.232	1.201	1.170	1.144	1.119	1.094
16	1.532	1.428	1.358	1.308	1.269	1.232	1.199	1.164	1.135	1.110	1.087
18	1.539	1.429	1.359	1.309	1.270	1.231	1.195	1.160	1.130	1.103	1.078
20	1.547	1.431	1.360	1.311	1.271	1.230	1.190	1.155	1.123	1.098	1.069

Spanwise station $y/(b/2) = 0.4$

A	0	0.1	0.2	0.3	0.4	0.5	0.6	0.7	0.8	0.9	1.0
2	1.217	1.190	1.178	1.172	1.172	1.171	1.170	1.169	1.169	1.168	1.168
3	1.220	1.191	1.176	1.166	1.161	1.160	1.159	1.158	1.157	1.156	1.155
4	1.223	1.192	1.173	1.162	1.156	1.151	1.149	1.148	1.147	1.146	1.145
5	1.226	1.193	1.172	1.159	1.149	1.142	1.140	1.138	1.136	1.134	1.133
6	1.229	1.193	1.171	1.155	1.145	1.138	1.132	1.128	1.127	1.126	1.125
7	1.229	1.193	1.170	1.152	1.140	1.131	1.124	1.121	1.120	1.119	1.118
8	1.229	1.192	1.168	1.150	1.138	1.128	1.120	1.116	1.113	1.111	1.110
10	1.228	1.192	1.167	1.148	1.132	1.121	1.113	1.108	1.104	1.102	1.100
12	1.228	1.192	1.166	1.145	1.125	1.111	1.107	1.102	1.099	1.094	1.090
14	1.228	1.191	1.161	1.136	1.116	1.104	1.100	1.096	1.090	1.087	1.082
16	1.228	1.189	1.158	1.131	1.112	1.101	1.097	1.091	1.086	1.081	1.075
18	1.228	1.186	1.152	1.129	1.111	1.100	1.092	1.087	1.080	1.076	1.070
20	1.228	1.182	1.149	1.127	1.110	1.098	1.089	1.083	1.078	1.071	1.065

Spanwise station $y/(b/2) = 0.6$

A	0	0.1	0.2	0.3	0.4	0.5	0.6	0.7	0.8	0.9	1.0
2	0.970	0.976	0.984	0.992	1.003	1.010	1.012	1.014	1.016	1.018	1.019
3	0.950	0.962	0.975	0.985	0.996	1.004	1.011	1.018	1.023	1.030	1.038
4	0.932	0.948	0.962	0.978	0.992	1.002	1.008	1.014	1.023	1.035	1.050
5	0.920	0.938	0.953	0.971	0.988	1.000	1.008	1.015	1.024	1.038	1.053
6	0.909	0.930	0.949	0.966	0.981	0.993	1.002	1.013	1.024	1.039	1.055
7	0.900	0.920	0.940	0.959	0.975	0.989	1.000	1.012	1.024	1.039	1.054
8	0.891	0.916	0.938	0.956	0.972	0.988	0.999	1.011	1.024	1.039	1.053
10	0.881	0.907	0.929	0.947	0.961	0.976	0.992	1.008	1.023	1.039	1.052
12	0.872	0.901	0.923	0.941	0.958	0.972	0.989	1.006	1.022	1.038	1.051
14	0.868	0.895	0.918	0.937	0.953	0.969	0.986	1.003	1.019	1.035	1.049
16	0.861	0.888	0.912	0.931	0.948	0.966	0.983	1.000	1.017	1.033	1.048
18	0.858	0.883	0.906	0.925	0.944	0.963	0.981	0.998	1.015	1.032	1.047
20	0.851	0.876	0.898	0.920	0.940	0.959	0.978	0.995	1.012	1.028	1.046

TABLE 2.—ADDITIONAL SPAN LIFT-DISTRIBUTION DATA.—(*Continued.*)

c_l/c_s A	0	0.1	0.2	0.3	0.4	0.5	0.6	0.7	0.8	0.9	1.0

Spanwise station $y/(b/2) = 0.8$

2	0.615	0.678	0.712	0.731	0.740	0.745	0.746	0.746	0.747	0.747	0.748
3	0.589	0.659	0.700	0.726	0.743	0.754	0.764	0.772	0.782	0.790	0.799
4	0.568	0.644	0.691	0.723	0.746	0.764	0.781	0.795	0.806	0.816	0.824
5	0.548	0.632	0.685	0.720	0.748	0.769	0.790	0.808	0.822	0.834	0.845
6	0.531	0.619	0.675	0.717	0.748	0.775	0.800	0.820	0.838	0.851	0.862
7	0.517	0.609	0.670	0.713	0.748	0.778	0.802	0.827	0.845	0.861	0.875
8	0.504	0.600	0.663	0.710	0.748	0.779	0.808	0.834	0.854	0.872	0.886
10	0.486	0.585	0.653	0.704	0.748	0.783	0.815	0.842	0.868	0.887	0.905
12	0.472	0.576	0.648	0.702	0.748	0.788	0.821	0.8~0	0.877	0.899	0.919
14	0.462	0.569	0.641	0.699	0.748	0.789	0.825	0.858	0.887	0.911	0.933
16	0.456	0.564	0.638	0.698	0.748	0.791	0.830	0.862	0.894	0.921	0.944
18	0.450	0.559	0.636	0.698	0.750	0.796	0.835	0.870	0.901	0.930	0.953
20	0.444	0.545	0.629	0.698	0.753	0.801	0.842	0.878	0.909	0.937	0.962

Spanwise station $y/(b/2) = 0.9$

2	0.378	0.465	0.508	0.525	0.531	0.534	0.535	0.536	0.537	0.538	0.539
3	0.352	0.447	0.500	0.528	0.543	0.552	0.559	0.564	0.568	0.571	0.575
4	0.331	0.435	0.495	0.532	0.554	0.569	0.581	0.590	0.598	0.603	0.609
5	0.314	0.424	0.490	0.531	0.560	0.583	0.600	0.613	0.622	0.630	0.636
6	0.300	0.416	0.487	0.531	0.565	0.595	0.615	0.631	0.643	0.652	0.659
7	0.290	0.410	0.484	0.535	0.572	0.603	0.628	0.646	0.660	0.671	0.678
8	0.282	0.403	0.481	0.536	0.579	0.612	0.638	0.658	0.673	0.686	0.696
10	0.266	0.383	0.472	0.541	0.590	0.628	0.656	0.679	0.698	0.712	0.723
12	0.253	0.376	0.469	0.542	0.597	0.639	0.669	0.698	0.718	0.736	0.751
14	0.245	0.370	0.468	0.545	0.602	0.648	0.684	0.715	0.739	0.759	0.776
16	0.239	0.366	0.468	0.547	0.609	0.659	0.698	0.729	0.756	0.780	0.801
18	0.234	0.367	0.470	0.552	0.618	0.669	0.710	0.743	0.773	0.800	0.822
20	0.231	0.368	0.473	0.560	0.625	0.679	0.722	0.759	0.791	0.819	0.846

Spanwise station $y/(b/2) = 0.95$

2	0.231	0.296	0.334	0.358	0.370	0.379	0.381	0.383	0.386	0.388	0.390
3	0.209	0.290	0.339	0.369	0.389	0.401	0.407	0.412	0.416	0.418	0.420
4	0.191	0.286	0.342	0.378	0.402	0.420	0.428	0.434	0.440	0.444	0.446
5	0.176	0.281	0.344	0.384	0.415	0.436	0.449	0.458	0.463	0.469	0.471
6	0.166	0.278	0.346	0.392	0.428	0.451	0.466	0.475	0.482	0.490	0.496
7	0.155	0.272	0.346	0.398	0.438	0.464	0.481	0.494	0.502	0.510	0.515
8	0.148	0.261	0.346	0.403	0.446	0.475	0.495	0.510	0.521	0.529	0.534
10	0.138	0.255	0.346	0.410	0.460	0.495	0.520	0.538	0.553	0.566	0.575
12	0.132	0.254	0.348	0.419	0.473	0.511	0.542	0.566	0.583	0.598	0.608
14	0.129	0.252	0.349	0.423	0.482	0.529	0.562	0.588	0.609	0.628	0.640
16	0.126	0.252	0.351	0.432	0.495	0.546	0.581	0.610	0.635	0.655	0.671
18	0.122	0.254	0.357	0.439	0.503	0.558	0.598	0.629	0.658	0.682	0.702
20	0.121	0.258	0.364	0.449	0.516	0.569	0.613	0.648	0.680	0.707	0.730

Spanwise station $y/(b/2) = 0.975$

2	0.132	0.172	0.207	0.239	0.263	0.272	0.274	0.277	0.279	0.281	0.282
3	0.119	0.166	0.210	0.250	0.278	0.289	0.291	0.294	0.298	0.300	0.301
4	0.107	0.163	0.214	0.258	0.288	0.304	0.308	0.311	0.315	0.319	0.322
5	0.098	0.158	0.217	0.269	0.304	0.320	0.322	0.328	0.333	0.338	0.342
6	0.089	0.158	0.219	0.272	0.314	0.332	0.340	0.344	0.350	0.357	0.361
7	0.081	0.158	0.222	0.278	0.320	0.342	0.351	0.359	0.366	0.373	0.381
8	0.077	0.158	0.228	0.283	0.328	0.352	0.363	0.374	0.383	0.391	0.400
10	0.069	0.158	0.233	0.295	0.343	0.373	0.390	0.403	0.415	0.428	0.438
12	0.068	0.161	0.242	0.308	0.360	0.395	0.413	0.430	0.448	0.461	0.473
14	0.066	0.163	0.248	0.320	0.376	0.413	0.438	0.458	0.478	0.495	0.510
16	0.064	0.166	0.255	0.331	0.394	0.435	0.463	0.488	0.510	0.529	0.546
18	0.063	0.169	0.263	0.346	0.412	0.461	0.492	0.518	0.539	0.560	0.580
20	0.062	0.171	0.271	0.363	0.435	0.483	0.515	0.544	0.570	0.593	0.615

where α_s = wing angle of attack measured from chord of root section

$\alpha_{l_{0_s}}$ = angle of zero lift of root section

J = a factor presented in Fig. 9

The angle of zero lift for the wing $\alpha_{s(L=0)}$ is obtained from

$$\alpha_{s(L=0)} = \alpha_{l_{0_s}} + J\epsilon$$

The induced-drag coefficient C_{D_i} is given by

$$C_{D_i} = \frac{C_L^2}{\pi A u} + C_L \epsilon a_e v + (\epsilon a_e)^2 w$$

where the effective section lift-curve slope a_e is taken as the average for

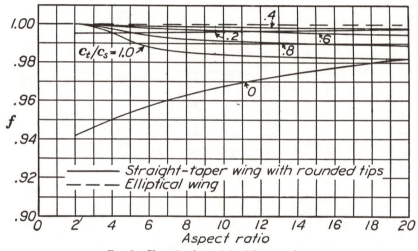

FIG. 8.   Chart for determining lift-curve slope.

$$a = f \frac{a_e}{1 + (57.3 a_e / \pi A)}$$

the sections composing the wing and u, v, and w are factors presented in Figs. 10, 11, and 12, respectively.

The drag coefficient C_D of the wing is obtained from

$$C_D = \frac{2}{S} \int_0^{b/2} c_d c \, dy + C_{D_i}$$

where y = distance along span

c_d = local section drag coefficient corresponding to local section lift coefficient c_l

The wing pitching-moment coefficient $C_{m_{ac}}$ about an axis through the aerodynamic center is given by

$$C_{m_{ac}} = E c_{m_{ac}} - G \epsilon a_e A \tan \beta$$

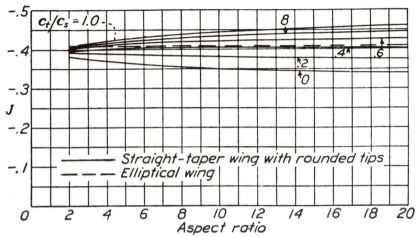

FIG. 9.   Chart for determining angle of attack.

$$\alpha_s = \frac{C_L}{a} + \alpha_{l_{0_s}} + J\epsilon \qquad \alpha_{s(L=0)} = \alpha_{l_{0_s}} + J\epsilon$$

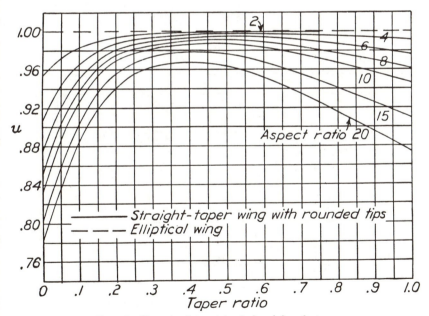

FIG. 10.   Chart for determining induced-drag factor u.

$$C_{D_i} = \frac{C_L{}^2}{\pi A u} + C_L \epsilon a_e v + (\epsilon a_e)^2 w$$

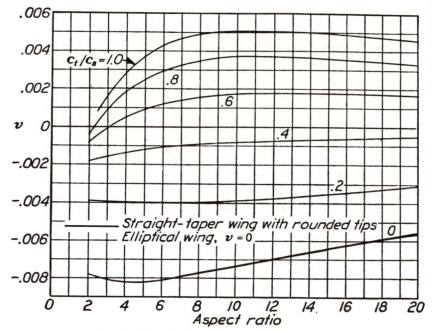

FIG. 11.   Chart for determining induced-drag factor v.

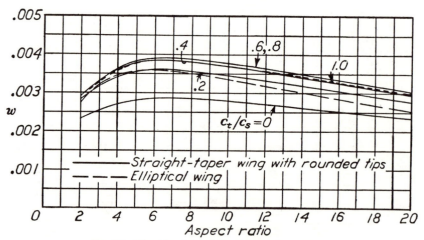

FIG. 12.   Chart for determining induced-drag factor w.

where the section pitching-moment coefficient $c_{m_{ac}}$ is constant along the span.   If the sections composing the wing have different pitching-moment coefficients, a weighted average may be used. E and G are factors presented in Figs. 13 and 14, respectively.

The longitudinal position of the aerodynamic center of the wing is given by

$$\frac{X_{ac}}{S/b} = HA \tan \beta$$

where X_{ac} = longitudinal distance between aerodynamic center of root section and aerodynamic center of wing, positive to the rear

H = a factor presented in Fig. 15

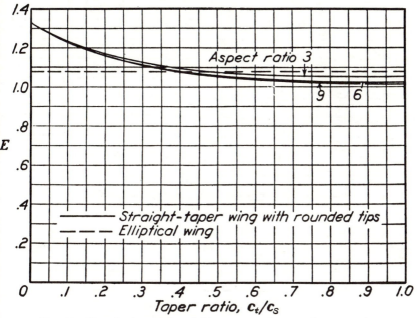

FIG. 13. Chart for determining pitching moment due to section moment.

The maximum lift coefficient for the wing may be estimated from the assumption that this coefficient is reached when the local section lift coefficient at any position along the span is equal to the local maximum lift coefficient for the corresponding section. This value may be found conveniently by the process indicated by Fig. 16. Spanwise variations of the local maximum lift coefficient $c_{l_{max}}$, and of the additional $c_{l_{a1}}$ and basic c_{l_b} lift distributions are plotted. The spanwise variation of $c_{l_{max}} - c_{l_b}$ is then plotted. The minimum value of the ratio

$$\frac{c_{l_{max}} - c_{l_b}}{c_{l_{a1}}}$$

is then found. The minimum value of this ratio is considered to be the maximum lift coefficient of the wing.

c. *Generalized Solution.* Although the Glauert[37] method used by Anderson[10] requires the assumption of a linear variation of lift with angle of attack, methods of successive approximation have been developed by Sherman[109] and Tani[118] that permit application of the actual wing section data. Boshar[16] applied Tani's method to the solution of this problem. Sivells and Neely[111] have developed procedures based on this method which

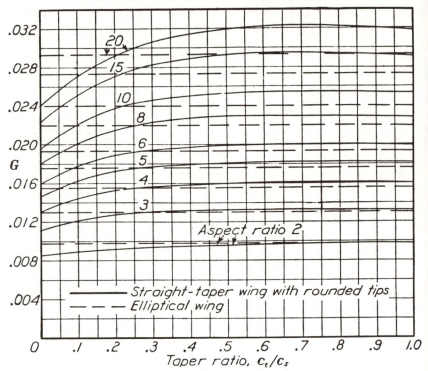

Fig. 14. Chart for determining pitching moment due to basic lift forces.
$$Cm_{l_b} = - Ge a_0 A \tan \beta$$

can be carried out by an experienced computer and which yield highly satisfactory and complete lift, drag, and pitching-moment characteristics of wings. This method will be presented in sufficient detail to permit such calculations to be made.

The basic problem is the determination of the downwash at each point along the span from summation of the effects of the trailing vortices. The strengths of the trailing vortices, however, are unknown, but they are intimately associated with the spanwise lift distribution which, in turn, is dependent upon the downwash. The relation between the downwash and the lift distribution as given by Prandtl[88] is

$$\alpha_i = \frac{180}{\pi} \frac{b}{2\pi} \int_{-b/2}^{b/2} \frac{(d/dy)(c_l c/4b)dy}{y_1 - y} \tag{1.8}$$

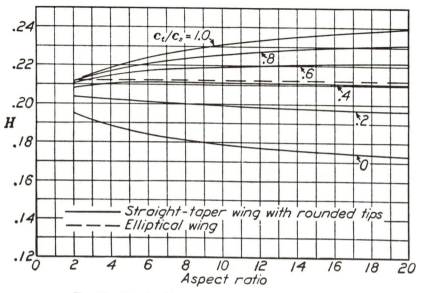

FIG. 15. Chart for determining aerodynamic-center position.

$$\frac{X_{ac}}{S/b} = HA \tan \beta$$

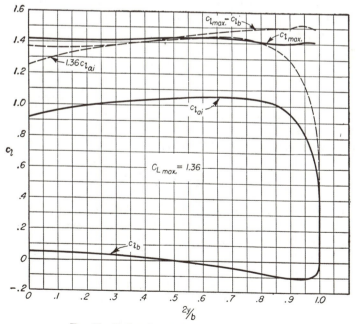

FIG. 16. Estimation of $C_{L\max}$ for example wing.

where α_i = angle of downwash (in degrees) at spanwise position y_1

y = spanwise position, variable in the integration and measured from center line

The effective angle of attack of each section of the wing is

$$\alpha_e = \alpha - \alpha_i \qquad (1.9)$$

where α_e = effective angle of attack of any section (angle between chord line and local wind direction)

α = geometric angle of attack of any section (angle between chord line and direction of free stream)

The effective angle of attack is a function of the lift coefficient of that section $f(c_l)$ so that the equation for the effective angle of attack becomes

$$\alpha_e = f(c_l) = \alpha - \frac{180}{\pi}\frac{b}{2\pi}\int_{-b/2}^{b/2}\frac{(d/dy)(c_l c/4b)dy}{y_1 - y} \qquad (1.10)$$

This general integral equation must be solved to determine the downwash and span-load distribution. Equation (1.10) may be solved by assuming a span-load distribution and solving for the corresponding downwash. The load distribution corresponding to this calculated downwash is then found using Eq. (1.9) and the section data. This load distribution is compared with the assumed distribution, and a second approximation is made. The process is continued until the load and downwash distributions are compatible.

The span-load distribution is expressed, following Glauert,[37] as the Fourier series

$$\frac{c_l c}{4b} = \Sigma A_n \sin n\theta \qquad (1.11)$$

where $\cos \theta = -2y/b$

The development of this method will be limited to load distributions symmetrical about the center line. Only odd values of n are therefore used. The induced angle from Eq. (1.8) becomes

$$\alpha_i = \frac{180}{\pi}\frac{\Sigma n A_n \sin n\theta}{\sin \theta} \qquad (1.12)$$

If it is assumed that values of $c_l c/4b$ are known at an even number r of stations equally spaced with respect to θ in the range $0 \le \theta \le \pi$, the coefficients A_n of the Fourier series

$$\left(\frac{c_l c}{4b}\right)_m = \sum_{n=1,3\ldots}^{r-1} A_n \sin n\frac{m\pi}{r}$$

where $m = 1, 2, 3, \ldots, r - 1$ may be found by harmonic analysis as

$$A_n = \sum_{m=1}^{r-1} \left(\frac{c_l c}{4b}\right)_m \eta_{mn}$$

where

$$\eta_{mn} = \frac{2}{r} \sin n \frac{m\pi}{r}$$

Because of the symmetrical relations

$$\left(\frac{c_l c}{4b}\right)_m = \left(\frac{c_l c}{4b}\right)_{r-m}$$

and

$$\sin n \frac{m\pi}{r} = \sin n \frac{(r-m)\pi}{r}$$

for odd values of n, the summation of A_n needs to be made only for values of m from 1 to $r/2$. Therefore

$$A_n = \sum_{m=1}^{r/2} \left(\frac{c_l c}{4b}\right)_m \eta_{mn} \qquad (1.13)$$

where

$$\eta_{mn} = \frac{4}{r} \sin n \frac{m\pi}{r} \qquad \text{for } m = 1, 2, 3, \ldots, (r/2) - 1$$

and

$$\eta_{mn} = \frac{2}{r} \sin n \frac{\pi}{2} \qquad \text{for } m = r/2$$

From these relations Sivells and Neely obtained multipliers η_{mn} which are presented in Table 3 for $r = 20$. The coefficients A_n may be obtained by multiplying the known value of $c_l c/4b$ by the appropriate multipliers and adding the resulting products.

By the substitution of the values of A_n from Eq. (1.13) into Eq. (1.12), Sivells and Neely[111] obtained the following expression for the downwash at the same stations at which $c_l c/b$ is known:

$$\alpha_{i_k} = \sum_{m=1}^{r/2} \left(\frac{c_l c}{b}\right)_m \lambda_{mk} \qquad (1.14)$$

where k designates the station at which α_i is to be evaluated, and m is also a station designation that is variable in the summation. Values of λ_{mk} for $r = 20$ are given in Table 4.[111] A similar table of values of $(4\pi/180)\lambda_{mk}$ was also given by Munk.[72] The induced-angle multipliers are independent of the aspect ratio and taper ratio of the wing and thus may be used for any wing whose load distribution is symmetrical.

Equation (1.14) permits the solution of Eq. (1.10) by successive approximations. For one geometric angle of attack of the wing, a lift dis-

TABLE 3.—MULTIPLIERS η_{mn} FOR A_n COEFFICIENTS OF GENERAL FOURIER SERIES

$$\sum_{n=1,3}^{19} A_n \sin n \frac{m\pi}{20} \qquad\qquad A_n = \sum_{m=1}^{10} \left(\frac{c_l c}{4b}\right)_m \eta_{mn}$$

$\frac{2y}{b}$	m \\ n	1	3	5	7	9	11	13	15	17	19
0	10	0.10000	-0.10000	0.10000	-0.10000	0.10000	-0.10000	0.10000	-0.10000	0.10000	-0.10000
0.1564	9	0.19754	-0.17820	0.14142	-0.09080	0.03129	0.03129	-0.09080	0.14142	-0.17820	0.19754
0.3090	8	0.19021	-0.11756	0	0.11756	-0.19021	0.19021	-0.11756	0	0.11756	-0.19021
0.4540	7	0.17820	-0.03129	-0.14142	0.19754	-0.09080	-0.09080	0.19754	-0.14142	-0.03129	0.17820
0.5878	6	0.16180	0.06180	-0.20000	0.06180	0.16180	-0.16180	-0.06180	0.20000	-0.06180	-0.16180
0.7071	5	0.14142	0.14142	-0.14142	-0.14142	0.14142	0.14142	-0.14142	-0.14142	0.14142	0.14142
0.8090	4	0.11756	0.19021	0	-0.19021	-0.11756	0.11756	0.19021	0	-0.19021	-0.11756
0.8910	3	0.09080	0.19754	0.14142	-0.03129	-0.17820	-0.17820	-0.03129	0.14142	0.19754	0.09080
0.9511	2	0.06180	0.16180	0.20000	0.16180	0.06180	-0.06180	-0.16180	-0.20000	-0.16180	-0.06180
0.9877	1	0.03129	0.09080	0.14142	0.17820	0.19754	0.19754	0.17820	0.14142	0.09080	0.03129

TABLE 3a. MULTIPLERS $\dfrac{\pi}{4}\eta_{m1}$ FOR DETERMINATION OF WING COEFFICIENTS

$\dfrac{2y}{b}$	$\dfrac{\pi}{4}\eta_{m1}$
0	0.07854
0.1564	0.15515
0.3090	0.14939
0.4540	0.13996
0.5878	0.12708
0.7071	0.11107
0.8090	0.09233
0.8910	0.07131
0.9511	0.04854
0.9877	0.02457

tribution is assumed from which the load distribution $c_l c/b$ is obtained. The corresponding downwash is calculated from Eq. (1.14) using values of λ_{mk} from Table 4. The downwash is subtracted from the geometric angle of attack at each station to give the effective angle of attack of the section [Eq (1.9)]. The section lift coefficients are obtained from curves of the section data plotted against the effective angle of attack α_e where

$$\alpha_e = E(\alpha_0 - \alpha_{l_0}) + \alpha_{l_0}$$

and E is the Jones[50] edge-velocity correction.

If the resulting lift distribution does not agree with that assumed, a second approximation to the lift distribution is assumed and the process is repeated until the assumed and calculated lift distributions agree. The entire process must be repeated for each angle of attack for which the wing characteristics are desired. The amount of labor required to obtain the complete characteristics of a wing may be reduced by using Anderson's method[10] for the range of lift coefficients where the section lift curves are substantially linear.

The wing lift coefficient is obtained by spanwise integration of the lift distribution. This process[111] has been reduced to the following summation:

$$C_L = A \sum_{m=1}^{r/2} \left(\frac{c_l c}{b}\right)_m \frac{\pi}{4}\eta_{m1} \tag{1.15}$$

The multipliers $\pi/4\ \eta_{m1}$ are equal to $\pi/4$ times the multipliers for A and are presented in Table 3a for $r = 20$.

The section induced-drag coefficient is equal to the product of the section lift coefficient and the downwash in radians.

$$c_{d_i} = \frac{\pi c_l \alpha_i}{180}$$

TABLE 4.—MULTIPLIERS λ_{mk} FOR INDUCED ANGLE OF ATTACK

$$\alpha_{ik} = \sum_{m=1}^{10} \left(\frac{c_l c}{b}\right)_m \lambda_{mk}$$

$\frac{2y}{b}$		0.9877	0.9511	0.8910	0.8090	0.7071	0.5878	0.4540	0.3090	0.1564	0
$\frac{2y}{b}$ \ k	m \ k	1	2	3	4	5	6	7	8	9	10
0	10	−1.468	0	−1.804	0	−2.865	0	−6.950	0	−58.533	143.239
0.1564	9	0	−3.394	0	−4.840	0	−10.158	0	−67.298	145.025	−115.624
0.3090	8	−3.768	0	−4.968	0	−9.916	0	−67.157	150.611	−64.802	0
0.4540	7	0	−5.812	0	−10.926	0	−72.472	160.761	−62.917	0	−12.384
0.5878	6	−7.713	0	−13.134	0	−82.083	177.054	−65.803	0	−8.320	0
0.7071	5	0	−17.388	0	−97.965	202.571	−71.743	0	−7.372	0	−4.051
0.8090	4	−26.635	0	−125.537	243.694	−81.434	0	−7.208	0	−2.880	0
0.8910	3	0	−180.528	315.512	−96.962	0	−7.370	0	−2.371	0	−1.638
0.9511	2	−329.976	463.533	−122.880	0	−7.599	0	−2.016	0	−1.062	0
0.9877	1	915.651	−167.045	0	−7.089	0	−1.491	0	−0.620	0	−0.459

The wing induced-drag coefficient may be obtained by means of a spanwise integration of the section induced-drag coefficients multiplied by the local chord. As in the case of the lift coefficient, this process has been reduced to the following summation:

$$C_{D_i} = \frac{\pi A}{180} \sum_{m=1}^{r/2} \left(\frac{c_l c}{b} \, \alpha_i \right)_m \frac{\pi}{4} \, \eta_{m1} \tag{1.16}$$

The section profile drag coefficient can be obtained from the section data for the appropriate wing section and lift coefficient. The wing profile drag coefficient may be obtained by means of a spanwise integration of the section profile drag coefficient multiplied by the local chord. This process has again been reduced to the following summation:

$$C_{D_0} = \sum_{m=1}^{r/2} \left(c_{d_0} \frac{c}{\bar{c}} \right)_m \frac{\pi}{4} \, \eta_{m1} \tag{1.17}$$

where $\bar{c}$ = mean geometric chord S/b

The section pitching-moment coefficient can be obtained from the section data for the appropriate wing section and lift coefficient. For each spanwise station the pitching-moment coefficient is transferred to the wing reference point by the equation

$$c_m = c_{m_{ac}} - \frac{x}{c} \left[c_l \cos \left(\alpha_s - \alpha_i \right) + c_{d_0} \sin \left(\alpha_s - \alpha_i \right) \right]$$
$$- \frac{z}{c} \left[c_l \sin \left(\alpha_s - \alpha_i \right) - c_{d_0} \cos \left(\alpha_s - \alpha_i \right) \right]$$

where c_m = section pitching-moment coefficient about wing reference point
$\quad \alpha_s$ = geometric angle of attack of root section
$\quad x$ = projected distance in plane of symmetry from wing reference point to aerodynamic center of wing section, measured parallel to chord of root section, positive to the rear
$\quad z$ = projected distance in plane of symmetry from wing reference point to aerodynamic center of wing section, measured perpendicular to root chord, positive upward.

The wing pitching-moment coefficient may be obtained by spanwise integration by use of the multipliers previously used.

$$C_m = \sum_{m=1}^{r/2} \left(\frac{c_m c^2}{\bar{c} c'} \right)_m \frac{\pi}{4} \, \eta_{m1} \tag{1.18}$$

where c' = mean aerodynamic chord, $2/S \int_0^{b/2} c^2 \, dy$

1.5. Applicability of Section Data. The applicability of section data to the prediction of the aerodynamic characteristics of wings is limited by the simplifying assumptions made in the development of wing theory. It is assumed, for instance, that each section acts independently of its neighboring sections except for the induced downwash. Strict compliance with this assumption would require two-dimensional flow, that is, no variation of section, chord, or lift along the span.

Such spanwise variations result in spanwise components of flow. If these components are small, the sections act nearly independently. Comparatively small spanwise variations of pressure, however, tend to produce large crossflows in the boundary layer. The air adjacent to the surface has lost most of its momentum and therefore tends to flow directly toward the region of lowest pressure rather than in the stream direction. These crossflows become particularly marked under conditions approaching separation. The flow of this low-energy air from one section to another tends to delay separation in some places and to promote it in others, with the result that the wing characteristics may depart seriously from the calculated ones.

In the development of the lifting-line theory, it was assumed that the effect of the trailing vortices was to change the local angle of attack, neglecting any change of downwash along the chord of the section. This variation of downwash is not negligible for sections close to a strong trailing vortex. Strong concentrations of the trailing vortices are obtained whenever a large spanwise variation of lift occurs. Consequently the sections are operating in a curved as well as a rotated flow field whenever the spanwise variation of lift is not small. This curved flow field may be interpreted as an effective change in camber which is not considered in ordinary lifting-line theories. This difficulty is avoided by lifting-surface theory, the treatment of which is outside the scope of this volume.

Experience has shown that usable results are obtained from lifting-line theory if no spanwise discontinuities or rapid changes of section, chord, or twist are present, and if the wing has no pronounced sweep. These conditions are obviously not satisfied near the wing tips, near the extremities of partial span flaps or deflected ailerons, near cutouts and large fillets, or if the wing is partly stalled. Since the assumed conditions are not satisfied near the wing tips, it is obvious that section data are not applicable to wings of low aspect ratio. In fact, an entirely different theory[51] applies to wings of very low aspect ratio. The ordinary three-dimensional wing theories are obviously not applicable at supercritical speeds when the velocity of sound is exceeded anywhere in the field of flow.

Despite these limitations, wing-section data have a wide field of application to wings of the aspect ratios and plan forms customarily used for airplanes flying at subcritical speeds. The lift, drag, and moment characteristics of such wings may be estimated from the section data with a fair degree of accuracy by Anderson's methods[10] and with a high degree of accuracy by the method of Sivells and Neely.[111] The order of accuracy

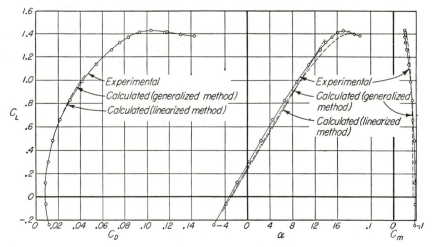

FIG. 17. Calculated and measured characteristics of a wing of aspect ratio 10 with NACA 44-series sections.

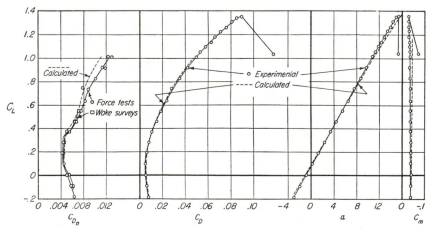

FIG. 18. Experimental and calculated characteristics of a wing having NACA 64-210 airfoil sections, 2-deg washout, aspect ratio of 9, and taper ratio of 2.5. $R \approx 4,400,000$; $M \approx 0.17$.

to be expected from predictions made by the latter method[74, 110] is indicated by Figs. 17 to 20. The three-dimensional wing characteristics were obtained from tests in the NACA 19-foot pressure tunnel. The predicted characteristics were based on section data obtained in the NACA two-dimensional low-turbulence pressure tunnel.[133] The agreement between the ex-

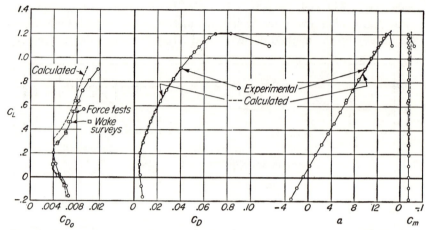

Fig. 19. Experimental and calculated characteristics of a wing having NACA 65-210 airfoil sections, 2-deg washout, aspect ratio of 9, and taper ratio of 2.5. $R \approx 4,400,000; M \approx 0.17$.

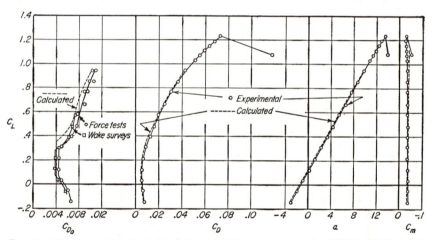

Fig. 20. Experimental and calculated characteristics of a wing having NACA 65-210 airfoil sections, 0-deg washout, aspect ratio of 9, and taper ratio of 2.5. $R \approx 4,400,000; M \approx 0.17$.

perimental and predicted characteristics is so good that the utmost care is required in model construction and experimental technique to obtain empirical wing characteristics as reliable as those predicted from the section data.

CHAPTER 2

SIMPLE TWO-DIMENSIONAL FLOWS

2.1. Symbols.

H total pressure

K nondimensional circulation, $\Gamma/2\pi a V$

S pressure coefficient, $(H - p)/\frac{1}{2}\rho V^2$

V resultant or free-stream velocity

X component of force in the x direction

Y component of force in the y direction

Z component of force in the z direction

a constant

a $\sqrt{\dfrac{\mu}{2\pi V}}$

b constant

c constant

c' constant

c'' constant

e base of Naperian logarithm, 2.71828

l section lift

$\ln$ logarithm to the base e

m source strength per unit length

p pressure

r radial coordinate

s position of source on x axis

t time

u component of velocity in the x direction

u' radial component of velocity

v component of velocity in the y direction

v' tangential component of velocity, positive counterclockwise

w component of velocity in the z direction

x length in Cartesian coordinates

y length in Cartesian coordinates

z length in Cartesian coordinates

Γ circulation, positive clockwise

θ angular coordinate, positive counterclockwise

ϕ velocity potential

ψ stream function

μ doublet strength, $2ms$

π ratio of the circumference of a circle to its diameter

ρ mass density of air

ω angular velocity, positive clockwise

2.2. Introduction. A considerable body of aerodynamic theory has been developed with which it is possible to calculate some of the important characteristics of wing sections. Conversely it is possible to design wing sections to have certain desirable aerodynamic characteristics. The purpose of this chapter is to review the basic fluid mechanics necessary for understanding the theory of wing sections.

2.3. Concept of a Perfect Fluid. The concept of a perfect fluid is an important simplification in fluid mechanics. The perfect fluid is considered to be a continuous homogeneous medium within which no shearing stresses can exist. For the purpose of this chapter the perfect fluid is also considered to be incompressible.

The assumption of zero shearing stresses, or zero viscosity, eliminates the possibility of obtaining any information about the drag of wing sections or about the separation of the flow from the surface. This assumption is very useful, nevertheless, because it simplifies the equations of motion that otherwise cannot generally be solved and because the resulting solutions represent reasonable approximations to many actual flows. In all cases under consideration, the viscous forces are small compared with the inertia forces except in the layer of fluid adjacent to the surface. The direct effects of viscosity are negligible except in this layer, and viscosity has little effect on the general flow pattern unless the local effects are such as to make the flow separate from the surface.

The assumption of incompressibility also results in simplified solutions that are reasonable approximations to actual flows except at high speeds. Although gases such as air are compressible, the relative change in density occurring in a field of flow is small if the variation of pressure is small compared with the absolute pressure. This assumption leads to increasingly important discrepancies as the local velocity anywhere in the field of flow approaches the velocity of sound. The effects of compressibility will be discussed in Chap. 9.

2.4. Equations of Motion. One of the fundamental conditions that must be satisfied is that no fluid can be created or destroyed within the field of flow considered. This condition means that the amount of fluid entering any small element of volume must equal the amount of fluid leaving the element. For an incompressible fluid the amount of fluid may be measured by its volume. The equation of continuity, expressing this condition, may be derived from the following considerations.

Figure 21 shows a small element of volume having dimensions dx, dy, and dz. The components of velocity along each of the three axes x, y, and z are u, v, and w. The volume of fluid entering across each of the faces of the element of volume, perpendicular to the x, y, and z axes, respectively, is

$$u \; dy \; dz \quad \text{entering } yz \text{ face}$$
$$v \; dz \; dx \quad \text{entering } zx \text{ face}$$
$$w \; dx \; dy \quad \text{entering } xy \text{ face}$$

To the first order of small quantities, the amount of fluid leaving the element of volume across the corresponding opposing faces is

$$\left(u + \frac{\partial u}{\partial x}\, dx\right) dy\, dz \qquad \text{leaving } yz \text{ face}$$

$$\left(v + \frac{\partial v}{\partial y}\, dy\right) dz\, dx \qquad \text{leaving } zx \text{ face}$$

$$\left(w + \frac{\partial w}{\partial z}\, dz\right) dx\, dy \qquad \text{leaving } xy \text{ face}$$

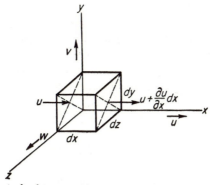

FIG. 21. Element of volume considered in derivation of equation of continuity.

The condition of continuity requires that the volume of fluid leaving the element must be equal to the volume entering. Hence the equation of continuity for an incompressible fluid is

$$\frac{\partial u}{\partial x} + \frac{\partial v}{\partial y} + \frac{\partial w}{\partial z} = 0 \qquad (2.1)$$

The other fundamental condition to be satisfied is that the motion of the fluid must be in accordance with Newton's laws of motion. This condition may be stated as follows:

$$\left. \begin{aligned} X &= \rho\, \frac{Du}{Dt}\, dx\, dy\, dz \\[2mm] Y &= \rho\, \frac{Dv}{Dt}\, dx\, dy\, dz \\[2mm] Z &= \rho\, \frac{Dw}{Dt}\, dx\, dy\, dz \end{aligned} \right\} \qquad (2.2)$$

where X, Y, Z = components of force on the element of fluid $dx\, dy\, dz$ in the x, y, and z directions, respectively

t = time

and the differentiations are performed following the motion of the element.

For axes stationary with respect to the observer, the components of velocity u, v, and w are, in general, functions of t, x, y, and z. The total derivatives given in Eqs. (2.2) may therefore be written

$$\frac{\partial u}{\partial t} + u\frac{\partial u}{\partial x} + v\frac{\partial u}{\partial y} + w\frac{\partial u}{\partial z} = \frac{Du}{Dt}$$

$$\frac{\partial v}{\partial t} + u\frac{\partial v}{\partial x} + v\frac{\partial v}{\partial y} + w\frac{\partial v}{\partial z} = \frac{Dv}{Dt}$$

$$\frac{\partial w}{\partial t} + u\frac{\partial w}{\partial x} + v\frac{\partial w}{\partial y} + w\frac{\partial w}{\partial z} = \frac{Dw}{Dt}$$

If gravitational forces are neglected, the only forces acting on the element of volume are normal pressure forces. Let $p\,dy\,dz$ be the force acting on the face of the element (Fig. 21) in the yz plane. Then to the first order $[p + (\partial p/\partial x)\,dx]\,dy\,dz$ will be the force acting on the opposite face. The resultant force on the element in the x direction is therefore

$$X = -\frac{\partial p}{\partial x}\,dx\,dy\,dz$$

Similarly the resultant forces in the y and z directions are

$$Y = -\frac{\partial p}{\partial y}\,dy\,dz\,dx$$

$$Z = -\frac{\partial p}{\partial z}\,dz\,dx\,dy$$

The equations stating that the motion of the fluid is in accordance with Newton's laws may be written in the following form by substituting the foregoing expressions for the forces and accelerations in Eqs. (2.2)

$$\left.\begin{aligned}
-\frac{\partial p}{\partial x} &= \rho\left(\frac{\partial u}{\partial t} + u\frac{\partial u}{\partial x} + v\frac{\partial u}{\partial y} + w\frac{\partial u}{\partial z}\right)\\
-\frac{\partial p}{\partial y} &= \rho\left(\frac{\partial v}{\partial t} + u\frac{\partial v}{\partial x} + v\frac{\partial v}{\partial y} + w\frac{\partial v}{\partial z}\right)\\
-\frac{\partial p}{\partial z} &= \rho\left(\frac{\partial w}{\partial t} + u\frac{\partial w}{\partial x} + v\frac{\partial w}{\partial y} + w\frac{\partial w}{\partial z}\right)
\end{aligned}\right\} \tag{2.3}$$

If we consider only steady motion, the derivatives with respect to time are equal to zero. In this case the differential equations for the path of an element are

$$\frac{dx}{u} = \frac{dy}{v} = \frac{dz}{w} \tag{2.4}$$

These equations merely state that the displacement along any axis is proportional to the component of velocity along that axis. Preparatory to integrating Eqs. (2.3) along the path of motion for steady flow, we shall multiply the equations for the x, y, and z components by dx, dy, and dz, respectively. For example,

$$-\frac{\partial p}{\partial x}\,dx = \rho\left(u\,\frac{\partial u}{\partial x}\,dx + v\,\frac{\partial u}{\partial y}\,dx + w\,\frac{\partial u}{\partial z}\,dx\right)$$

According to Eqs. (2.4), we have

$$v\,dx = u\,dy$$
$$w\,dx = u\,dz$$

Hence

$$-\frac{\partial p}{\partial x}\,dx = \rho\left(u\,\frac{\partial u}{\partial x}\,dx + u\,\frac{\partial u}{\partial y}\,dy + u\,\frac{\partial u}{\partial z}\,dz\right) = \rho\,d\!\left(\frac{1}{2}u^2\right)$$

Similarly

$$-\frac{\partial p}{\partial y}\,dy = \rho\,d\!\left(\frac{1}{2}v^2\right)$$

$$-\frac{\partial p}{\partial z}\,dz = \rho\,d\!\left(\frac{1}{2}w^2\right)$$

Adding the three foregoing relations, we have

$$-dp = \frac{\rho}{2}\,d(u^2 + v^2 + w^2) = \frac{\rho}{2}\,d(V^2)$$

where V = magnitude of velocity.

Integrating this equation gives Bernoulli's equation

$$p + \tfrac{1}{2}\,\rho V^2 = H \tag{2.5}$$

where H, the constant of integration, is the total pressure.

The application of Eq. (2.5) is limited by the assumptions made in its derivation, namely,

1. Perfect incompressible fluid.

2. Steady motion.

3. Integration along path of motion (streamline). It follows that this equation may be applied only along streamlines in unaccelerated flow where the effects of viscosity and compressibility are negligible.

2.5. Description of Flow Patterns. It is evident from the foregoing equations that the flow pattern would be completely determined if the values of u, v, and w were known at every point. The theory of wing sections assumes that the flow is two-dimensional, that is, $w = 0$. Even with this simplification, it is still necessary to specify both the components of velocity u and v to define the flow. By the use of the equation of continuity (2.1) it is possible to simplify the problem further so that it is necessary to specify only a single quantity at each point to determine the flow pattern.

Consider the flow across the arbitrary line *oba* that connects the origin *o* with a point *a* (Fig. 22). The amount of flow across this line will be the same as that across any other line connecting the two points, for example, line *oca*, provided that the equation of continuity holds for all points within the region enclosed by the two lines. With a fixed origin *o* the amount of fluid crossing any line joining the points *o* and *a* is therefore a function only of the position of the point *a* for a given flow pattern. The amount of fluid passing between the points *o* and *a* is given by the following expression

$$\psi = \int_o^a u \, dy - v \, dx$$

or

$$d\psi = u \, dy - v \, dx \qquad (2.6)$$

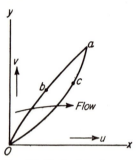

Because the value of ψ is independent of the path of integration, Eq. (2.6) remains valid even though dx and dy are varied independently. The general expression for the total differential of ψ is

$$d\psi = \frac{\partial \psi}{\partial x} \, dx + \frac{\partial \psi}{\partial y} \, dy \qquad (2.7)$$

Fig. 22. Derivation of stream function.

This equation is also valid for independent variations of dx and dy. The coefficients of dx and dy in Eqs. (2.6) and (2.7) must therefore be equal, or

$$\left. \begin{aligned} u &= \frac{\partial \psi}{\partial y} \\ v &= -\frac{\partial \psi}{\partial x} \end{aligned} \right\} \qquad (2.8)$$

A single function ψ has therefore been found by means of which it is possible to define both components of velocity at all points in the field of flow. This function ψ is called the "stream function." Lines in the flow along which ψ is constant are called "streamlines," and these lines are the paths of motion of fluid elements for steady motion.

In general, the component of velocity in any direction may be obtained by differentiating the stream function in a direction 90° counterclockwise to the component desired. In polar coordinates, therefore, the expressions for the radial and tangential components of velocity are

$$\left. \begin{aligned} u' &= \frac{1}{r}\frac{\partial \psi}{\partial \theta} \qquad \text{radial} \\ v' &= -\frac{\partial \psi}{\partial r} \qquad \text{tangential} \end{aligned} \right\} \qquad (2.9)$$

Most flows of a perfect fluid are of the type known as "irrotational motion." In irrotational motion the fluid elements move with translation only. Their angular velocity is zero. This absence of angular velocity does not, of course, prevent the elements from moving in curved paths. To obtain expressions for the angular velocity of an element in terms of the velocity derivatives, consider the motion of a solid disk. If we assume clockwise motion to be positive, the angular velocity of the disk may be expressed as (Fig. 23)

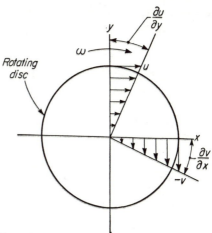

$$\omega = \frac{1}{2}\left(\frac{\partial u}{\partial y} - \frac{\partial v}{\partial x}\right)$$

The expression

$$\left(\frac{\partial u}{\partial y} - \frac{\partial v}{\partial x}\right)$$

Fig. 23. Velocity derivatives for rotating disk.

is called "vorticity." The vorticity is seen to be twice the angular velocity.

The integral of the tangential component of velocity around any closed curve is defined as the circulation. A simple relation exists between the circulation around a curve and the vorticity over the area enclosed by the curve. Consider the small element shown in Fig. 24.

$$d\Gamma = v\,dy + \left(u + \frac{\partial u}{\partial y}\,dy\right)dx - \left(v + \frac{\partial v}{\partial x}\,dx\right)dy - u\,dx$$

or

$$d\Gamma = \left(\frac{\partial u}{\partial y} - \frac{\partial v}{\partial x}\right)dx\,dy$$

That is, the element of circulation is equal to the vorticity multiplied by the element of area. For any finite area, the circulation is given by the following expression:

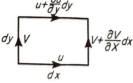

Fig. 24. Calculation of circulation about element of area.

$$\Gamma = \int\int\left(\frac{\partial u}{\partial y} - \frac{\partial v}{\partial x}\right)dx\,dy \qquad (2.10)$$

The angular velocity of any element can be changed only by the application of tangential (shearing) forces which, by definition, are absent in a perfect fluid. It follows that, if the flow is once irrotational, it will remain irrotational and the circulation around any closed path will equal zero. If the flow is not irrotational, the circulation around any path moving with the fluid will remain constant.

If the flow is irrotational, it is possible to derive a second quantity which, like the stream function, can be used to describe the flow pattern completely. Consider the flow field indicated in Fig. 25. The line integral of the velocity over the path *oap* must be equal to the line integral of the velocity over the path *obp* if the motion is irrotational in the region between the two paths. The value of this integral, called the "velocity potential ϕ," therefore depends only on the position of the point *p* relative to the origin *o*. The value of the velocity potential is

$$\phi = \int_o^p u\, dx + v\, dy$$

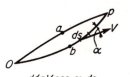

$d\phi = V \cos \alpha\, ds$

Fig. 25. Definition of velocity potential.

or

$$d\phi = u\, dx + v\, dy \qquad (2.11)$$

By a process of reasoning similar to that previously given for the stream function, we obtain

$$\left. \begin{aligned} u &= \frac{\partial \phi}{\partial x} \\ v &= \frac{\partial \phi}{\partial y} \end{aligned} \right\} \qquad (2.12)$$

In general, the component of velocity in any direction may be obtained by differentiating the velocity potential in the direction of the desired component. This property of the velocity potential makes it particularly useful for the study of three-dimensional flows. In polar coordinates, the expressions for the radial and tangential components of velocity are

$$\left. \begin{aligned} u' &= \frac{\partial \phi}{\partial r} \qquad \text{radial} \\ v' &= \frac{1}{r}\frac{\partial \phi}{\partial \theta} \qquad \text{tangential} \end{aligned} \right\} \qquad (2.13)$$

The equation of continuity for two-dimensional flow

$$\frac{\partial u}{\partial x} + \frac{\partial v}{\partial y} = 0 \qquad (2.14)$$

assumes a particularly simple form when expressed in terms of the velocity potential

$$\frac{\partial^2 \phi}{\partial x^2} + \frac{\partial^2 \phi}{\partial y^2} = 0 \qquad (2.15)$$

This equation is Laplace's equation in two dimensions.

The equation stating that the flow is irrotational is

$$\frac{\partial u}{\partial y} - \frac{\partial v}{\partial x} = 0$$

When written in terms of the stream function, this equation becomes

$$\frac{\partial^2 \psi}{\partial y^2} + \frac{\partial^2 \psi}{\partial x^2} = 0 \qquad (2.16)$$

In writing Eq. (2.15), it is implicit in the definition of ϕ that the motion is irrotational and the equation itself states that no fluid is being created or destroyed. In writing Eq. (2.16), it is implicit in the definition of ψ that no fluid is being created or destroyed, and the equation itself states that the flow is irrotational. Equations (2.15) and (2.16) impose rather general conditions on functions chosen to represent actual flow patterns. Any function of ϕ or ψ satisfying Eqs. (2.15) or (2.16) represents a possible flow.

It can be seen from Eqs. (2.8) and (2.12) that

$$u = \frac{\partial \phi}{\partial x} = \frac{\partial \psi}{\partial y}$$

$$v = \frac{\partial \phi}{\partial y} = -\frac{\partial \psi}{\partial x}$$

These equations indicate that lines along which ϕ is constant intersect lines along which ψ is constant at right angles.

Because Eqs. (2.15) and (2.16) are linear, functions of ϕ or ψ representing possible flow patterns may be added to obtain new flow patterns. This method of obtaining new flow patterns by superposition of known flows is fundamental to the theory of wing sections because it leads to simple solutions of complicated problems.

2.6. Simple Two-dimensional Flows. A few simple flows upon which the theory of wing sections is based are described in this section.

a. Uniform Stream. Consider the functions

$$\left.\begin{array}{l} \psi = by - ax \\[2mm] \phi = bx + ay \end{array}\right\} \qquad (2.17)$$

and

which satisfy Eqs. (2.15) and (2.16). The component of velocity along the x axis is

$$u = \frac{\partial \psi}{\partial y} = \frac{\partial \phi}{\partial x} = b$$

The component of velocity along the y axis is

$$v = -\frac{\partial \psi}{\partial x} = \frac{\partial \phi}{\partial y} = a$$

Equations (2.17) therefore represent a uniform field of flow having a velocity V equal to $\sqrt{a^2 + b^2}$ inclined to the x axis at an angle whose tangent is a/b.

b. Sources and Sinks. The concept of sources and sinks is one of the building blocks used to construct desired flow patterns. A source is considered to be a point at which fluid is being created at a given rate. A sink is a point at which fluid is being destroyed. The flow about a point source or sink is assumed to be uniform in all directions and to obey the equations of continuity and of irrotational motion everywhere except at the point itself. In two-dimensional motion, the flow is assumed to be the same in all planes perpendicular to the z axis. The point source in two-dimensional motion is therefore a line parallel to the z axis from which fluid emanates at a uniform rate throughout its infinite length. In accordance with these assumptions, the radial component of velocity u' at any point at a distance r from a source creating fluid at a rate m per unit length perpendicular to the plane of flow may be expressed as

$$u' = \frac{m}{2\pi r} \tag{2.18}$$

Because the sink is simply a negative source, the equation for the radial component of velocity is the same as Eq. (2.18) except that the velocity is directed toward the sink. The tangential component of velocity resulting from a source or sink is zero.

Taking the origin at the source we have

$$u = u' \cos \theta$$
$$v = u' \sin \theta$$
$$r = \sqrt{x^2 + y^2}$$

where

$$\cos \theta = \frac{x}{\sqrt{x^2 + y^2}}$$

$$\sin \theta = \frac{y}{\sqrt{x^2 + y^2}}$$

Substituting in Eq. (2.8), we have

$$u = \frac{\partial \psi}{\partial y} = \frac{mx}{2\pi(x^2 + y^2)}$$

Integration of this expression with respect to y results in the expression

$$\psi = \frac{m}{2\pi} \tan^{-1} \frac{y}{x} \tag{2.19}$$

The same expression may be obtained by a corresponding process using the expression for v and integrating with respect to x. It is seen that ψ is a multiple-valued function of x and y. This peculiarity results from the fact that the equation of continuity is not valid at the source itself.

The expression for the flow from a point source in terms of the velocity potential ϕ may be obtained as follows:

$$u = \frac{\partial \phi}{\partial x} = \frac{mx}{2\pi(x^2 + y^2)}$$

$$\phi = \frac{m}{2\pi} \ln \sqrt{x^2 + y^2} \tag{2.20}$$

From Eqs. (2.19) and (2.20) it is seen that the streamlines for a source situated at the origin are radial lines and that the equipotential lines are circles about the origin.

c. *Doublets.* A doublet is the limiting case of the flow about a single source and a sink of equal strength that approach each other in such a manner that the product of the source strength and the distance between the source and sink remains constant. The axis of the doublet is the line joining the source and sink. To obtain expressions for the flow about a doubtlet, consider the arrangement of a source and a sink shown in Fig. 26. The source and sink are situated on the x axis at $-s$ and s, respectively. The flow resulting from the combination is

$$\psi = \frac{m}{2\pi}\left(\tan^{-1}\frac{y}{x+s} - \tan^{-1}\frac{y}{x-s}\right)$$

Since

$$\tan^{-1} a - \tan^{-1} b = \tan^{-1}\frac{a-b}{1+ab}$$

we have

$$\psi = \frac{m}{2\pi}\tan^{-1}\frac{-2ys}{x^2+y^2-s^2}$$

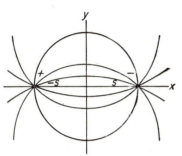

Fig. 26. Streamlines of flow for a source and sink.

For values of s small with respect to x and y,

$$\psi = \frac{m}{2\pi}\left(\frac{-2ys}{x^2+y^2}\right)$$

If the product of m and s remains constant as s approaches zero,

$$\psi = -\frac{\mu y}{2\pi(x^2+y^2)} \tag{2.21}$$

where $\mu = 2ms$

d. *Circular Cylinder in a Uniform Stream.* The flow about a circular cylinder in a uniform stream is obtained by superposing a uniform stream [Eq. (2.17)] on the flow about a doublet [Eq. (2.21)] with the uniform stream flowing from the source to the sink.

$$\psi = y\left(V - \frac{\mu}{2\pi(x^2+y^2)}\right)$$

This equation may be written as follows in terms of polar coordinates:

$$\psi = Vr\left(1 - \frac{a^2}{r^2}\right)\sin\theta \tag{2.22}$$

where
$$a^2 = \mu/2\pi V$$
$$r^2 = x^2 + y^2$$

It is obvious from Eq. (2.22) that part of the streamline $\psi = 0$ is the circle $r = a$. At large distances from the origin the flow is uniform and parallel to the x axis. Equation (2.22) therefore represents the flow about a circular cylinder in a uniform stream. The streamlines for this flow pattern are shown in Fig. 27.

The velocity distribution about the cylinder may be obtained by finding the tangential component of velocity on the circle $r = a$.

$$v = -\frac{\partial\psi}{\partial r} = -V\left(1 + \frac{a^2}{r^2}\right)\sin\theta = -2V\sin\theta \qquad (2.23)$$

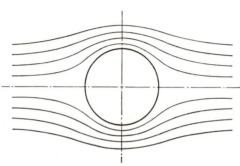

FIG. 27.   Streamlines for the flow about a circular cylinder in a uniform stream.

Because the circle $r = a$ is a streamline, we may apply Bernoulli's equation (2.5) to obtain the pressure distribution about the cylinder.

$$p + \tfrac{1}{2}\rho(4V^2\sin^2\theta) = H$$

If we define a pressure coefficient S as

$$S = \frac{H - p}{\tfrac{1}{2}\rho V^2} \qquad (2.24)$$

the distribution of S over the surface of the cylinder is given by

$$S = 4\sin^2\theta \qquad (2.25)$$

Integration of the pressure distribution over the surface of the cylinder will show that the resultant force is zero.

e. Vortex. A vortex is a flow pattern in which the elements of fluid follow circular paths about a point, and the flow is irrotational at all points except the center. The equations defining the flow pattern may be derived directly from Laplace's equation for the stream function. Equation (2.16) may be written as follows in terms of polar coordinates:[80]

$$\frac{\partial^2\psi}{\partial r^2} + \frac{1}{r}\frac{\partial\psi}{\partial r} + \frac{1}{r^2}\frac{\partial^2\psi}{\partial\theta^2} = 0 \qquad (2.26)$$

Because the radial component of velocity is zero,

$$\frac{1}{r}\frac{\partial \psi}{\partial \theta} = 0$$

and hence

$$\frac{1}{r^2}\frac{\partial^2 \psi}{\partial \theta^2} = 0$$

Equation (2.26) therefore becomes

$$\frac{d^2 \psi}{dr^2} + \frac{1}{r}\frac{d\psi}{dr} = 0$$

Integrating once with respect to r we obtain

$$\ln \frac{d\psi}{dr} + \ln r = c$$

or

$$\frac{d\psi}{dr} = \frac{c'}{r} = \frac{\Gamma}{2\pi r} = -v' \tag{2.27}$$

where the constant of integration c' is assigned the value $\Gamma/2\pi$. The velocity therefore varies inversely with the dis-
tance from the center of the vortex. The
circulation Γ about the vortex is equal to
$-2\pi r v'$. The circulation is considered to be
positive in the clockwise direction, whereas v'
is positive in the counterclockwise direction.
The stream function may be obtained by integration of Eq. (2.27)

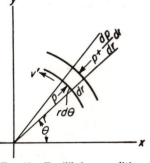

$$\psi = \frac{\Gamma}{2\pi} \ln r + c'' \tag{2.28}$$

The pressure in the field of flow may be
calculated from the condition that the pressure

Fig. 28. Equilibrium conditions for a vortex.

gradients must maintain the fluid elements in equilibrium with the accelerations (d'Alembert's principle). The centrifugal force acting on each element must be (Fig. 28)

$$-\frac{\rho v'^2}{r} r\, d\theta\, dr$$

The pressure force acting on the element is

$$\left[p - \left(p + \frac{dp}{dr} dr \right) \right] r\, d\theta$$

Equating the pressure and acceleration forces, we have

$$\frac{\rho v'^2}{r} = \frac{dp}{dr}$$

Substituting the value of v' from Eq. (2.27) gives

$$\frac{\rho\Gamma^2}{4\pi^2 r^3} = \frac{dp}{dr}$$

$$p = -\frac{\rho\Gamma^2}{8\pi^2 r^2} + H$$

where H = constant of integration

This equation may be written in the form

$$p + \tfrac{1}{2}\rho v'^2 = H$$

which shows that Bernoulli's equation may be applied throughout the field of flow.

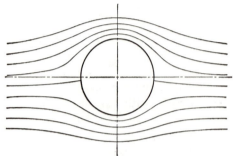

FIG. 29.   Streamlines for the flow about a circular cylinder in a uniform stream with circulation corresponding to a wing-section lift coefficient of 0.6.

f. Circular Cylinder with Circulation.   The flow pattern represented by a circular cylinder with circulation is the basic flow pattern from which the flow about wing sections of arbitrary shape at various angles of attack is calculated. Such a flow pattern is obtained by superposing the flow produced by a point vortex upon the flow about a circular cylinder. Adding the flows corresponding to Eqs. (2.22) and (2.28), we obtain

$$\psi = Vr\left(1 - \frac{a^2}{r^2}\right)\sin\theta + \frac{\Gamma}{2\pi}\ln\frac{r}{a} \tag{2.29}$$

where the constant c'' in Eq. (2.28) has been taken equal to

$$\frac{-\Gamma}{2\pi}\ln a$$

A typical flow pattern for a moderate value of the circulation Γ is given in Fig. 29.

The velocity distribution about the cylinder is found by differentiating the expression for the stream function [Eq. (2.29)] as follows:

$$\frac{\partial\psi}{\partial r} = V\left(1 + \frac{a^2}{r^2}\right)\sin\theta + \frac{\Gamma}{2\pi r}$$

The tangential component of velocity v' (positive counterclockwise) at the surface of the cylinder is obtained from Eq. (2.9) and the substitution of $r = a$.

$$v' = -2V \sin \theta + \frac{\Gamma}{2\pi a} \tag{2.30}$$

It is seen that the addition of the circulation Γ moves the points of zero velocity (stagnation points) from the positions $\theta = 0$ and π to the positions

$$\theta = \sin^{-1} \frac{\Gamma}{4\pi a V}$$

The pressure distribution about the cylinder may be found by applying Bernoulli's equation (2.5) along the streamline $\psi = 0$.

$$p + \frac{1}{2}\rho\left(4V^2 \sin^2 \theta - \frac{2V\Gamma \sin \theta}{\pi a} + \frac{\Gamma^2}{4\pi^2 a^2}\right) = H \tag{2.31}$$

Setting

$$\frac{\Gamma}{2\pi a V} = K \tag{2.32}$$

the pressure coefficient S [Eq. (2.24)] becomes

$$S = 4 \sin^2 \theta - 4K \sin \theta + K^2 \tag{2.33}$$

The symmetry of Eq. (2.33) about the line $\theta = \pi/2$ shows that there can be no drag force. The lift on the cylinder can be obtained by integration, over the surface, of the components of pressure normal to the stream.

$$l = \frac{1}{2}\rho V^2 \int_0^{2\pi} Sa \sin \theta \, d\theta$$

$$= \frac{1}{2}\rho V^2 \int_0^{2\pi} (4a \sin^3 \theta - 4aK \sin^2 \theta + aK^2 \sin \theta) \, d\theta$$

$$l = \frac{1}{2}\rho V^2 ak[2\theta - \sin 2\theta]_0^{2\pi}$$

$$l = 2\rho V^2 ak\pi = \rho V \Gamma$$

It can be shown that the relation

$$l = \rho V \Gamma \tag{2.34}$$

is valid regardless of the shape of the body.[61]

CHAPTER 3

THEORY OF WING SECTIONS OF FINITE THICKNESS

3.1. Symbols.

A_n, B_n coefficients of the transformation from x' to x

H total pressure

S pressure coefficient $(H - p)/\frac{1}{2}\rho V^2$

V velocity of free stream

a radius of circular cylinder

c_l section lift coefficient

e base of Naperian logarithms, 2.71828

i $\sqrt{-1}$

$\ln$ logarithm to the base e

m source strength per unit length

p $1 - \left(\dfrac{x}{2a}\right)^2 - \left(\dfrac{y}{2a}\right)^2$

p local static pressure

r radius vector of z (modulus)

u component of velocity along x axis

v component of velocity along y axis

v local velocity at any point on the surface of the wing section

w complex variable

x Cartesian coordinate

x real part of the complex variable z or ζ

y Cartesian coordinate

y magnitude of the imaginary part of the complex variable z

z complex variable

z' complex variable in the near-circle plane

z^* complex variable for the flow about a circle whose center is shifted from the center of coordinates

Γ circulation, positive clockwise

α_0 section angle of attack

ϵ distance of the center of the circle from the center of coordinates

ζ complex variable

η magnitude of the imaginary part of the complex variable ζ

θ angular coordinate of the complex variable z (argument)

θ angular coordinate of z' (see Fig. 32)

λ ae^λ is the radius vector of z

μ doublet strength per unit length

ξ real part of the complex variable ζ

π ratio of the circumference of a circle to its diameter

ρ mass density of air

ϕ real part of the complex variable w

ϕ potential function

φ angular coordinate of z
ψ magnitude of the imaginary part of the complex variable w
ψ stream function
ψ ae^ψ is the radius vector of z' (see Fig. 32)
ψ_0 average value of ψ
∞ infinity

3.2. Introduction. It was shown in Chap. 2 that the field of flow about a circular cylinder with circulation in a uniform stream is known. It is possible to relate this field of flow to that about an arbitrary wing section by means of conformal mapping. In relating these fields of flow, the circulation is selected to satisfy the Kutta condition that the velocity at the trailing edge of the section must be finite. Such characteristics as the lift and pressure distribution may then be determined from the known flow about the circular cylinder. The resulting theory permits the approximate calculation of the angle of zero lift, the moment coefficient, the pressure distribution, and the field of flow about the section under conditions where the flow adheres closely to the surface.

3.3. Complex Variables. Conformal mapping, which makes the calculation of wing-section characteristics possible, depends on the use of complex variables. A complex number is a number composed of two parts, a real part and a so-called imaginary part. The real part is just an ordinary number that may have any value. The imaginary part contains the factor $\sqrt{-1}$, which is given the symbol i. A complex variable z may therefore be written in the form

$$z = x + iy$$

where the symbols in the expression $x + iy$ obey all the usual rules of algebra. It should be remembered that i^2 is equal to -1. Such a variable z can be represented conveniently by plotting the real part x as the abscissa of a point and the magnitude y of the imaginary part as the ordinate. By De Moivre's theorem, z may also be written in the form

$$z = re^{i\theta}$$

where r and θ are the polar coordinates.

Let us consider the complex variable

$$w = \phi + i\psi$$

and let $w = f(z)$, that is,

$$\phi + i\psi = f(x + iy)$$

Then

$$\frac{\partial^2\phi}{\partial x^2} + i\frac{\partial^2\psi}{\partial x^2} = f_{xx}(x + iy) = f''(z)$$

and

$$\frac{\partial^2\phi}{\partial y^2} + i\frac{\partial^2\psi}{\partial y^2} = f_{yy}(x + iy) = -f''(z)$$

adding, we have

$$\frac{\partial^2 \phi}{\partial x^2} + \frac{\partial^2 \phi}{\partial y^2} + i\left(\frac{\partial^2 \psi}{\partial x^2} + \frac{\partial^2 \psi}{\partial y^2}\right) = 0$$

In any equation involving complex variables, the real and imaginary parts must be equal to each other independently. Therefore

$$\frac{\partial^2 \phi}{\partial x^2} + \frac{\partial^2 \phi}{\partial y^2} = 0$$

and

$$\frac{\partial^2 \psi}{\partial x^2} + \frac{\partial^2 \psi}{\partial y^2} = 0$$

Because these equations are the same as those of Eqs. (2.15) and (2.16), any differentiable function

$$w = f(z)$$

where

$$w = \phi + i\psi$$

and

$$z = x + iy$$

may be interpreted as a possible case of irrotational fluid motion by giving ϕ and ψ the meaning of velocity potential and stream function, respectively.

The derivative dw/dz has a simple meaning in terms of the velocities in the field of flow.

$$dw = d\phi + i\,d\psi$$
$$dz = dx + i\,dy$$

Further

$$d\phi = \frac{\partial \phi}{\partial x}\,dx + \frac{\partial \phi}{\partial y}\,dy$$

$$d\psi = \frac{\partial \psi}{\partial x}\,dx + \frac{\partial \psi}{\partial y}\,dy$$

Therefore

$$\frac{dw}{dz} = \frac{[(\partial\phi/\partial x)dx + (\partial\phi/\partial y)dy] + i\,[(\partial\psi/\partial x)dx + (\partial\psi/\partial y)dy]}{dx + i\,dy}$$

In order for dw/dz to have a definite meaning, it is necessary that the value of dw/dz be independent of the manner with which dz approaches zero. If dy is assumed to be zero, the value of the differential quotient dw/dz is

$$\frac{\partial \phi}{\partial x} + i\frac{\partial \psi}{\partial x} = u - iv = \frac{dw}{dz} \tag{3.1}$$

according to Eqs. (2.8) and (2.12). Similarly, if dx is assumed to be zero, the value of the differential quotient dw/dz is

$$\frac{1}{i}\frac{\partial \phi}{\partial y} + \frac{\partial \psi}{\partial y} = -iv + u$$

The expressions for simple two-dimensional flows given in Sec. 2.6 may be expressed conveniently in terms of complex variables.
Uniform stream parallel to x axis

$$w = Vz \tag{3.2}$$

Source at the origin

$$w = \frac{m}{2\pi} \ln z \tag{3.3}$$

Doublet at origin with axis along x axis

$$w = \frac{\mu}{2\pi z} \tag{3.4}$$

Circular cylinder of radius a in a uniform stream

$$w = V\left(z + \frac{a^2}{z}\right) \tag{3.5}$$

Vortex at origin

$$w = \frac{i\Gamma}{2\pi} \ln z \tag{3.6}$$

Circular cylinder with circulation

$$w = V\left(z + \frac{a^2}{z}\right) + \frac{i\Gamma}{2\pi} \ln \frac{z}{a} \tag{3.7}$$

3.4. Conformal Transformations. A conformal transformation consists in mapping a region of one plane on another plane in such a manner that the detailed shape of infinitesimal elements of area is not changed. This restriction does not mean that the shape of finite areas cannot be considerably altered. It has been shown previously that the equipotential lines and streamlines intersect at right angles, thus dividing the field of flow into a large number of small rectangles. It has also been shown that the equation

$$w = f(z)$$

represents a possible flow pattern. The equation

$$w = g(\zeta)$$

represents another flow pattern where ζ is the complex variable

$$\xi + i\eta$$

The coordinates in the z plane are considered to be x and y, and those in the ζ plane are ξ and η. If the equipotential lines and streamlines are plotted in either of the planes, they will divide the plane into a large number of small rectangles. These rectangles will be similar at corresponding points in both planes. The corresponding points are found from the relation

$$f(z) = g(\zeta)$$

This equation accordingly represents the conformal transformation from the z plane to the ζ plane or the converse. In practical use, the flow function in the z plane is known and the corresponding flow function in the ζ plane is desired. To plot the flow known on the z plane on the ζ plane, it is necessary to solve this relation for ζ and to obtain an equation in the form

$$\zeta = h(z)$$

In the transformations we shall consider, this relation will be given. The velocities in the z plane, by Eq. (3.1), are

$$\frac{dw}{dz} = u - iv$$

The corresponding velocities in the ζ plane are

$$\frac{dw}{d\zeta} = \frac{dw}{dz}\frac{dz}{d\zeta} \tag{3.8}$$

As a simple example of a conformal transformation, consider the relations

$$w = V\left(z + \frac{a^2}{z}\right) = V\zeta$$

These relations transform the flow about a circular cylinder on the z plane [Eq. (3.5)] to uniform flow parallel to the ξ axis on the ζ plane [Eq. (3.2)]. Corresponding points of both planes are obtained from the relation

$$\zeta = z + \frac{a^2}{z}$$

and are indicated in Fig. 30.

3.5. Transformation of a Circle into a Wing Section. A circle can be transformed into a shape resembling that of a wing section by substitution of the variable

$$\zeta = z + \frac{a^2}{z} \tag{3.9}$$

into the expression for the flow about a circular cylinder having a radius slightly larger than a, and so placed that the circumference passes through the point $x = a$. If, in addition, the center of the larger cylinder is placed on the x axis, the transformed curve will be that of a symmetrical wing section (Fig. 31). In the present example, let the center of the larger cylinder be placed at the point $x = -\epsilon$ where ϵ is a real quantity. The radius of this cylinder will then be $a + \epsilon$. The equation of flow about the larger cylinder with circulation is then

$$w = V\left(z^* + \epsilon + \frac{(a + \epsilon)^2}{z^* + \epsilon}\right) + \frac{i\Gamma}{2\pi}\ln\frac{z^* + \epsilon}{a + \epsilon}$$

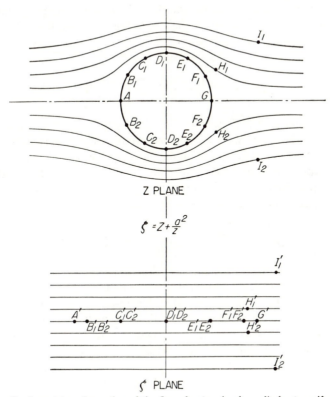

FIG. 30. Conformal transformation of the flow about a circular cylinder to uniform flow.

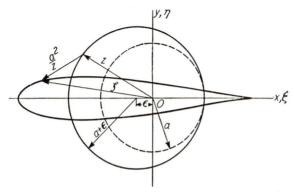

FIG. 31. Conformal transformation of a circle into a symmetrical wing section.

The more general expression for the flow about the circular cylinder with the flow at infinity inclined at an angle α_0 to the x axis is found by substituting the expression

$$z + \epsilon = (z^* + \epsilon)e^{i\alpha_0}$$

$$w = V\left[(z + \epsilon)e^{-i\alpha_0} + \frac{(a + \epsilon)^2 e^{i\alpha_0}}{z + \epsilon}\right] + \frac{i\Gamma}{2\pi}\ln\frac{(z + \epsilon)e^{-i\alpha_0}}{a + \epsilon} \qquad (3.10)$$

Substitution of Eq. (3.9) into Eq. (3.10) would result in the equation for the flow about the wing section but would lead to a complicated expression. A simple way of obtaining the shape of the wing section is to select values of z corresponding to points on the larger cylinder and find the corresponding points in the ζ plane by the use of Eq. (3.9). The velocity of any point on the wing section can be found from Eq. (3.8).

$$\frac{dw}{d\zeta} = \frac{dw}{dz}\frac{dz}{d\zeta} = \left[V\left(e^{-i\alpha_0} - \frac{(a + \epsilon)^2 e^{i\alpha_0}}{(z + \epsilon)^2}\right) + \frac{i\Gamma}{2\pi(z + \epsilon)}\right]\left(\frac{z^2}{z^2 - a^2}\right)$$

It can be seen from this equation that the velocity at the point $z = a$ is infinite unless the first factor is zero. The point $z = a$ corresponds to the trailing edge of the wing section. The Kutta-Joukowsky condition states that the value of the circulation is such as to make the first factor equal to zero, which is the condition that ensures smooth flow at the trailing edge. The value of the circulation satisfying this condition is found as follows:

$$V\left[e^{-i\alpha_0} - \frac{(a + \epsilon)^2 e^{i\alpha_0}}{(a + \epsilon)^2}\right] + \frac{i\Gamma}{2\pi(a + \epsilon)} = 0$$

$$i\Gamma = 2\pi(a + \epsilon)V(e^{i\alpha_0} - e^{-i\alpha_0})$$

Since

$$\frac{e^{i\alpha_0} - e^{-i\alpha_0}}{2} = \sinh i\alpha_0 = i\sin\alpha_0$$

$$i\Gamma = 4\pi(a + \epsilon)Vi\sin\alpha_0$$

or

$$\Gamma = 4\pi(a + \epsilon)V\sin\alpha_0$$

The leading edge of the wing section corresponds to the point

$$\zeta = -a - 2\epsilon - \frac{a^2}{a + 2\epsilon}$$

neglecting powers of ϵ greater than one

$$\zeta = -2a$$

Because the trailing edge corresponds to the point

$$\zeta = 2a$$

the chord of the wing section is $4a$. The lift on the wing section is $\rho V \Gamma$, and the lift coefficient is

$$c_l = 2\pi \left(1 + \frac{\epsilon}{a}\right) \sin \alpha_0$$

In the limiting case as ϵ/a approaches zero, the slope of the lift curve $dc_l/d\alpha_0$ is 2π per radian for small angles of attack. Detailed computations[37]

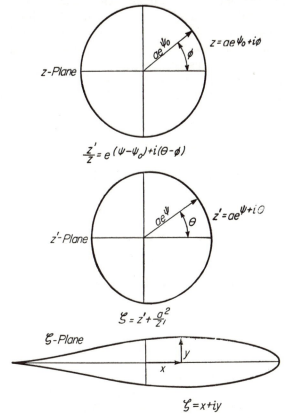

Fig. 32. Illustration of transformations used to derive airfoils and calculate pressure distributions.

will show that the thickness ratio of the wing section is nearly equal to $(3\sqrt{3}/4)(\epsilon/a)$. For wing sections approximately 12 per cent thick, the theoretical lift-curve slope is about nine per cent greater than its limiting value for thin sections.

3.6. Flow about Arbitrary Wing Sections. It was seen in the previous section that the transformation

$$\zeta = z + \frac{a^2}{z}$$

transforms a circle in the z plane into a curve resembling a wing section in the ζ plane. Most wing sections have a general resemblance to each other. If the aforementioned transformation is applied to a wing section, the resulting curve in the z plane will therefore be nearly circular in shape. Theodorsen recognized this fact and showed that the flow about the nearly circular curve, and hence about the wing section, can be derived from the flow about the true circle by a rapidly converging process. The basic method is presented in reference 122, and a detailed discussion of the method is given in reference 125. The derivation of Theodorsen's relation for the velocity distribution about the wing is divided into three parts (Fig. 32).

1. Derivation of relations between the flow in the plane of the wing section (ζ plane) and in the plane of the near circle (z' plane).

2. Derivation of relations between the flow in the z' plane and in the plane of the true circle (z plane).

3. Combination of the foregoing relations to obtain the final expression for the velocity distribution in the ζ plane in terms of the ordinates of the wing section.

The basic relation between the z' plane and the ζ plane is

$$\zeta = z' + \frac{a^2}{z'} \tag{3.11}$$

The coordinates of ζ are defined by the relation

$$\zeta = x + iy$$

and the coordinates of z' are defined by the relation

$$z' = ae^{\psi + i\theta}$$

By Eq. (3.11),

$$\zeta = ae^{\psi + i\theta} + ae^{-\psi - i\theta}$$

Since

$$e^{i\theta} = \cos \theta + i \sin \theta$$
$$\zeta = a(e^{\psi} + e^{-\psi}) \cos \theta + ia(e^{\psi} - e^{-\psi}) \sin \theta$$
$$= 2a \cosh \psi \cos \theta + 2ia \sinh \psi \sin \theta$$

and since

$$\left.\begin{aligned}
\zeta &= x + iy \\
x &= 2a \cosh \psi \cos \theta \\
y &= 2a \sinh \psi \sin \theta
\end{aligned}\right\} \tag{3.12}$$

Expressions for ψ and θ in terms of x and y are desired and may be obtained as follows:

$$\cosh \psi = \frac{x}{2a \cos \theta}$$

$$\sinh \psi = \frac{y}{2a \sin \theta}$$

Since

$$\cosh^2 \psi - \sinh^2 \psi = 1$$

$$\left(\frac{x}{2a \cos \theta}\right)^2 - \left(\frac{y}{2a \sin \theta}\right)^2 = 1$$

or

$$2 \sin^2 \theta = p + \sqrt{p^2 + \left(\frac{y}{a}\right)^2} \tag{3.13}$$

where

$$p = 1 - \left(\frac{x}{2a}\right)^2 - \left(\frac{y}{2a}\right)^2$$

Because sin θ, and hence cos θ, are known in terms of the ordinates x and y of the wing section, the values of sinh ψ and cosh ψ can be found from Eq. (3.12).

The factor relating velocities in the z' plane to those in the ζ plane is $dz'/d\zeta$. From Eq. (3.11),

$$\frac{d\zeta}{dz'} = 1 - \frac{a^2}{z'^2} = \frac{1}{z'}\left(z' - \frac{a^2}{z'}\right)$$

$$= \frac{1}{z'}\left(ae^{\psi + i\theta} - ae^{-\psi - i\theta}\right)$$

$$= \frac{1}{z'}\, a(e^\psi - e^{-\psi}) \cos \theta + ia(e^\psi + e^{-\psi}) \sin \theta$$

$$= \frac{1}{z'}\, (2a \sinh \psi \cos \theta + 2ia \cosh \psi \sin \theta) \tag{3.14}$$

The second step is to find the relation between the flows in the z' plane and those in the z plane. The coordinates of z are defined by the relation

$$z = ae^{\lambda + i\varphi}$$

The transformation relating the z' to the z plane is the general transformation

$$z' = ze^{\sum\limits_{1}^{\infty}(A_n + iB_n)(1/z^n)}$$

By definition

$$z' = ze^{\psi - \lambda + i(\theta - \varphi)}$$

Consequently

$$\psi - \lambda + i(\theta - \varphi) = \sum_{1}^{\infty} (A_n + iB_n)\frac{1}{z^n}$$

or

$$\psi - \lambda + i(\theta - \varphi) = \sum_{1}^{\infty} (A_n + iB_n)\frac{1}{r^n} (\cos n\varphi - i \sin n\varphi)$$

where z has been expressed in polar form

$$z = r(\cos \varphi + i \sin \varphi)$$

Equating the real and imaginary parts, we obtain the two Fourier expansions

$$\psi - \lambda = \sum_{1}^{\infty} \left(\frac{A_n}{r^n} \cos n\varphi + \frac{B_n}{r^n} \sin n\varphi \right)$$

and

$$\theta - \varphi = \sum_{1}^{\infty} \left(\frac{B_n}{r^n} \cos n\varphi - \frac{A_n}{r^n} \sin n\varphi \right)$$

(3.15)

These relations show that $\psi - \lambda$ and $\theta - \varphi$ are conjugate functions. In order for the deviation of the near circle from the true circle to be a minimum, the value of λ corresponding to the radius of the true circle is taken to be

$$\lambda = \psi_0 = \frac{1}{2\pi} \int_0^{2\pi} \psi \, d\varphi$$

where the values of ψ correspond to points on the near circle. The required transformation from the near circle to the true circle is then

$$\frac{z'}{z} = a e^{\psi - \psi_0 + i(\theta - \varphi)}$$

(3.16)

If ψ is known as a function of φ, $\theta - \varphi$ can be found by a method developed by Naiman.[73] The ψ function is assumed to be approximated by a finite trigonometric series of the form

$$\psi(\varphi) = \psi_0 + A_1 \cos \varphi + \cdots + A_{n-1} \cos (n - 1)\varphi$$
$$+ A_n \cos \varphi + B_1 \sin \varphi + \cdots + B_{n-1} \sin (n - 1)\varphi$$

$$\psi(\varphi) = \psi_0 + \sum_{m=1}^{n-1} (A_m \cos m\varphi + B_m \sin m\varphi) + A_n \cos n\varphi$$

If ψ is specified at $2n$ equally spaced intervals in the range $0 \leq \varphi \leq 2\pi$, that is, $0, \pi/n, 2\pi/n, \ldots, [(2n - 1)\pi]/n$ then

$$\psi_0 = \frac{1}{2n} \sum_{r=0}^{2n-1} \psi_r$$

where ψ_r = value of ψ at $\varphi = r\pi/n$

$$A_m = \frac{1}{n} \sum_{r=0}^{2n-1} \psi_r \cos m \frac{r\pi}{n}$$

$$B_m = \frac{1}{n} \sum_{r=0}^{2n-1} \psi_r \sin m \frac{r\pi}{n}$$

$$A_n = \frac{1}{n} \sum_{r=0}^{2n-1} (-1)^r \psi_r \qquad B_n = 0$$

Now, by Eqs. (3.15),

$$\varphi - \theta = \epsilon(\varphi) = \sum_{m=1}^{n-1} (A_m \sin m\varphi - B_m \cos m\varphi) + A_n \sin n\varphi$$

$$= \frac{1}{n} \sum_{m=1}^{n-1} \left(\sin m\varphi \sum_{r=0}^{2n-1} \psi_r \cos m \frac{r\pi}{n} \right.$$

$$\left. - \cos m\varphi \sum_{r=0}^{2n-1} \psi_r \sin m \frac{r\pi}{n} \right) + \frac{1}{2n} \sin n\varphi \sum_{r=0}^{2n-1} (-1)^r \psi_r$$

$$= \frac{1}{n} \sum_{m=1}^{n-1} \sum_{r=0}^{2n-1} \psi_r \sin m \left(\varphi - \frac{r\pi}{n} \right) + \frac{1}{2n} \sin n\varphi \sum_{r=0}^{2n-1} (-1)^r \psi_r$$

Upon interchanging the order of summation, there is obtained

$$\epsilon(\varphi) = \frac{1}{n} \sum_{r=0}^{2n-1} \psi_r \sum_{m=1}^{n-1} \sin m \left(\varphi - \frac{r\pi}{n} \right) + \frac{1}{2n} \sin n\varphi \sum_{r=0}^{2n-1} (-1)^r \psi_r$$

If ϵ is evaluated at the same points φ at which ψ is given, that is, at the points $\varphi = r'\pi/n$, the variable $\varphi - (r\pi/n)$ becomes $[(r' - r)\pi]/n = -(K\pi/n)$, and the last term becomes zero.

$$\psi_r = \psi \frac{r\pi}{n} = \psi \left(\frac{r'\pi}{n} + \frac{K\pi}{n} \right) = \psi \left(\varphi + \frac{K\pi}{n} \right) \equiv \psi_K$$

$$\epsilon(\varphi) = -\frac{1}{n} \sum_{K=0}^{2n-1} \psi_K \sum_{m=1}^{n-1} \sin m \frac{K\pi}{n}$$

The summation over K may also be taken from 0 to $2n - 1$ because of the periodicity of ψ_K. A simple expression can be obtained for the coefficients of ψ_K

$$\sum_{m=1}^{n-1} \sin m \frac{K\pi}{n} = \begin{cases} \cot \dfrac{K\pi}{2n} & K \text{ odd} \\ 0 & K \text{ even} \end{cases}$$

and therefore

$$\epsilon(\varphi) = -\frac{1}{n} \sum_{K=1}^{2n-1} \psi_K \cot \frac{K\pi}{2n} \qquad K \text{ odd}$$

or

$$\epsilon(\varphi) = \frac{1}{n} \sum_{K=1}^{n} (\psi_{-K} - \psi_K) \cot \frac{K\pi}{2n} \qquad K \text{ odd} \qquad (3.17)$$

$$\psi_K \equiv \psi \left(\varphi + \frac{K\pi}{n} \right)$$

In most cases a value of n equal to 40 gives sufficiently accurate results. Ordinarily ψ is known as a function of θ, and first approximations to the values of ψ_0 and $\theta - \varphi$ are obtained by substitution of θ for φ in the relations for $\theta - \varphi$ and ψ_0.

The factor relating the velocities in the true-circle plane to those in the near-circle plane is dz'/dz. From Eq. (3.16),

$$\frac{dz'}{dz} = z' \left\{ \frac{1}{z} + \frac{d}{dz} [(\psi - \psi_0) + i(\theta - \varphi)] \right\}$$

$$= z' \frac{d}{dz} [\psi + i(\theta - \varphi) + \ln z]$$

But, on the true circle,

$$z = ae^{\psi_0 + i\varphi}$$

from which

$$\frac{1}{z} = \frac{d}{dz} \ln z = \frac{d}{dz} (\ln a + \psi_0 + i\varphi) = \frac{d}{dz} (i\varphi)$$

Therefore

$$\frac{dz'}{dz} = z' \frac{d}{dz} [\psi + i(\theta - \varphi) + i\varphi]$$

$$\frac{dz'}{dz} = z' \frac{d}{dz} (\psi + i\theta)$$

This expression may be written

$$\frac{dz'}{dz} = z' \frac{d}{d\theta} (\psi + i\theta) \frac{d\theta}{dz}$$

But

$$\frac{1}{z} = i \frac{d\varphi}{dz}$$

or

$$\frac{dz}{z} = i\, d\varphi = i\, d(\varphi - \theta) + i\, d\theta$$

and

$$\frac{dz}{d\theta} = iz \left[1 + \frac{d(\varphi - \theta)}{d\theta} \right]$$

Hence

$$\frac{dz'}{dz} = \frac{z'}{z} \frac{d}{d\theta} (-i\psi + \theta) \frac{1}{1 + (d\epsilon/d\theta)}$$

where

$$\epsilon = \varphi - \theta$$

or

$$\frac{dz'}{dz} = \frac{z'}{z} \frac{1 - i(d\psi/d\theta)}{1 + (d\epsilon/d\theta)} \qquad (3.18)$$

The flow in the plane of the true circle of radius ae^{ψ_0} is described by the following equation:

$$w = V\left(z + \frac{a^2 e^{2\psi_0}}{z}\right) + \frac{i\Gamma}{2\pi} \ln \frac{z}{ae^{\psi_0}} \tag{3.19}$$

where the value of a in Eq. (3.7) is here taken to be ae^{ψ_0}. The velocities in the true-circle plane are

$$\frac{dw}{dz} = V\left(1 - \frac{a^2 e^{2\psi_0}}{z^2}\right) + \frac{i\Gamma}{2\pi z} \tag{3.20}$$

In order for the rear stagnation point of the circular cylinder to correspond to the trailing edge of the wing section, the flow about the cylinder is rotated through an angle α_0 equal to the angle of attack of the wing section. This process corresponds to the application of the Kutta-Joukowski condition. At zero angle of attack, the trailing edge corresponds to the point

$$z = ae^{\psi_0 + i\epsilon_T}$$

It is therefore necessary for the circulation Γ to have a value corresponding to a rotation of the stagnation point by an amount $\alpha_0 + \epsilon_T$. This value may be seen from Eq. (2.30) to be

$$\Gamma = 4\pi ae^{\psi_0}V \sin (\alpha_0 + \epsilon_T)$$

Substituting the general value of z on the surface of the cylinder into Eq. (3.20)

$$z = ae^{\psi_0 + i(\alpha_0 + \varphi)}$$

$$\frac{dw}{dz} = V[1 - e^{-2i(\alpha_0 + \varphi)} + 2i \sin (\alpha_0 + \epsilon_T)e^{-i(\alpha_0 + \varphi)}]$$

from which the absolute value of dw/dz may be obtained

$$\left|\frac{dw}{dz}\right| = 2V[\sin (\alpha_0 + \varphi) + \sin (\alpha_0 + \epsilon_T)] \tag{3.21}$$

The third step is to obtain the velocities in the plane of the wing section from the expression for the velocities in the plane of the true circle. The expression for the velocity in the plane of the wing section is

$$\frac{dw}{d\zeta} = \frac{dw}{dz}\frac{dz}{dz'}\frac{dz'}{d\zeta}$$

Multiplying Eqs. (3.14) and (3.18), we obtain

$$\frac{d\zeta}{dz'}\frac{dz'}{dz} = \frac{1}{z}\left[\frac{1 - i(d\psi/d\theta)}{1 + (d\epsilon/d\theta)}\right](2a \sinh \psi \cos \theta + 2ia \cosh \psi \sin \theta)$$

The absolute value of

$$\left|\frac{d\zeta}{dz}\right| = \frac{2\sqrt{1 + (d\psi/d\theta)^2}}{e^{\psi_0}[1 + (de/d\theta)]}\sqrt{\sinh^2 \psi + \sin^2 \theta}$$

Dividing Eq. (3.21) by this expression,

$$\left|\frac{dw}{d\zeta}\right| = v = V \frac{[\sin(\alpha_0 + \varphi) + \sin(\alpha_0 + \epsilon_T)][1 + (d\epsilon/d\theta)]e^{\psi_0}}{\sqrt{(\sinh^2\psi + \sin^2\theta)[1 + (d\psi/d\theta)^2]}} \tag{3.22}$$

where v = local velocity at any point on surface of wing section

V = free-stream velocity

The necessary calculations to obtain the pressure distribution for a given wing section are as follows:

1. The coordinates of the wing section are found with respect to a line joining a point midway between the nose of the section and its center of curvature and its trailing edge for sections having sharp trailing edges.[122] The coordinates of these points are taken as $(-2a, 0)$ and $(2a, 0)$, respectively. It is usually convenient to let a have the value of unity.

2. $\sin\theta$ is found from Eq. (3.13).

3. $\sinh\psi$ is found from Eqs. (3.12).

4. ψ is found from tables of hyperbolic functions.

5. ψ is plotted as a function of θ.

6. ψ_0 is determined from the relation

$$\psi_0 = \frac{1}{2\pi}\int_0^{2\pi}\psi\,d\theta$$

7. A first approximation to ϵ is obtained by conjugating the curve of ψ plotted against θ using Eq. (3.17).

8. From the curves of ϵ and ψ plotted against θ, $d\epsilon/d\theta$ and $d\psi/d\theta$ are found.

9. Determine F by the relation

$$F = \frac{[1 + (d\epsilon/d\theta)]e^{\psi_0}}{\sqrt{(\sinh^2\psi + \sin^2\theta)[1 + (d\psi/d\theta)^2]}}$$

10. $\dfrac{v}{V} = F[\sin(\theta + \alpha_0 + \epsilon) + \sin(\alpha_0 + \epsilon_T)]$

11. The pressure coefficient $S = (v/V)^2$.

For most purposes the first approximation to ϵ is sufficiently accurate. If greater accuracy is desired, a second approximation may be found by plotting ψ against $\theta + \epsilon$ and repeating steps 6 and 7 before proceeding.

3.7. Empirical Modification of the Theory. Although the foregoing theory for computing the pressure distribution about arbitary wing sections is exact for perfect fluid flow, the presence of viscous effects in actual flows leads to discrepancies. Even when the flow is not separated from the surface, the thickness of the boundary layer effectively distorts the shape of the section. One result of this distortion is that the theoretical slope of the lift curve is not realized. A comparison of the actual lift and moment characteristics with the theoretical ones is made in Fig. 33 for the NACA 4412 wing section.[82] This comparison is typical in that the experimental lift coefficient is lower than the theoretical value at a given angle of attack and the theoretical value of the pitching-moment coefficient is not realized.

Several methods suggest themselves for bringing the theory into closer agreement with experiment. It is obvious that the lift coefficient can be

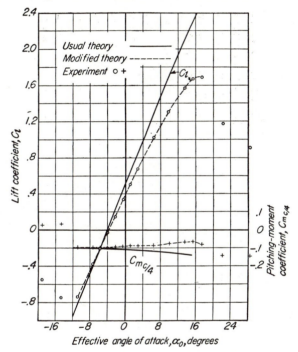

FIG. 33. Comparison between theoretical and measured lift and pitching-moment coefficients, NACA 4412 wing section.

brought into agreement by reducing the theoretical angle of attack. This method seriously alters the flow about the leading edge, leading to discrepancies in the pressure distribution over the forward portion of the wing section. Another method tried by Pinkerton[82] is to reduce the circulation to the required value by disregarding the Kutta condition. This method leads to fair agreement over the greater part of the wing section (Fig. 34) but, as would be expected, leads to infinite velocities at the trailing edge.

Pinkerton[82] found that fairly satisfactory agreement with experiment could be obtained by effectively distorting the shape of the section (Fig. 34). This distortion is affected by finding an increment $\Delta \epsilon_T$ required to avoid infinite velocities at the trailing edge when the circulation is adjusted to produce the experimentally observed lift coefficient. The altered ϵ function ϵ_α is arbitrarily assumed to be given by

$$\epsilon_\alpha = \epsilon + \frac{\Delta\epsilon_T}{2}(1 - \cos\theta) \qquad (3.23)$$

where ϵ = original value computed by methods of preceding section
        ϵ_α = modified value used to compute flow about distorted section

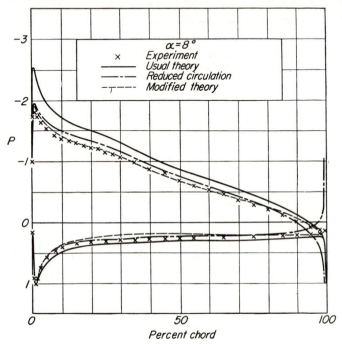

Fig. 34. Comparison between measured and various theoretical pressure distributions NACA 4412 wing section.

The agreement obtained by this method is indicated by Fig. 34 and the moment coefficients of Fig. 33.

3.8. Design of Wing Sections. Section 3.6 presents a method for obtaining the pressure distribution for arbitrary wing sections. A knowledge of the pressure distribution is desirable for structural design and for the estimation of the critical Mach number and the moment coefficient if tests are not available. The pressure distribution also exerts a strong or predominant influence on the boundary-layer flow and, hence, on the section characteristics. It is therefore usually advisable to relate the aerodynamic characteristics to the pressure distribution rather than directly to the geometry of the wing section. In the experimental development of wing sections, it is accordingly desirable to have a method of determining changes of shape of the section corresponding to desired changes in the pressure distribution. A method that has been used successfully for the development of the newer NACA wing sections is presented in reference 3.

Equation (3.22) gives the relation between the velocity distribution over the wing section and the shape parameters ψ and θ. It is shown by Theodorsen and Garrick[125] that an alternate expression is

$$\frac{v}{V} = \frac{\sin (\alpha_0 + \varphi) + \sin (\alpha_0 + \epsilon_T)e^{\psi_0}}{\sqrt{(\sinh^2 \psi + \sin^2 \theta)\{[1 - (d\epsilon/d\varphi)]^2 + [(d\psi/d\varphi)]^2\}}} \tag{3.24}$$

Basic symmetrical shapes are derived by assuming suitable values of $d\epsilon/d\varphi$ as a function of φ. These values are chosen on the basis of previous experience and are subject to the conditions that

$$\int_0^\pi \frac{d\epsilon}{d\varphi} = 0$$

and $d\epsilon/d\varphi$ at φ is equal to $d\epsilon/d\varphi$ at $-\varphi$. These conditions are necessary for obtaining closed symmetrical shapes. Values of $\epsilon(\varphi)$ are obtained by integrating $(d\epsilon/d\varphi)\ d\varphi$. Values of $\psi(\varphi)$ are found by obtaining the conjugate curve of $\epsilon(\varphi)$ by means of Eq. (3.17) and adding an arbitrary value ψ_0 sufficient to make the value of ψ equal to or slightly greater than zero at $\varphi = \pi$. This condition assures a sharp trailing-edge shape. If ψ equals zero at $\varphi = \pi$, the wing section will have a cusplike trailing edge. Slightly positive values of ψ result in more conventional trailing-edge shapes. Theodorsen[123] has shown that small changes in the velocity distribution at any point on the surface are approximately proportional to $1 + (d\epsilon/d\varphi)$. The initially assumed values of $d\epsilon/d\varphi$ are accordingly altered by a process of successive approximations until the desired type of velocity distribution is obtained. After the final values of ψ and ϵ are obtained, the ordinates of the symmetrical section are computed by Eq. (3.12).

Although a similar procedure may be used to design cambered sections, it has been found convenient to design symmetrical sections which may then be cambered by a method to be described in Sec. 4.5. This method is satisfactory for thin or moderately thick sections. Very thick sections must be designed directly with the desired camber. Goldstein[37a] has also developed useful approximate methods for the design of cambered wing sections.

CHAPTER 4

THEORY OF THIN WING SECTIONS

4.1. Symbols.

A coefficients of Fourier series

A multiplier for obtaining the angle of zero lift

B multiplier for obtaining the pitching-moment coefficient

L lower surface ordinate in fraction of the chord

P_R local load coefficient

S pressure coefficient $(H - p)/\frac{1}{2}\rho V^2$

U upper surface ordinate in fraction of the chord

V velocity of the free stream

a mean-line designation; fraction of the chord from leading edge over which loading is uniform at the ideal angle of attack

c chord

c_l section lift coefficient

c_{l_0} value of c_l corresponding to calculated value of $\Delta v_a/V$

c_{l_i} ideal lift coefficient

$c_{m_{LE}}$ section pitching-moment coefficient about the leading edge

$c_{m_{c/4}}$ section pitching-moment coefficient about the quarter-chord point

e base of Naperian logarithm, 2.71828

k multiplier for obtaining the angle of zero lift

l section lift

$\ln$ logarithm to the base e

m_{LE} section pitching moment about the leading edge

p local static pressure

v local velocity over the surface of a symmetrical wing section at zero lift

v_n component of velocity normal to the chord

Δv increment of local velocity over the surface of a wing section associated with camber

Δv_a increment of local velocity over the surface of a wing section associated with angle of attack

x distance along chord

x_1 fixed point on the x axis

y ordinate of the mean line

Γ circulation

α_0 section angle of attack

α_i ideal angle of attack

$\alpha_{L=0}$ angle of zero lift

β_0 quantity defined by Eq. (4.13)

γ circulation per unit length along chord

θ angle whose cosine is $1 - (2x/c)$

θ_1 angle whose cosine is $1 - (2x_1/c)$

μ_0 quantity defined by Eq. (4.14)

π ratio of the circumference of a circle to its diameter

ρ mass density of air

∞ infinity

4.2. Basic Concepts. Many of the properties of wing sections are primarily functions of the shape of the mean line. The mean line is considered to be the locus of points situated halfway between the upper and lower surfaces of the section, these distances being measured normal to the mean line. Although the mean line of an arbitrary wing section is rather difficult to obtain, the construction of cambered wing sections from given mean lines and symmetrical thickness distributions is relatively easy. Among the properties mainly associated with the shape of the mean line are

1. The chordwise load distribution.
2. The angle of zero lift.
3. The pitching-moment coefficient.

Other important results obtained from the theory of thin wing sections

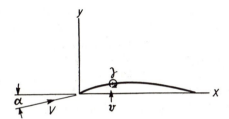

FIG. 35. Configuration for analysis of mean lines.

are a value of the slope of the lift curve and the approximate position of the aerodynamic center.

The theory of thin wing sections was developed by Munk.[71] Later contributions to the theory were made by Birnbaum, Glauert,[37] Theodorsen,[121] and Allen.[7] The present treatment follows that given by Glauert.[37]

For the purposes of the theory, the wing section is considered to be replaced by its mean line. The chordwise distribution of load is assumed to be connected with a chordwise distribution of vortices.

Let γ be the difference in velocity between the upper and lower surfaces. The total circulation around the section is then given by the relation

$$\Gamma = \int_0^c \gamma \, dx \tag{4.1}$$

where c is the chord and the camber is assumed to be sufficiently small for distances along the chord line to be substantially equal to those along the mean line. The configuration for the analysis is given in Fig. 35.

The vertical components of velocity caused by an element of the vortex distribution along the mean line have the magnitudes [see Eq. (2.27)]

$$\frac{\gamma \, dx}{2\pi(x - x_1)}$$

where x is the position of the element and x_1 is the point where velocity is being calculated. The total vertical component of velocity at any point x_1 is then

$$v_n(x_1) = \int_0^c \frac{\gamma \, dx}{2\pi(x - x_1)} \tag{4.2}$$

where v_n = component of velocity normal to chord line

The thin wing section must of course be a streamline. There is no flow through it. The equation expressing this condition is

$$\alpha_0 + \frac{v_n}{V} = \frac{dy}{dx} \tag{4.3}$$

for small angles. Equations (4.2) and (4.3) are the fundamental relations between the shape of the mean line and its aerodynamic characteristics.

It is possible to obtain the characteristics of arbitary thin wing sections from these relations. The pressure distribution or the chordwise load distribution may, however, be obtained more accurately and just as conveniently by the method presented in Sec. 3.6.

4.3. Angle of Zero Lift and Pitching Moment. Simple relations for the angle of zero lift and the pitching moment of arbitrary thin wing sections may be obtained by the application of a Fourier method of analysis to Eqs. (4.2) and (4.3).

Let

$$x = \frac{c}{2}(1 - \cos \theta) \tag{4.4}$$

then

$$dx = \frac{c}{2} \sin \theta \, d\theta$$

Assume that γ may be represented by a trignometric series as follows:

$$\gamma = 2V\left(A_0 \cot \frac{\theta}{2} + \sum_{n=1}^{\infty} A_n \sin n\theta\right) \tag{4.5}$$

When γ is expressed in this manner, it will be shown later that the coefficient A_0 depends only on the angle of attack, and the coefficients A_n depend only on the shape of the mean line. Since

$$\cot \frac{\theta}{2} = \frac{1 + \cos \theta}{\sin \theta}$$

$$\gamma \, dx = cV[A_0(1 + \cos \theta) + \sum_1^{\infty} A_n \sin n\theta \sin \theta] d\theta$$

The lift is

$$l = \int_0^c \rho V \gamma \, dx$$

$$= \int_0^\pi c \rho V^2 [A_0(1 + \cos \theta) + \sum_1^\infty A_n \sin n\theta \sin \theta] \, d\theta$$

$$= \pi c \rho V^2 (A_0 + \tfrac{1}{2} A_1) \tag{4.6}$$

and the lift coefficient

$$c_l = \frac{l}{\tfrac{1}{2}\rho V^2 c} = 2\pi \left(A_0 + \frac{1}{2} A_1 \right) \tag{4.7}$$

The expression for the pitching moment about the leading edge is given by the relation

$$m_{LE} = -\int_0^c \rho V \gamma x \, dx$$

Since

$$\sin \theta (1 - \cos \theta) = \sin \theta - \sin \theta \cos \theta$$
$$= \sin \theta - \tfrac{1}{2} \sin 2\theta$$

$$m_{LE} = -\int_0^\pi \frac{c^2}{2} \rho V^2 \left[A_0(1 - \cos^2 \theta) + \sum_1^\infty A_n \sin n\theta \left(\sin \theta - \frac{1}{2} \sin 2\theta \right) \right] d\theta$$

Upon integration

$$m_{LE} = -\frac{\pi}{4} c^2 \rho V^2 \left(A_0 + A_1 - \frac{1}{2} A_2 \right)$$

$$c_{m_{LE}} = \frac{\pi}{4} (A_2 - A_1) - \frac{1}{4} c_l \tag{4.8}$$

Equations (4.7) and (4.8) provide simple relations between the lift and moment and the first few coefficients of the Fourier series. The next step is to determine the relations between the coefficients of the Fourier series and the shape of the mean line.

Substituting Eqs. (4.4) and (4.5) into Eq. (4.2), the expression for the vertical component of velocity becomes

$$v_n(\theta_1) = \frac{V}{\pi} \int_0^\pi \left\{ \frac{A_0(1 + \cos \theta)}{\cos \theta_1 - \cos \theta} + \frac{\tfrac{1}{2} \sum_1^\infty A_n [\cos (n - 1)\theta - \cos (n + 1)\theta]}{\cos \theta_1 - \cos \theta} \right\} d\theta$$

since

$$\tfrac{1}{2}[\cos (n - 1)\theta - \cos (n + 1)\theta] = \sin n\theta \sin \theta$$

It is shown by Glauert[37] that

$$\int_0^\pi \frac{\cos n\theta}{\cos \theta - \cos \phi} \, d\theta = \pi \frac{\sin n\phi}{\sin \phi} \tag{4.9}$$

Consequently

$$v_n(\theta_1) = V\left[-A_0 + \frac{1}{2}\sum_1^\infty A_n \frac{\sin(n+1)\theta_1 - \sin(n-1)\theta_1}{\sin\theta_1}\right]$$

and, dropping the subscript,

$$\frac{v_n}{V} = -A_0 + \sum_1^\infty A_n \cos n\theta \qquad (4.10)$$

The required relation between the Fourier coefficients and the shape of the mean line is obtained by substitution of Eq. (4.10) in Eq. (4.3)

$$\frac{dy}{dx} = \alpha_0 - A_0 + \sum_1^\infty A_n \cos n\theta \qquad (4.11)$$

The standard expressions for the Fourier series coefficients are then

$$\left. \begin{array}{l} \alpha_0 - A_0 = \dfrac{1}{\pi}\displaystyle\int_0^\pi \dfrac{dy}{dx}\,d\theta \\[3mm] A_n = \dfrac{2}{\pi}\displaystyle\int_0^\pi \dfrac{dy}{dx}\cos n\theta\,d\theta \end{array} \right\} \qquad (4.12)$$

Equations (4.12) for A_0, A_1, and A_2 can be transformed in such a manner that the angle of zero lift and the pitching moment are expressed in terms of the ordinates of the mean line rather than in terms of its slope. The quantities β_0 and μ_0 defined as follows are convenient in effecting the change.

$$\beta_0 = \frac{2}{\pi}\int_0^\pi \frac{y}{c}\frac{d\theta}{1+\cos\theta} \qquad (4.13)$$

and

$$\mu_0 = \int_0^\pi \frac{y}{c}\cos\theta\,d\theta \qquad (4.14)$$

The value of β_0 in terms of the coefficients of the original Fourier series may be found by the following process.

Integrating Eq. (4.13) by parts we obtain

$$\beta_0 = \frac{2}{\pi}\left[\frac{y}{c}\sqrt{\frac{1-\cos\theta}{1+\cos\theta}}\right]_0^\pi - \frac{2}{\pi}\int_0^\pi \frac{1}{c}\frac{dy}{dx}\frac{dx}{d\theta}\sqrt{\frac{1-\cos\theta}{1+\cos\theta}}\,d\theta$$

(reference 80, formulas 296 and 578). The first term of this expression equals zero if y approaches zero at $\theta = \pi$ faster than $\sqrt{c-x}$. This is true for nearly all wing sections. From Eq. (4.4),

$$\frac{dx}{d\theta} = \frac{c}{2}\sin\theta = \frac{c}{2}\sqrt{(1+\cos\theta)(1-\cos\theta)}$$

Hence

$$\beta_0 = -\frac{1}{\pi}\int_0^\pi \frac{dy}{dx}(1-\cos\theta)d\theta$$

and, from Eq. (4.12),

$$\beta_0 = A_0 - \alpha_0 + \tfrac{1}{2}A_1 \tag{4.15}$$

The value of μ_0 in terms of the Fourier coefficients A_n is found as follows: Integrating Eq. (4.14) by parts,

$$\mu_0 = \left[\frac{y}{c}\,sin\,\theta\right]_0^\pi - \int_0^\pi \frac{1}{c}\frac{dy}{dx}\frac{dx}{d\theta}\sin\theta\,d\theta$$

$$= -\int_0^\pi \frac{1}{4}\frac{dy}{dx}(1-\cos 2\theta)d\theta$$

$$= -\frac{\pi}{4}\left(\alpha_0 - A_0 - \frac{1}{2}A_2\right) \tag{4.16}$$

Comparing Eq. (4.7) with Eq. (4.15)

$$c_l = 2\pi(\alpha_0 + \beta_0) \tag{4.17}$$

The angle of zero lift $\alpha_{L=0}$ is therefore $-\beta_0$, and the slope of the lift curve according to Eq. (4.17) is 2π per radian.

The equation for the pitching-moment coefficient about the leading edge is found by substituting Eqs. (4.15) and (4.16) into Eq. (4.8)

$$c_{m_{LE}} = \left(2\mu_0 - \frac{\pi}{2}\beta_0\right) - \frac{1}{4}c_l \tag{4.18}$$

The pitching-moment coefficient about a point one-quarter of the chord behind the leading edge $c_{m_{c/4}}$ equals $c_{m_{LE}} + (\tfrac{1}{4})c_l$. From Eq. (4.18),

$$c_{m_{c/4}} = 2\mu_0 - \frac{\pi}{2}\beta_0 \tag{4.19}$$

It should be noted that the pitching-moment coefficient about the quarter-chord point is independent of the lift coefficient. The quarter-chord point is therefore the aerodynamic center of thin wing sections. The aerodynamic center for most commonly used wing sections is found to be approximately the quarter-chord point at speeds where the velocity of sound is not reached in the field of flow.

By substituting the expression

$$x = \frac{c}{2}(1-\cos\theta)$$

into Eqs. (4.13) and (4.14)

$$\beta_0 = \int_0^1 \frac{y}{c}f_1\left(\frac{x}{c}\right)\frac{dx}{c} \tag{4.20}$$

and

$$\mu_0 = \int_0^1 \frac{y}{c} f_2\left(\frac{x}{c}\right) \frac{dx}{c} \tag{4.21}$$

where

$$f_1\left(\frac{x}{c}\right) = \frac{1}{\pi[1 - (x/c)]\sqrt{(x/c)[1 - (x/c)]}}$$

$$f_2\left(\frac{x}{c}\right) = \frac{1 - (2x/c)}{\sqrt{(x/c)[1 - (x/c)]}}$$

In order to avoid infinite velocities at the leading edge, it is necessary that the coefficient A_0 of Eq. (4.5) be equal to zero. From Eq. (4.12), A_0 equals zero when

$$\alpha_0 = \frac{1}{\pi} \int_0^\pi \frac{dy}{dx} \, d\theta$$

Substituting the value of dx from Eq. (4.4)

$$\alpha_0 = \frac{1}{\pi} \int_0^\pi \frac{dy \, d\theta}{(c/2) \sin \theta \, d\theta}$$

and integrating by parts

$$\alpha_0 = \frac{1}{\pi} \left[\frac{2y/c}{\sin \theta}\right]_0^\pi + \frac{1}{\pi} \int_0^\pi \frac{2(y/c) \cos \theta \, d\theta}{\sin^2 \theta}$$

As previously discussed in connection with the integration of Eq. (4.13) the first term of this expression equals zero for most wing sections. Substitution of the expression

$$x = \frac{c}{2}(1 - \cos \theta)$$

gives

$$\alpha_i = \int_0^1 \frac{y}{c} f_3\left(\frac{x}{c}\right) \frac{dx}{c} \tag{4.22}$$

where

$$f_3\left(\frac{x}{c}\right) = \frac{1 - (2x/c)}{2\pi\{(x/c)[1 - (x/c)]\}^{3/2}}$$

This angle of attack is called the "ideal angle of attack" by Theodorsen.[121] The lift coefficient corresponding to this angle of attack [obtained from Eq. (4.17)] is called the "ideal" or "design" lift coefficient.

In general the angle of zero lift, the pitching moment about the quarter-chord point, and the ideal angle of attack may be calculated by graphical integration of Eqs. (4.20), (4.21), and (4.22). For convenience in such calculations, values of the functions f_1, f_2, and f_3 corresponding to several values of x/c are presented in Table 5. Although the functions f_1, f_2, and f_3 become infinite at both the leading and trailing edges, the integrands of

Eqs. (4.20) and (4.21) approach zero at the leading edge for mean lines that approach zero faster than $\sqrt{x/c}$. Most mean lines satisfy this condition. Similarly, the integrand of Eq. (4.21) approaches zero at the trailing edge in most cases. In order to avoid the difficulties at the trailing edge in Eq. (4.20) and at the leading and trailing edges in Eq. (4.22), parts of these integrals are evaluated analytically.[49] The analytical determina-

TABLE 5.—VALUES OF FUNCTIONS f_1, f_2, AND f_3

x/c	$f_1(x/c)$	$f_2(x/c)$	$f_3(x/c)$
0	∞	∞	∞
0.0125	2.901	8.774	113.15
0.0250	2.091	6.085	39.73
0.0500	1.537	4.131	13.84
0.0750	1.306	3.226	7.403
0.1000	1.179	2.667	4.716
0.15	1.049	1.960	2.447
0.20	0.995	1.502	1.492
0.25	0.980	1.156	0.980
0.30	0.992	0.873	0.662
0.40	1.083	0.408	0.271
0.50	1.273	0	0
0.60	1.624	−0.408	− 0.271
0.70	2.315	−0.873	− 0.662
0.80	3.979	−1.502	− 1.492
0.90	10.61	−2.667	− 4.716
0.95	29.21	−4.131	− 13.84
1.00	∞	∞	− ∞

tion of the increments at the critical points is accomplished by assuming the mean line near the ends to be of the form

$$\frac{y}{c} = A + B\frac{x}{c} + C\left(\frac{x}{c}\right)^2$$

Evaluation of the increment of the angle of zero lift gives

$$\Delta\beta_0 = 0.964 y_{0.95} - 0.0954 \frac{dy}{dx_1} \qquad \left(\frac{x}{c} = 0.95 \text{ to } 1.0\right)$$

Where $y_{0.95}$ is the ordinate of the mean line at $x/c = 0.95$ and dy/dx_1 is the slope of the mean line at $x/c = 1.0$. This increment is added to the value of β_0 obtained by graphical integration from $x/c = 0$ to $x/c = 0.95$. The increments for the ideal angle of attack are

$$\Delta\alpha_i = \begin{cases} + 0.467 y_{0.05} + 0.0472 \dfrac{dy}{dx_0} & \left(\dfrac{x}{c} = 0 \text{ to } 0.05\right) \\[2ex] - 0.467 y_{0.95} + 0.0472 \dfrac{dy}{dx_1} & \left(\dfrac{x}{c} = 0.95 \text{ to } 1.0\right) \end{cases}$$

These increments are added to the values of α_i obtained by graphical integration from $x/c = 0.05$ to $x/c = 0.95$.

By application of Gauss's rules for numerical integration to the expression for the angle of zero lift, Munk[70] obtained a simple approximate solution.

$$- \alpha_0 = k_1 y_1 + k_2 y_2 + k_3 y_3 + k_4 y_4 + k_5 y_5$$

where y_1, y_2, etc., are the ordinates of the mean line expressed as fractions of the chord at the points x_1, x_2, etc., as tabulated together with corresponding values of the constants k_1, k_2, etc., calculated to give the angle of zero lift in degrees.

x_1	0.99458	k_1	1,252.24
x_2	0.87426	k_2	100.048
x_3	0.50000	k_3	32.5959
x_4	0.12574	k_4	15.6838
x_5	0.00542	k_5	5.97817

This method is useful for quick calculation of the angle of zero lift when a high order of accuracy is not required.

Approximate solutions for the angle of zero lift and the moment coefficient were also obtained by Pankhurst[73] in the following form:

$$\alpha_{L=0} = \Sigma A (U + L)$$

$$c_{m_{c/4}} = \Sigma B (U + L)$$

where U,L = upper and lower ordinates of wing section in fractions of chord

A,B = constants given in following table

x	A	B
0	1.45	− 0.119
0.025	2.11	− 0.156
0.05	1.56	− 0.104
0.1	2.41	− 0.124
0.2	2.94	− 0.074
0.3	2.88	− 0.009
0.4	3.13	0.045
0.5	3.67	0.101
0.6	4.69	0.170
0.7	6.72	0.273
0.8	11.75	0.477
0.9	21.72	0.786
0.95	99.85	3.026
1.00	− 164.90	− 4.289

Pankhurst's solution for the angle of zero lift is often more convenient than Munk's because the constants are given for stations at which the ordinates are usually specified. The solution is based on the assumptions that the mean-line ordinates are represented satisfactorily by $(U + L)/2$ and that the tangent of the mean line at the trailing edge coincides with the last two points. The accuracy is accordingly expected to be limited for sections with large curvatures of the mean line near the trailing edge, or for thick, highly cambered sections.

4.4. Design of Mean Lines. It is possible to design mean lines to have certain desired load distributions by means of the theory of thin wing sections. The fundamental relations are Eqs. (4.1), (4.2), and (4.3). A simple example of the manner in which a mean line is designed may be demonstrated by the design of a mean line to have uniform load.[113]

For a uniform chordwise load distribution, the value of γ in Eq. (4.2) is a constant over the chord. Equation (4.2) may therefore be written

$$v_n(x_1) = \frac{\gamma}{2\pi} \int_0^c \frac{dx}{x - x_1} \tag{4.23}$$

The lift is

$$l = \int_0^c \rho V \gamma \, dx = \rho V \gamma c$$

and c_{l_i} equals

$$c_{l_i} = \frac{2l}{\rho V^2 c} = \frac{2\gamma}{V}$$

or

$$\gamma = \frac{V c_{l_i}}{2}$$

By substitution, Eq. (4.23) becomes

$$\frac{v_n}{V}\left(\frac{x_1}{c}\right) = \frac{c_{l_i}}{4\pi} \int_0^1 \frac{d(x/c)}{(x/c) - (x_1/c)} \tag{4.24}$$

Special means must be used to integrate Eq. (4.24) because the integrand becomes infinite as x approaches x_1. The integration is performed as follows:

$$\frac{v_n}{V}\left(\frac{x_1}{c}\right) = \frac{c_{l_i}}{4\pi} \lim_{\epsilon \to 0} \left[\int_0^{(x_1/c)-\epsilon} \frac{d(x/c)}{(x/c) - (x_1/c)} + \int_{(x_1/c)+\epsilon}^1 \frac{d(x/c)}{(x/c) - (x_1/c)} \right]$$

$$= \frac{c_{l_i}}{4\pi} \lim_{\epsilon \to 0} \left\{ \left[\ln\left(\frac{x}{c} - \frac{x_1}{c}\right) \right]_0^{x_1 - \epsilon} + \left[\ln\left(\frac{x}{c} - \frac{x_1}{c}\right) \right]_{x_1 + \epsilon}^1 \right\}$$

$$= \frac{c_{l_i}}{4\pi} \left[\ln\left(1 - \frac{x_1}{c}\right) - \ln\left(\frac{x_1}{c}\right) \right]$$

The angle of attack at which the uniform load distribution is realized is zero because the assumed load distribution, and accordingly the mean line, is symmetrical about the mid-chord point. Equation (4.3) becomes

$$\frac{dy}{dx} = \frac{c_{l_i}}{4\pi}\left[\ln\left(1 - \frac{x_1}{c}\right) - \ln\left(\frac{x_1}{c}\right)\right] \tag{4.25}$$

Integration of this expression gives

$$y = -\frac{c_{l_i}}{4\pi}\left[\left(1 - \frac{x}{c}\right)\ln\left(1 - \frac{x}{c}\right) + \frac{x}{c}\ln\frac{x}{c}\right] \tag{4.26}$$

where the constant of integration has been selected to make the ordinates zero at the leading and trailing edges.

By similar but more complicated processes, the mean lines corresponding to other types of load distribution may be calculated. In this connection, it is interesting to note that the distribution of γ and the slope of the mean line are conjugate functions at the ideal angle of attack [see Eqs. (4.5) and (4.11)]. The slope of the mean line can therefore be obtained from the assumed load distribution by the proper application of Naiman's coefficients[8,73] as in Eq. (3.17). The mean-line ordinates can then be obtained by graphical or analytical methods.

Mean lines have been calculated[45] corresponding to a load distribution that is uniform from $x/c = 0$ to $x/c = a$ and decreases linearly to zero at $x/c = 1$. The ordinates are given by the expression

$$\frac{y}{c} = \frac{c_{l_i}}{2\pi(a+1)}\left\{\frac{1}{1-a}\left[\frac{1}{2}\left(a - \frac{x}{c}\right)^2\ln\left|a - \frac{x}{c}\right| - \frac{1}{2}\left(1 - \frac{x}{c}\right)^2\ln\left(1 - \frac{x}{c}\right)\right.\right.$$
$$\left.\left. + \frac{1}{4}\left(1 - \frac{x}{c}\right)^2 - \frac{1}{4}\left(a - \frac{x}{c}\right)^2\right] - \frac{x}{c}\ln\frac{x}{c} + g - h\,\frac{x}{c}\right\} \tag{4.27}$$

where

$$g = \frac{-1}{1-a}\left[a^2\left(\frac{1}{2}\ln a - \frac{1}{4}\right) + \frac{1}{4}\right]$$

$$h = \frac{1}{1-a}\left[\frac{1}{2}\left(1 - a\right)^2\ln\left(1 - a\right) - \frac{1}{4}\left(1 - a\right)^2\right] + g$$

The ideal angle of attack for these mean lines is

$$\alpha_i = \frac{c_{l_i}h}{2\pi(a+1)}$$

It will be noted that the ordinates of the mean lines are directly proportional to the design lift coefficient. The fact that the load distributions and the ordinates of the corresponding mean lines are additive may be seen from examination of Eqs. (4.5) and (4.11). This property is, of course, a direct consequence of the linearizing assumptions made in writing the fundamental expressions (4.1), (4.2), and (4.3). This linearity is convenient because ordinates of a mean line having a load distribution made up of the sums of load distributions for which the corresponding ordinates are known can be obtained by simply adding the ordinates of the component mean lines.

The linearizing assumptions lead to increasing errors as the slope of the mean line increases. For geometrically similar mean lines, the slope increases with the design lift coefficient, and the errors are consequently greater for the larger design lift coefficients. In the case of the uniform-load mean line [Eq. (4.26)], the slope is infinite at the leading and trailing edges, and it is not to be expected that the assumed load will be maintained

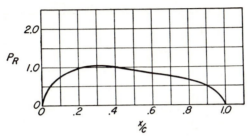

Fig. 36. Typical basic load distribution at ideal angle of attack.

near these points. The region in which the slope is large is extremely small, however, because of the manner in which the slope approaches infinity [Eq. (4.25)].

4.5. Engineering Applications of Section Theory. The theory of wing sections presented in Sec. 3.6 permits the calculation of the pressure distribution and certain other characteristics of arbitrary sections with considerable accuracy. Although this method is not unduly laborious, the computations required are too long to permit quick and easy calculations for large numbers of wing sections. The need for a simple method of quickly obtaining pressure distributions with engineering accuracy has led to the development of a method[3] combining features of thin and thick wing-section theory. This simple method makes use of previously calculated characteristics of a limited number of mean lines and thickness forms that may be combined to form large numbers of related wing sections.

The theory of thin wing sections as presented in Sec. 4.3 and reference 121 shows that the load distribution of a thin section may be considered to consist of

1. A basic distribution at the ideal angle of attack (Fig. 36).

2. An additional distribution proportional to the angle of attack as measured from the ideal angle of attack (Fig. 37).

The first load distribution is a function only of the shape of the thin wing section or (if the thin wing section is considered to be a mean line) of the mean-line geometry. Integration of this load distribution along the chord results in a normal force coefficient which, at small angles of attack, is substantially equal to the design lift coefficient. If, moreover, the camber of the mean line is changed by multiplying the mean-line ordinates by a constant factor, the resulting load distribution, the ideal or design angle of attack α_i, and the design lift coefficient c_{l_i} may be obtained simply by multiplying the original values by the same factor.

The second load distribution, which results from changing the angle of attack, is designated the "additional load distribution," and the corresponding lift coefficient is designated the "additional lift coefficient."

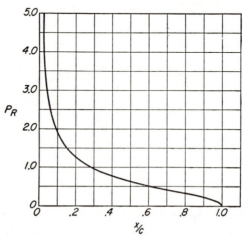

FIG. 37. Additional load distribution associated with angle of attack.

This additional load distribution contributes no moment about the quarter-chord point and, according to the theory of thin wing sections, is independent of the geometry of the wing section except for angle of attack. The additional load distribution obtained from the theory of thin wing sections is of limited practical application, however, because this simple theory leads to infinite values of the velocity at the leading edge. This difficulty is obviated by the exact theory of wing sections presented in Sec. 3.6, which also shows that the additional load distribution is neither completely independent of the airfoil shape nor exactly a linear function of the angle of attack. Suitable additional load distributions can be calculated by the methods of Sec. 3.6 for each thickness form (symmetrical section) and a typical value of the additional lift coefficient. The variation of the additional load distribution with the additional lift coefficient is arbitrarily assumed to be linear.

In addition to the pressure distributions associated with these two load distributions, another pressure distribution exists which is associated with the thickness form, or basic symmetrical section, at zero angle of attack, and which is calculated by the methods of Sec. 3.6.

The velocity distribution about the wing section is thus considered to be composed of three separate and independent components as follows:
1. The distribution corresponding to the velocity distribution over the basic thickness form at zero angle of attack.
2. The distribution corresponding to the load distribution of the mean line at its ideal angle of attack.
3. The distribution corresponding to the additional load distribution associated with angle of attack.

The local load at any chordwise position is caused by a difference of velocity between the upper and the lower surfaces. It is assumed[7] that the velocity increment on one surface is equal to the velocity decrement on the other surface. This assumption is in accord with the basic concept of the distribution of circulation used in the theory of thin wing sections. It can be shown that the relation between the local load coefficient P_R and the velocity increment ratios $\Delta v/V$ or $\Delta v_a/V$ is

$$\frac{\Delta v}{V} \quad \text{or} \quad \frac{\Delta v_a}{V} = \frac{1}{4}\frac{P_R}{v/V}$$

where v/V is the velocity ratio at the corresponding point on the surface of the basic thickness form. In actual practice the value of $\Delta v/V$ used is that corresponding to the thin wing section or mean line where v/V equals unity:

The velocity increment ratios $\Delta v/V$ and $\Delta v_a/V$ corresponding to components 2 and 3 are added to the velocity ratio corresponding to component 1 to obtain the total velocity at one point from which the pressure coefficient S is obtained, thus

$$S = \left(\frac{v}{V} \pm \frac{\Delta v}{V} \pm \frac{\Delta v_a}{V}\right)^2 \tag{4.28}$$

This procedure is illustrated in Fig. 38. When this formula is used, values of the ratios corresponding to one value of x/c are added together and the resulting value of the pressure coefficient S is assigned to the surface of the wing section at the same value of x/c. The values of $\Delta v/V$ and $\Delta v_a/V$ are positive on the upper surface and negative on the lower surface for positive values of the corresponding local loads.

When the ratio $\Delta v_a/V$ has the value of zero, integration of the resulting pressure-distribution diagram will give approximately the design lift coefficient c_{l_i} of the mean line. In general, however, the value of c_l will be greater than c_{l_i} by an amount dependent on the thickness ratio of the basic thickness form. This discrepancy is caused by applying the values of $\Delta v/V$ obtained for the mean line to the sections of finite thickness where v/V is greater than unity over most of the surface.

The pressure distribution will usually be desired at some specified lift coefficient not corresponding to c_{l_i}. For this purpose, the ratio $\Delta v_a/V$ must

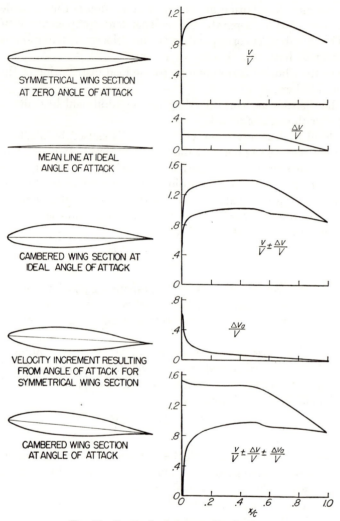

FIG. 38. Synthesis of pressure distribution.

be assigned some value by multiplying the originally calculated values of this ratio by a factor $f(\alpha)$. For a first approximation, this factor may be assigned the value

$$f(\alpha) = \frac{c_l - c_{l_i}}{c_{l_0}}$$

where c_l is the lift coefficient for which the pressure distribution is desired and c_{l_0} is the lift coefficient for which the values of $\Delta v_a/V$ were originally calculated. If greater accuracy is desired, the value of $f(\alpha)$ may be adjusted by trial and error to produce the actual desired life coefficient as determined by integration of the pressure-distribution diagram.

Although this method of superposition of velocities has inadequate theoretical justification, experience has shown that the results obtained are

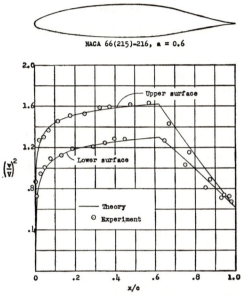

FIG. 39. Comparison of theoretical and experimental pressure distributions for the NACA 66(215)-216 airfoil; $c_l = 0.23$

adequate for many engineering uses. In fact the results of even the first approximation agree well with experimental data and are adequate for at least preliminary consideration and selection of wing sections. A comparison of a first-approximation theoretical pressure distribution with an experimental distribution is shown in Fig. 39. Some discrepancy naturally occurs between the results of experiment and the results of any theoretical method based on potential flow because of the presence of the boundary layer. These effects are small, however, over the range of lift coefficients for which the boundary layer is thin and the drag coefficient is low.

CHAPTER 5

THE EFFECTS OF VISCOSITY

5.1. Symbols.

A constant
F velocity gradient
F skin friction
H total pressure outside boundary layer
H ratio of the displacement thickness to the momentum thickness
K constant
L characteristic length
$\bar{L}_f$ length of mean free path of molecules
R Reynolds number $\rho VL/\mu$
R_c Reynolds number $\rho Vc/\mu$
R_r Reynolds number $\rho \bar{u}r/\mu$
R_x Reynolds number $\rho Vx/\mu$
$R_\delta{}^*$ Reynolds number $\rho V\delta^*/\mu$ or $\rho U\delta^*/\mu$
R_θ Reynolds number $\rho U\theta/\mu$
U velocity outside boundary layer
U_0 maximum velocity
U_v velocity outside boundary layer at the point x_0
U_s velocity outside boundary layer at the laminar separation point
U_x value of U at the point x
V velocity of the free stream
c chord
c_d section drag coefficient
c_F skin-friction coefficient
c_f local skin-friction coefficient
d diameter of pipe
e base of Naperian logarithms, 2.71828
l length
l wave length
ln   logarithm to the base e
log   logarithm to the base 10
n exponent
p local static pressure
q dynamic pressure
q_0 dynamic pressure of the free stream $\frac{1}{2}\rho V^2$
r radius of pipe
u component of velocity in boundary layer along the x axis
$\bar{u}$ average velocity of flow in pipe
v component of velocity in boundary layer along the y axis
$\bar{v}$ mean molecular velocity
v^* friction velocity $\sqrt{\tau/\rho}$
v_s velocity of slip

x Cartesian coordinate (distance along surface)
x_e effective length of flat plate
x_0 position on surface at start of integration
y Cartesian coordinate (distance normal to surface)
α $2\pi/l$
β_i time rate of amplification of disturbance
δ boundary-layer thickness
δ^* displacement thickness of the boundary layer
η nondimensional variable $(y/x)\sqrt{R_x}$
η distance parameter $\rho v^* y/\mu$
θ momentum thickness of the boundary layer
θ_0 momentum thickness at the point x_0
λ pressure-loss coefficient
λ laminar boundary layer shape parameter
μ viscosity of the fluid
ξ coefficient of slip
π ratio of the circumference of a circle to its diameter
ρ mass density of the fluid
τ shearing stress at the wall (skin friction)
ϕ velocity parameter u/v^*
ψ stream function
ψ_1 nondimensional stream function
∞ infinity

5.2. Concept of Reynolds Number and Boundary Layer. In order to be able to compare directly the forces acting on geometrically similar bodies of various sizes at various air speeds, it is customary to express the forces in terms of nondimensional coefficients, as explained in Chap. 1. For geometrically similar configurations, these coefficients tend to remain constant, and they would remain exactly constant if all factors influencing the flow about the body were properly accounted for. It is well known, however, that some characteristics such as the drag and maximum lift coefficients vary with the size of the wing for a given air speed, and with the air speed for a given size of wing.

The two most important factors that are neglected in defining the coefficients are effects associated with the compressibility and viscosity of air. An examination of the complete equations of motion[89] shows the significant parameters to be the ratio of the air speed to the speed of sound in air (the Mach number) and the Reynolds number $\rho V L/\mu$ where ρ is the mass density of the fluid, V is the velocity of the free stream, L is the characteristic length, and, μ is the viscosity of the fluid. It can be shown that similarity of the flows about different bodies is obtained only if the bodies are geometrically similar and if the Reynolds number and the Mach number are the same. At speeds where the pressure variations around the body are small compared with the absolute pressure (low Mach numbers) the effects of compressibility are negligible, and the viscous effects may be considered independently.

Some concept of the physical significance of the Reynolds number may be obtained by expressing this number in terms of the mean velocity and the mean free path of the molecules, as suggested by von Kármán.[137] The kinetic theory of gases indicates that the coefficient of viscosity μ of a homogeneous gas may be expressed by a formula of the type[56]

$$\mu = K\rho\bar{v}\bar{L}_f$$

where K = constant

$\bar{v}$ = mean molecular velocity

$\bar{L}_f$ = mean free path of the molecules

Substituting this value of μ into the formula for the Reynolds number, we obtain

$$R = \frac{1}{K} \frac{V}{\bar{v}} \frac{L}{\bar{L}_f}$$

where R = Reynolds number

The Reynolds number is thus proportional to the product of the ratio of the speed of the body to the mean speed of the molecules and of the ratio of the size of the body to the mean free path of the molecules. For bodies of ordinary size moving at low Mach numbers in air of ordinary density, the ratio $V/\bar{v}$ is small and the ratio $L/\bar{L}_f$ is very large. Under these conditions, the flow around the body corresponds to the conditions under which this formula was derived, namely, that the velocities are not large compared with the mean velocity and that the spatial dimensions of the phenomenon are large compared with the mean free path.

The mean velocity $\bar{v}$ is of the same order of magnitude as the velocity of sound, and consequently the ratio $V/\bar{v}$ is of the same order as the Mach number. It may accordingly be inferred that the Reynolds and Mach numbers are not independent parameters and that the scale effects experienced at high Mach numbers would be different from those at low Mach numbers even though the Reynolds numbers were the same in each case. Moreover, in a rarefied gas, as at extremely high altitudes, the mean free path may be of the same order as, or even much larger than, the characteristic length of the body. The concept of Reynolds number as presented here presupposes a mean free path that is small compared with the thickness of the boundary layer. Consequently the significance of this concept as applied to flows in very rarefied gases or at high Mach numbers is uncertain. Under such conditions, it is preferable to compare flows on the basis of the independent parameters $V/\bar{v}$ and $L/\bar{L}_f$.

The effects of viscosity are of primary importance in a thin region near the surface of the wing called the "boundary layer." Boundary layers, in general, are of two types, namely, laminar and turbulent. The flow in the laminar layer is smooth and free from any eddying motion. The flow in the turbulent layer is characterized by the presence of a large number of relatively small eddies. The eddies in the turbulent layer produce a transfer of momentum from the relatively fast moving outer parts of the boundary layer to the portions closer to the surface. Consequently the distribution of average velocity is characterized by relatively higher velocities near the surface and a greater total boundary-layer thickness in a turbulent than in a laminar boundary layer developed under otherwise identical conditions. Skin friction is therefore higher for turbulent boundary-layer flow than for laminar flow.

Aerodynamically, the concept of the boundary layer is based on the premise of a continuous homogeneous viscous fluid. According to this concept the velocity within the boundary layer varies from zero at the surface to the full local stream value at the outer edge of the layer. Such a concept is valid only if the density of the gas is sufficiently great to limit the mean free path of the molecules to a length very small compared with the thickness of the boundary layer. The kinetic theory of gases[56] indicates that the velocity at the surface is not exactly zero but that there is a velocity of slip proportional to the velocity gradient.

$$v_s = \xi \, \frac{du}{dy}$$

The coefficient of slip ξ has the dimension of length and may be considered as a backward displacement of the wall with the velocity gradient extending effectively right up to the displaced wall where the velocity is zero. It has been shown by Maxwell, Millikan, and others that, for most surfaces, the coefficient of slip is very nearly equal to the mean free path of the molecules. At ordinary altitudes, this distance is so small that it may properly be neglected. At very high altitudes, this slip velocity may have large effects even though the mean free path is still only a fraction of the boundary-layer thickness. At still higher altitudes where the mean free path is long, the entire concept of the boundary layer and viscosity as presented here becomes invalid.

5.3. Flow around Wing Sections. When the pressures along the wing surfaces are increasing in the direction of flow, a general deceleration takes place. At the outer limits of the boundary layer, this deceleration takes place in accordance with Bernoulli's law. Closer to the surface, no such simple law can be given because of the action of the viscous forces within the boundary layer. In general, however, the relative loss of speed is somewhat greater for the particles of fluid within the boundary layer than for those at the outer limits of the layer, because the reduced kinetic energy of the boundary-layer air limits its ability to flow against the adverse pressure gradient. If the rise in pressure is sufficiently great, portions of the fluid within the boundary layer may actually have their direction of motion reversed and may start moving upstream. When this reversal occurs, the boundary layer is said to be "separated." Because of the increased interchange of momentum from different parts of the layer, turbulent boundary layers are much more resistant to separation than are laminar layers. Except under very special circumstances laminar boundary layers can exist for only a relatively short distance in a region in which the pressure increases in the direction of flow.

After laminar separation occurs, the flow may either leave the surface permanently or reattach itself in the form of a turbulent layer. Not much is known concerning the factors controlling this phenomenon. Laminar separation on wings is usually not permanent at flight values of the Reynolds number except when it occurs on some wing sections near the leading edge under conditions corresponding to maximum lift. The size of the locally separated region that is formed when the laminar boundary layer separates and the flow returns to the surface decreases with increasing Reynolds number at a given angle of attack.

The flow over aerodynamically smooth wings at low and moderate lift coefficients is characterized by laminar boundary layers from the leading edge back to approximately the location of the first minimum-pressure point on both upper and lower surfaces. If the region of laminar flow is extensive, separation occurs immediately downstream from the location of minimum pressure and the flow returns to the surface almost immediately at flight Reynolds numbers as a turbulent layer. This turbulent boundary layer extends to the trailing edge. If the surfaces are not sufficiently smooth and fair, if the air stream is turbulent, or perhaps if the Reynolds number is sufficiently large, transition from laminar to turbulent flow may occur anywhere upstream of the calculated laminar separation point.

For low and moderate lift coefficients where inappreciable separation occurs, the wing profile drag is largely caused by skin friction, and the value of the drag coefficient depends mostly on the relative amounts of laminar and turbulent flow. If the location of transition is known or assumed, the drag coefficient may be calculated with reasonable accuracy from boundary-layer theory.

As the lift coefficient of the wing is increased by changing the angle of attack, the resulting application of the additional type of lift distribution moves the minimum-pressure point upstream on the upper surface, and the possible extent of laminar flow is thus reduced. The resulting greater proportion of turbulent flow, together with the larger average velocity of flow over the surfaces, causes the drag to increase with lift coefficient.

At high lift coefficients, a large part of the drag is contributed by pressure or form drag resulting from separation of the flow from the surface. The flow over the upper surface is characterized by a negative pressure peak near the leading edge, which causes laminar separation. The onset of turbulence causes the flow to return to the surface as a turbulent boundary layer. High Reynolds numbers are favorable to the development of turbulence and aid in this process. If the lift coefficient is sufficiently high or if the reestablishment of the flow following laminar separation is unduly delayed by low Reynolds numbers, the turbulent layer will separate from the surface near the trailing edge with resulting large drag increases. The eventual loss of lift with increasing angle of attack may result either from relatively sudden failure of the boundary layer to reattach itself to the surface following separation of the laminar boundary layer near the leading edge or from progressive forward movement of turbulent separation. Under the latter condition, the flow over a relatively large portion of the surface may be separated prior to maximum lift.

5.4. Characteristics of the Laminar Layer. The characteristics of the laminar boundary layer may be deduced from detailed consideration of the general equations of motion. For two-dimensional steady motion neglecting the effects of compressibility, these equations are[28]

$$\left. \begin{aligned} u\frac{\partial u}{\partial x} + v\frac{\partial u}{\partial y} &= -\frac{1}{\rho}\frac{\partial p}{\partial x} + \frac{\mu}{\rho}\left(\frac{\partial^2 u}{\partial x^2} + \frac{\partial^2 u}{\partial y^2}\right) \\ u\frac{\partial v}{\partial x} + v\frac{\partial v}{\partial y} &= -\frac{1}{\rho}\frac{\partial p}{\partial y} + \frac{\mu}{\rho}\left(\frac{\partial^2 v}{\partial x^2} + \frac{\partial^2 v}{\partial y^2}\right) \end{aligned} \right\} \tag{5.1}$$

These equations are known as the "Navier-Stokes equations." Their general solution is difficult. A simple integration can be performed only in special cases. These equations, however, may be simplified by the use of dimensionless variables and consideration of the order of magnitude of each term. Prandtl[37] has shown that, if the viscous effects are assumed to be confined to a thin layer over the surface, the following approximate relation applies in this region

$$u\frac{\partial u}{\partial x} + v\frac{\partial u}{\partial y} = -\frac{1}{\rho}\frac{\partial p}{\partial x} + \frac{\mu}{\rho}\frac{\partial^2 u}{\partial y^2} \tag{5.2}$$

This equation is a simplified form of the first relation of Eq. (5.1). Consideration of the order of magnitude of the various terms of the second relation of Eq. (5.1) indicates that $\partial p/\partial y$ is small and hence that pressures are transmitted unchanged through the boundary layer.

An interesting conclusion is reached by substitution of the following nondimensional variables in Eq. (5.2).

$$x_1 = \frac{x}{c} \qquad y_1 = \frac{y}{c}\sqrt{R}$$

$$u_1 = \frac{u}{V} \qquad v_1 = \frac{v}{V}\sqrt{R}$$

$$p_1 = \frac{p}{\rho V^2}$$

Equation (5.2) becomes

$$u_1 V \frac{\partial(u_1 V)}{\partial(x_1 c)} + \frac{v_1 V}{\sqrt{R}} \frac{\partial(u_1 V)}{\partial(y_1 c/\sqrt{R})} = -\frac{1}{\rho}\frac{\partial(p_1 \rho V^2)}{\partial(x_1 c)} + \frac{\partial^2(u_1 V)}{\partial(y_1 c/\sqrt{R})^2}$$

and, upon simplification,

$$u_1 \frac{\partial u_1}{\partial x_1} + v_1 \frac{\partial u_1}{\partial y_1} = -\frac{\partial p_1}{\partial x_1} + \frac{\partial^2 u_1}{\partial y_1^2} \qquad (5.3)$$

It will be noted that the viscosity does not appear directly in Eq. (5 3). Solution of this equation will give u_1 and v_1 in terms of x_1 and y_1. Because y_1 equals $(y/c)\sqrt{R}$, the effect of variations of the Reynolds number will be merely to change y in a manner inversely proportional to the square root

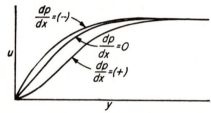

FIG. 40.   Effect of pressure gradient on velocity distribution in laminar boundary layer.

of the Reynolds number. It follows directly that the thickness of the laminar boundary layer varies inversely with the square root of the Reynolds number. It may also be concluded that the shape of the velocity distribution through the boundary layer is independent of the Reynolds number, unless, of course, the pressure distribution is a function of the Reynolds number. It also follows that the position of the laminar separation point, characterized by the condition that $\partial u/\partial y$ is zero at the surface, is independent of the Reynolds number.

A considerable amount of information concerning the velocity distribution in the laminar boundary layer can be derived from Eq. (5.3) without actually attempting a solution. At the surface u_1 and v_1 must both equal zero. At the outer limit of the boundary layer u_1 approaches a constant value. Consequently near the outer edge of the layer $\partial^2 u/\partial y^2$ must be negative and approach zero with increasing distance from the surface, as shown in Fig. 40. At the surface, Eq. (5.3) simplifies to the form

$$\frac{\partial^2 u_1}{\partial y_1^2} = \frac{\partial p_1}{\partial x_1}$$

When $\partial p_1/\partial x_1$ is positive (increasing pressure in the direction of flow), $\partial^2 u/\partial y^2$ must be positive at the surface, and the shape of the velocity distribution will resemble that shown in Fig. 40 for this condition. Similarly the velocity distribution must be linear near the surface for zero pressure gradient, and the velocity profile will be continuously convex upward as shown in Fig. 40 for decreasing pressures in the direction of flow.

5.5. Laminar Skin Friction. Blasius[14] obtained a solution for Eq. (5.2) for the case where $\partial p/\partial x$ is zero, that is, for a flat plate with uniform velocity outside the boundary layer. Blasius found that Eq. (5.2) could be greatly simplified by assuming a variable oi the form

$$\eta = \frac{y}{x}\sqrt{R_x}$$

where $R_x = \rho V x/\mu$

and by the introduction of the nondimensional stream function

$$\psi_1 = \int_0^\eta u_1 \, d\eta = f(\eta)$$

where $u_1 = u/V$

The true stream function is

$$\psi = \int_0^y u \, dy$$

$$\psi = \frac{Vx}{\sqrt{R_x}}\int_0^\eta u_1 \, d\eta = \frac{Vx}{\sqrt{R_x}} f(\eta) = \sqrt{\frac{\mu}{\rho}} x V f(\eta)$$

$$u = \frac{\partial \psi}{\partial y} = \frac{\partial \psi}{\partial \eta}\frac{\partial \eta}{\partial y} = \left[\sqrt{\frac{\mu}{\rho}} x V f'(\eta)\right]\left(\frac{\sqrt{R_x}}{x}\right) = V f'(\eta)$$

The velocity v perpendicular to the flat plate is given by

$$v = -\frac{\partial \psi}{\partial x} = -\frac{\partial \psi}{\partial \eta}\frac{\partial \eta}{\partial x}$$

$$v = \frac{1}{2}\sqrt{\frac{\mu V}{\rho x}}\left[\eta f'(\eta) - f(\eta)\right]$$

Making similar calculations for $\partial u/dx$, $\partial u/dy$, and $\partial^2 u/\partial y^2$ and substituting in Eq. (5.2) with $\partial p/\partial x$ zero, we obtain

$$-\frac{V^2}{2x}\eta f'(\eta)f''(\eta) + \frac{V^2}{2x}\left[\eta f'(\eta) - f(\eta)\right]f''(\eta) = \frac{\mu}{\rho}\frac{V^2}{x}\frac{\rho}{\mu}f'''(\eta)$$

or

$$f(\eta)f''(\eta) + 2f'''(\eta) = 0 \tag{5.4}$$

The fact that neither x nor y appears explicitly in Eq. (5.4) indicates that the initial assumption that ψ_1 is a function of η alone is correct, that is, u_1 is a function of η alone.

The boundary conditions for the solution of Eq. (5.4) are that

$$\begin{aligned}
f(\eta) &= 0 &\quad \text{for } \eta = 0\\
f'(\eta) &= 1 &\quad \text{for } \eta \to \infty\\
f''(\eta) &\to 0 &\quad \text{for } \eta \to \infty
\end{aligned}$$

Blasius obtained the solution of Eq. (5.4) by an approximate integration in a Taylor series.[12] The solution has not been given in explicit form but is presented in the following table.

η	$f'(\eta)$	η	$f'(\eta)$	η	$f'(\eta)$
0	0	2.0	0.6298	4.0	0.9555
0.2	0.0664	2.2	0.6813	4.2	0.9670
0.4	0.1328	2.4	0.7290	4.4	0.9759
0.6	0.1990	2.6	0.7725	4.6	0.9827
0.8	0.2647	2.8	0.8115	4.8	0.9878
1.0	0.3298	3.0	0.8461	5.0	0.9916
1.2	0.3938	3.2	0.8761	5.2	0.9943
1.4	0.4563	3.4	0.9018	5.4	0.9962
1.6	0.5168	3.6	0.9233	5.6	0.9975
1.8	0.5778	3.8	0.9411	5.8	0.9984
				6.0	0.9990

A plot of these values is given in Fig. 41. These values define the velocity distribution in a laminar boundary layer in the absence of a pres-

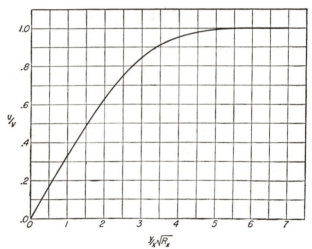

Fig. 41. Velocity distribution in the boundary layer over a flat plate according to Blasius.

sure gradient. The form of the variables in which this velocity distribution is specified shows that the boundary-layer thickness increases as the square root of the distance from the leading edge and inversely as the square root of the Reynolds number and that the velocity distribution is similar at all points along a flat plate.

In addition to the velocity distribution through the boundary layer, the solution of Eq. (5.4) gives a value of $f''(\eta)$ at the surface equal to 0.332, that is,

$$0.332 = f''(\eta) = \frac{d(u/V)}{\partial \eta} = \frac{d(u/V)}{d[(y/x)\sqrt{R_x}]}$$

For the skin friction, we have

$$\left.\begin{array}{l} \tau = \mu\left(\dfrac{\partial u}{dy}\right)_{y=0} = 0.332\,\dfrac{\mu V}{x}\,\sqrt{R_x} \\[3mm] c_f = \dfrac{\tau}{\frac{1}{2}\rho V^2} = 0.664\,\dfrac{\mu}{\rho V x}\,\sqrt{R_x} = \dfrac{0.664}{\sqrt{R_x}} \end{array}\right\} \tag{5.5}$$

The total drag coefficient for one side of a flat plate is then given by

$$c_F = \frac{1}{x}\int_0^x c_f\,dx = \frac{1}{x}\int_0^x \frac{0.664}{\sqrt{x}}\sqrt{\frac{\mu}{\rho V}}\,dx$$

$$= \frac{1}{x}\left(1.323\,\sqrt{\frac{\mu x}{\rho V}}\right) = \frac{1.328}{\sqrt{R_x}} \tag{5.6}$$

The obvious definition of the boundary-layer thickness is the distance from the surface to where the local velocity defect due to viscosity is zero, that is, $u/V = 1$. Practically, this distance is not well defined because u/V approaches unity asymptotically. A definition that has frequently been used is that the boundary-layer thickness is the distance from the surface to where $u/V = 0.995$. The preceding table indicates that, according to this definition, $\delta/x \simeq 5.3/\sqrt{R_x}$. A thickness parameter that is convenient for experimental work is the distance from the surface to the point where $(u/V)^2 = \frac{1}{2}$ or $\delta/x \simeq 2.3/\sqrt{R_x}$. For theoretical work it is frequently desirable to obtain a measure of the effect of the boundary layer on the outside flow. The reduced velocities within the boundary layer cause a displacement of the streamlines around the body. This displacement is called the "displacement thickness" of the boundary layer δ^* and is calculated from the relation

$$\delta^* = \frac{1}{V}\int_0^\delta (V - u)\,dy$$

or in the general case

$$\delta^* = \frac{1}{U}\int_0^\delta (U - u)\,dy \tag{5.7}$$

where U = velocity just outside boundary layer

For the laminar boundary layer over a flat plate with zero pressure gradient

$$\frac{\delta^*}{x} = \frac{1.73}{\sqrt{R_x}} \tag{5.8}$$

5.6. Momentum Relation. The Kármán integral relation[27, 136] provides a method for relating the boundary-layer thickness to the pressure distribution if the skin friction and the velocity distribution through the boundary layer are known or assumed. The relation is useful because good approximations to the skin friction and the velocity distribution can often be obtained from consideration of local conditions. This relation is applicable to both laminar and turbulent boundary layers. The Kármán integral relation is based on the principle of mechanics that, if any closed surface is described within a steady stream of fluid, the time rate of increase, within the surface, of momentum in any direction is equal to the sum of the components in that direction of all the forces acting on the fluid.

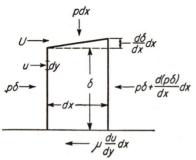

FIG. 42. Forces on an element of the boundary layer.

If body forces such as weight are absent, the only forces to be considered are those acting on the surface, that is, normal forces caused by pressures and tangential forces caused by viscous shearing.

Consider the equilibrium of the element shown in Fig. 42. The momentum from left to right entering the left face per unit time is

$$\int_0^\delta u(\rho u \; dy)$$

while the momentum leaving the right face per unit time is

$$\int_0^\delta u(\rho u \; dy) + \left[\frac{d}{dx}\int_0^\delta u(\rho u \; dy)\right]dx$$

The mass flow per unit time leaving the right face must be the sum of the mass flow entering the left face plus the mass flow entering the upper boundary. The mass flow per unit time through the upper boundary is

$$\frac{d}{dx}\left(\int_0^\delta \rho u \; dy\right)dx$$

This mass enters with the velocity U. The momentum entering the upper boundary per unit time is then

$$U\left(\frac{d}{dx}\int_0^\delta \rho u \; dy\right)dx$$

The total rate of change of momentum from left to right in the element is then

$$\left(\frac{d}{dx}\int_0^\delta \rho u^2 \; dy\right)dx - U\left(\frac{d}{dx}\int_0^\delta \rho u \; dy\right)dx$$

The forces acting on the element are

$$p\delta \qquad \text{on the left face}$$

$$-\left[p\delta + \frac{d}{dx}(p\delta)dx\right] \qquad \text{on the right face}$$

$$p\frac{d\delta}{dx}\,dx \qquad \text{on the upper boundary}$$

$$-\tau\,dx \qquad \text{on the lower boundary}$$

The resultant force is

$$-\frac{d}{dx}(p\delta)dx + p\frac{d\delta}{dx}\,dx - \tau\,dx$$

which reduces to

$$-\delta\frac{dp}{dx}\,dx - \tau\,dx$$

Equating the resultant force and the rate of change of momentum

$$\left(\frac{d}{dx}\int_0^\delta \rho u^2\,dy\right)dx - U\left(\frac{d}{dx}\int_0^\delta \rho u\,dy\right)dx = -\delta\frac{dp}{dx}\,dx - \tau\,dx$$

Dividing through by $\rho\,dx$, we obtain

$$\frac{d}{dx}\int_0^\delta u^2\,dy - U\frac{d}{dx}\int_0^\delta u\,dy = -\frac{\delta}{\rho}\frac{dp}{dx} - \frac{\tau}{\rho}$$

This equation is not very convenient because the values of the integrands do not approach zero at the upper limit of integration. To avoid this difficulty the integrands are expressed in terms of the variable $(U - u)$ by making the substitution

$$u = U - (U - u)$$

$$\frac{d}{dx}\int_0^\delta\left[U - (U - u)\right]^2 dy - U\frac{d}{dx}\int_0^\delta\left[U - (U - u)\right]dy = -\frac{\delta}{\rho}\frac{dp}{dx} - \frac{\tau}{\rho}$$

Expanding and collecting terms

$$U\delta\frac{dU}{dx} - \frac{d}{dx}U\int_0^\delta(U - u)dy + U\frac{d}{dx}\int_0^\delta(U - u)dy - \frac{d}{dx}\int_0^\delta(U - u)u\,dy$$

$$= -\frac{\delta}{\rho}\frac{dp}{dx} - \frac{\tau}{\rho}$$

At the outer limit of the boundary layer, Bernoulli's equation is valid; hence

$$\frac{d}{dx}\left(p + \frac{1}{2}\rho U^2\right) = \frac{dH}{dx} = 0$$

and

$$\frac{dp}{dx} = -\rho U\frac{dU}{dx}$$

Substitution of this relation in the previous equation gives

$$-\frac{dU}{dx}\int_0^\delta (U - u)dy - \frac{d}{dx}\int_0^\delta (U - u)u\, dy = -\frac{\tau}{\rho} \qquad (5.9)$$

This equation is known as the "Kármán integral relation." In this form the integrals approach zero as the upper limit is approached. This equation is suitable for the experimental determination of the skin friction, although the accuracy of the determination may be affected considerably by the form in which this equation is expressed.

Another form of Eq. (5.9) can be obtained by the use of the displacement and momentum thicknesses. The displacement thickness is defined by Eq. (5.7). The momentum thickness is related to the loss of momentum of the air in the boundary layer and is defined as

$$\theta = \int_0^\delta \left(1 - \frac{u}{U}\right)\frac{u}{U}\, dy = \frac{1}{U^2}\int_0^\delta (U - u)u\, dy \qquad (5.10)$$

Substitution of δ^* and θ in Eq. (5.9) gives

$$U\delta^* \frac{dU}{dx} + \frac{d}{dx} U^2\theta = \frac{\tau}{\rho}$$

$$U\delta^* \frac{dU}{dx} + 2U\theta \frac{dU}{dx} + U^2 \frac{d\theta}{dx} = \frac{\tau}{\rho}$$

$$\frac{d\theta}{dx} + \frac{\theta}{U}\frac{dU}{dx}\left(\frac{\delta^*}{\theta} + 2\right) = \frac{\tau}{\rho U^2}$$

Introducing the shape parameter $H = \delta^*/\theta$ and the dynamic pressure $q = \frac{1}{2}\rho U^2$

$$\frac{d\theta}{dx} + \left(\frac{H + 2}{2}\right)\frac{\theta}{q}\frac{dq}{dx} = \frac{\tau}{2q} \qquad (5.11)$$

Equation (5.11) is in a convenient form for the calculation of the thickness of the laminar boundary layer when the pressures are not constant, as was assumed in the Blasius solution.[132] In many cases of interest, the pressure gradients are sufficiently small so that the type of velocity distribution through the boundary layer will not differ greatly from that in the boundary layer of a flat plate. Application of Eq. (5.10) to the Blasius distribution gives a value of θ/x equal to $0.664/\sqrt{R_x}$. Consequently the value of H for the Blasius distribution is $1.73/0.664 = 2.605$. From Eq. (5.5),

$$\tau = 0.332 \frac{\mu U}{x} \sqrt{R_x}$$

Substituting the value of θ

$$\tau = 0.664 \times 0.332 \frac{\mu U}{\theta} = 0.220 \frac{\mu U}{\theta}$$

Substituting the values for H and τ in Eq. (5.11)

$$\frac{d\theta}{dx} + \left(\frac{4.605}{2}\right)\left(\frac{\theta}{\frac{1}{2}\rho U^2}\right)\rho U \frac{dU}{dx} = \frac{0.220\mu U}{\rho U^2\theta}$$

By use of the integrating factor $2\theta U^{9.210}$, this expression becomes

$$\theta^2 = \frac{0.440\mu}{\rho U_x{}^{9.210}} \int_0^x U^{8.210}\, dx \tag{5.12}$$

where U_x = velocity at limit x

If U is known as a function of x, that is, if the pressure distribution is known, Eq. (5.12) permits the boundary-layer thickness to be calculated approximately if the pressure gradients are sufficiently small for the actual distribution of velocity through the boundary layer to be approximated by the Blasius distribution. For most wing sections at low and moderate lift coefficients, the pressure gradients satisfy this condition back to the position of minimum pressure. The skin friction may be assumed to be that corresponding to the Blasius velocity distribution for the calculated thickness of the boundary layer. A convenient way of performing these calculations is to find an equivalent length of flat plate[132] which is given by

$$x_e = \frac{1}{U_x{}^{8.210}} \int_0^x U^{8.210}\, dx \tag{5.13}$$

where x_e is the length of flat plate in a stream of velocity U_x that will generate the same laminar boundary-layer thickness as actually exists on the body in question at the point x.

5.7. Laminar Separation. It was pointed out in Sec. 5.3 that the laminar boundary layer cannot ordinarily exist for long distances in a region where the pressures are increasing in the direction of flow, and the extent of the laminar boundary layer under such conditions is limited by separation. The Blasius solution does not, of course, give any information about laminar separation because this solution is premised on the absence of pressure gradients. Calculations of the location of laminar separation require a solution of the boundary-layer equations in more general form. One such attempt was made by Pohlhausen.[86] He assumed that the velocity distribution in the boundary layer could be expressed by a polynomial of the form

$$\frac{u}{U} = a_1 \frac{y}{\delta} + a_2 \left(\frac{y}{\delta}\right)^2 + a_3 \left(\frac{y}{\delta}\right)^3 + a_4 \left(\frac{y}{\delta}\right)^4$$

Using plausible assumptions regarding the physical characteristics of the layer, Pohlhausen developed a method for calculating the velocity distribution in the boundary layer and the position of the separation point. This method has only limited applicability to the calculation of the laminar separation point.[105] Kármán and Millikan[141] obtained more accurate solutions of Eq. (5.2). The solutions are involved and require much calculation, but the results appear to be satisfactory.

For most actual work on wing sections, the extent of the laminar boundary layer in regions of adverse pressure gradients is too short to require much detailed attention. It is frequently desirable, however, to know the possible extent of the laminar layer as determined by laminar separation. It has been possible to develop a method for applying the Kármán-Millikan solution to the rapid estimation of the laminar separation point[132] which is sufficiently accurate for most applications. More detailed studies may, however, be required for the condition associated with maximum lift.

The method for rapidly estimating the position of laminar separation is based on the assumption that it is possible to replace the actual velocity distribution effectively by a simplified velocity distribution consisting of a region of uniform velocity followed by a region of linearly decreasing velocity. The actual pressure distribution up to the point of minimum pressure is used to calculate the equivalent length of flat plate from Eq. (5.13). Because dp/dx is zero at the point of minimum pressure, the velocity distribution in the boundary layer at this point must approximate the Blasius distribution very closely. The replacement of the actual pressure distribution downstream of the point of minimum pressure by a linear velocity gradient is usually justified by the relatively small extent of the laminar boundary layer in this region. Consequently, the laminar separation point may be estimated from positions calculated for a family of such simplified pressure distributions.

The position of the laminar separation point was calculated by the Kármán-Millikan method for a series of pressure distributions of the following type.

$$\frac{U}{U_0} = 1 \qquad \text{for } 0 \leq \frac{x}{x_e} \leq 1$$

and

$$\frac{U}{U_0} = 1 + F\frac{x - x_e}{x_e} \qquad \text{for } 1 \leq \frac{x}{x_e}$$

where x = distance along surface from leading edge
  x_e = length of flat plate equivalent to length from leading edge to position of minimum pressure
  U = velocity outside boundary layer at any point x
  U_0 = maximum velocity
  F = velocity gradient $(x_e/U_0)(dU/dx)$ (constant for any given case)

The values of U_S/U_0, where U_S is the velocity at the laminar separation point, were calculated by the Kármán-Millikan method for a series of values of F, and the results are shown in Fig. 43.

In applying this method, the value of F is obtained from calculated values of x_e/U_0 and the value of dU/dx corresponding to the estimated equivalent adverse velocity gradient. The corresponding value of U_S/U_0 is found from Fig. 43 and is applied to the actual velocity distribution to locate the position of laminar separation.

5.8. Turbulent Flow in Pipes. The earliest information on the nature of turbulent flow was obtained from studies of the flow in pipes. Reynolds found that, if the diameter of the pipe and the velocity were sufficiently

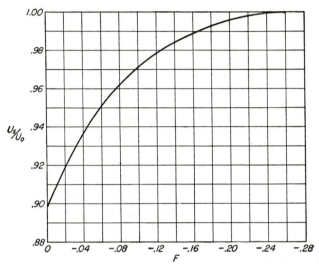

FIG. 43. Velocity ratio for laminar separation as a function of the nondimensional velocity gradient F.

small, the flow was rectilinear or laminar, as shown by the behavior of colored filaments of fluid which did not mix with the surrounding fluid. With larger diameters and greater velocities, the flow became turbulent, as shown by diffusion of the colored filaments. Reynolds found that, if the quantity $\rho V d/\mu$ was less than about 2,000, the flow was always laminar. The upper limit of this Reynolds number for laminar flow was indefinite and depended mainly on the care taken to provide steady conditions and smooth surfaces. Values of the upper limit as much as ten to twenty-five times the lower limit have been found by various observers.[27]

With laminar flow, the pressure gradient along the pipe increases linearly with the velocity; but, with turbulent flow, the gradient increases nearly as the square of the velocity. The velocity distributions across the diameter of the pipe for laminar and turbulent flows are shown in Fig. 44. For turbulent flow, the velocity distribution is more nearly uniform than for laminar flow, and the shearing stresses near the walls are considerably higher. These conditions result from the increased transport of momentum associated with the turbulent motion.

It has been observed that the nondimensional velocity distribution across the pipe is affected only slightly by the Reynolds number. On the basis of dimensional reasoning similar to that used for the formation of dimensionless coefficients for wings, Blasius[15] expressed the pressure loss in coefficient form as follows:

$$\lambda = \frac{p_1 - p_2}{l} \frac{r}{\frac{1}{2}\rho\overline{u}^2}$$

where $p_1 - p_2$ = pressure drop in a pipe of length l and radius r with a fluid of density ρ flowing at average velocity $\overline{u}$.

For smooth pipes or pipes having geometrically similar roughness, the

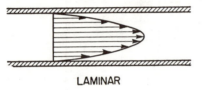

LAMINAR

coefficient λ is a function only of the Reynolds number $R_r = \rho\overline{u}r/\mu$. The formula given by Blasius for the coefficient λ for the case of turbulent flow in smooth pipes is

$$\lambda = \frac{0.133}{\sqrt[4]{R_r}}$$

TURBULENT

FIG. 44. Velocity distribution across pipe for laminar and turbulent flow.

Prandtl[89] arrived at the conclusion that there should be a relation between the velocity distribution across the pipe and the rate at which the pressure-loss coefficient changes with Reynolds number. The most important region in which to study the nature of the phenomena appeared to be in the vicinity of the wall of the pipe where the velocity gradient is high rather than in the middle of the pipe where the velocity is nearly uniform. It appeared reasonable to assume that the velocity distribution near the wall depended on the local quantities ρ and μ and the skin friction τ and was independent of the radius of the pipe. It was assumed, in approximate agreement with experimental evidence, that the velocity distribution across the entire pipe could be expressed as a function of y/r alone where y is the distance from the wall of the pipe; that is, the velocity distribution remains similar for various mean velocities and pipe sizes.

Equating the pressure loss to the skin friction

$$(p_1 - p_2)\pi r^2 = 2\pi r l \tau$$

$$p_1 - p_2 = \frac{2l\tau}{r}$$

From Blasius' formula

$$p_1 - p_2 = \frac{0.133}{\sqrt[4]{R_r}} \frac{l}{r} \frac{\rho}{2} \overline{u}^2$$

$$\tau = \frac{0.033}{\sqrt[4]{R_r}} \rho\overline{u}^2 = 0.033\rho^{3/4}\mu^{1/4}r^{-1/4}\overline{u}^{7/4} \tag{5.14}$$

If it is further assumed that the velocity distribution may be expressed as a power function of y/r,

$$u = V \left(\frac{y}{r} \right)^n$$

where V = velocity at center of pipe

$$\tau = K \rho^{3/4} \mu^{1/4} r^{-1/4} V^{1/4} \left(\frac{r}{y} \right)^{(7/4)n}$$

where K = a constant factor dependent on ratio of maximum velocity to mean velocity

Applying the assumption that the velocity distribution near the wall and hence the skin friction is independent of the radius of the pipe, the exponent of r in the foregoing expression must be zero,

$$-\tfrac{1}{4} + \tfrac{7}{4}n = 0$$
$$n = \tfrac{1}{7}$$

or

$$u = V \left(\frac{y}{r} \right)^{1/7} \tag{5.15}$$

The assumption made in the derivation of this equation that the velocity distribution is a function only of local parameters near the wall would indicate that this equation would also apply to the velocity distribution in a turbulent boundary layer on a flat plate. In this case, r is interpreted as the thickness of the boundary layer and V as the velocity just outside the layer. The validity of the analogy between the flow in pipes and that over a flat plate is shown by the data of reference 19, some of which are presented in Fig. 45.

5.9. Turbulent Skin Friction. It is possible to calculate the skin friction on a flat plate with turbulent flow by application of the one-seventh power law [Eq. (5.15)] and the Blasius skin-friction formula [Eq. (5.14)] to the local boundary-layer conditions. In the case of the flow of an incompressible fluid in a pipe, the Reynolds number along the pipe remains constant and the skin friction results in a loss of pressure. In the case of the flat plate, the pressure is constant along the plate and the skin friction causes a deceleration of the fluid particles, resulting in an increase of the thickness of the boundary layer and a corresponding increase of the boundary-layer Reynolds number. At any station along the plate, the skin-friction drag over that portion of the plate upstream of the station can be obtained from the rate of loss of momentum in the boundary layer. It is assumed that, for any station, the velocity distribution in the boundary layer corresponds to that in the pipe and that the local skin friction is the same as that in a pipe at the same Reynolds number.

The particles in the boundary layer having an original velocity V up-

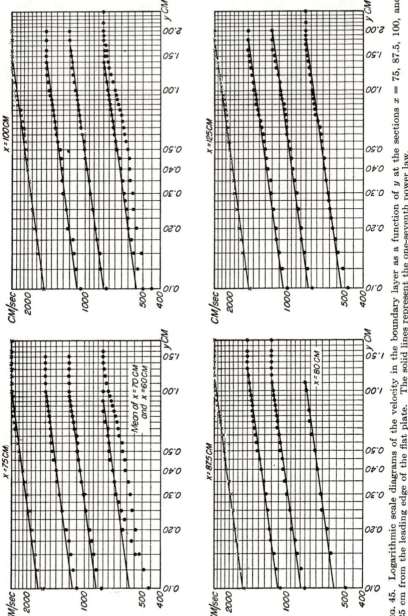

Fig. 45. Logarithmic scale diagrams of the velocity in the boundary layer as a function of y at the sections $x = 75$, 87.5, 100, and 125 cm from the leading edge of the flat plate. The solid lines represent the one-seventh power law.

stream of the plate are decelerated to a velocity u at station x. The rate of loss of momentum at station x is given by the following expression and is equal to the skin-friction drag of that portion of the plate upstream of station x.

$$F = \rho \int_0^\delta u(V - u)dy$$

From the assumption that

$$\frac{u}{V} = \left(\frac{y}{\delta}\right)^{\frac{1}{7}}$$

$$F = \frac{7}{72} \rho V^2 \delta$$

but

$$\frac{dF}{dx} = \tau$$

hence

$$\tau = \frac{7}{72} \rho V^2 \frac{d\delta}{dx}$$

A corresponding value of τ may be obtained from Eq. (5.14) if the relation between $\bar{u}$ and V is known. A commonly used experimental value of $\bar{u}/V$ is approximately 0.8. Using this value Eq. (5.14) may be written in the form

$$\tau = 0.0225\rho^{\frac{3}{4}}\mu^{\frac{1}{4}}\delta^{-\frac{1}{4}}V^{\frac{7}{4}} \tag{5.16}$$

Equating the two expressions for τ and integrating

$$\frac{4}{5} \delta^{\frac{5}{4}} = \frac{72}{7} 0.0225 \left(\frac{\mu}{\rho V}\right)^{\frac{1}{4}} x$$

Solving for δ

$$\delta = 0.37 \left(\frac{\mu}{\rho V}\right)^{\frac{1}{5}} x^{\frac{4}{5}} \tag{5.17}$$

Substituting this value of δ into the expression for the drag

$$F = 0.036\rho V^2 x \left(\frac{\mu}{\rho V x}\right)^{\frac{1}{5}}$$

In coefficient form

$$c_F = \frac{F}{\frac{1}{2}\rho V^2 x} = 0.072 R_x^{-\frac{1}{5}} \tag{5.18}$$

Equation (5.18) is a formula for the drag coefficient for one side of a smooth flat plate of length x with turbulent flow over the whole surface. Figure 46 shows the skin-friction coefficient for turbulent flow from Eq. (5.18) and for laminar flow from Eq. (5.6).

Nikuradse performed pipe experiments over a range of Reynolds numbers much larger than for previous investigations.[75, 75a] These experiments showed that the pressure losses departed from values predicted by the Blasius formula above Reynolds numbers of about 50,000, the experimental values decreasing less rapidly than the predicted values as the Reynolds number was increased. As expected from the simplified analysis of the relation between the pressure-loss coefficients and the velocity distributions, the velocity distributions at the higher Reynolds numbers are approximated better by one-eighth, one-ninth, or smaller powers than

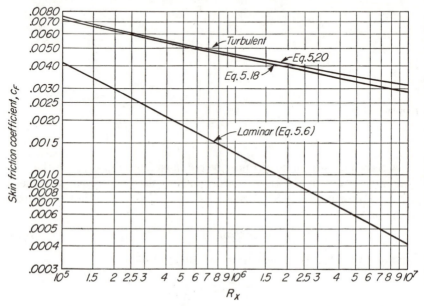

Fig. 46. Laminar and turbulent skin-friction coefficients for one side of a flat plate.

by the one-seventh power. On the basis of an elaborate and extended dimensional analysis, a logarithmic expression was obtained[28, 138] for the velocity distributions corresponding to Nikuradse's values for the pressure losses.

$$\phi = 2.5 \ln \eta + 5.5 \tag{5.19}$$

where

$$\phi = \frac{u}{v^*}$$

$$\eta = \frac{\rho v^* y}{\mu}$$

$$v^* = \sqrt{\frac{\tau}{\rho}}$$

and is called the "friction velocity." The skin friction on a flat plate may be calculated using this expression by a process using the same general principles as that for the derivation of Eq. (5.18).[28] The tabular solution to this problem may be approximated by the expression

$$c_F = \frac{0.472}{(\log R_x)^{2.58}} \tag{5.20}$$

Skin-friction coefficients calculated from this formula are shown in Fig. 46.

Formulas (5.18) and (5.20) are in convenient form for the calculation of skin friction on flat plates. For application to bodies having arbitrary pressure distributions, it is more convenient to relate the local skin-friction coefficient to a Reynolds number based on the momentum thickness of the boundary layer. For the one-seventh power velocity distribution, such an expression may be obtained from Eq. (5.16). For this case, according to Eq. (5.10),

$$\theta = \frac{7}{72} \delta$$

$$c_f = \frac{\tau}{\frac{1}{2}\rho U^2} = 0.0251 \left(\frac{\mu}{\rho U \theta}\right)^{\frac{1}{4}} = 0.0251 R_\theta^{-\frac{1}{4}} \tag{5.21}$$

For the case of the logarithmic velocity distribution, approximate formulas have been obtained relating the local skin-friction coefficient to the boundary-layer Reynolds number R_θ. That given by Squire and Young[112] is

$$\frac{\rho U^2}{\tau} = [5.890 \log (4.075 R_\theta)]^2 \tag{5.22}$$

It is known that Eq. (5.21) is valid only for moderately small Reynolds numbers, whereas Eq. (5.22) is a much better approximation at large Reynolds numbers. Equation (5.22) is, however, inconvenient for use in connection with the momentum equation (5.11). Interpolation formulas have therefore been developed in the form

$$c_f = A R_\theta^{-n} \tag{5.23}$$

where the coefficients and exponents are selected to correspond with the desired range of Reynolds numbers.

For any particular application, the two constants of Eq. (5.23) may be found by making Eq. (5.23) agree with the assumed skin-friction law at two values of R_θ including the range under consideration. If the subscript 1 corresponds to conditions at the low value of R_θ and the subscript 2 to conditions at the high value of R_θ, then

$$n = \frac{\log \dfrac{(\tau/\rho U^2)_1}{(\tau/\rho U^2)_2}}{\log (R_{\theta_2}/R_{\theta_1})}$$

$$A = c_{f_1} R_{\theta_1}{}^n$$

The assumed skin-friction law may be Eq. (5.22) or the following formula suggested by Tetervin[120]

$$\frac{\tau}{\rho U^2} = \frac{1}{\left[2.5 \ln \dfrac{R_\theta}{2.5(1 - 5\sqrt{\tau/\rho U^2})} + 5.5 \right]^2} \qquad (5.24)$$

The values of $\dot{c}_f$ corresponding to Eqs. (5.22) and (5.24) are plotted in Fig. 47.

5.10. Calculation of Thickness of the Turbulent Layer. A knowledge of the thickness of the turbulent boundary layer is required for several

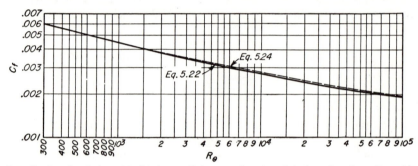

Fig. 47. Local turbulent skin-friction coefficient as a function of the boundary-layer Reynolds number.

purposes. The profile drag of the wing is, of course, intimately associated with the thickness of the boundary layer at the trailing edge. It is also desirable to know the thickness of the boundary layer at various points on the wing surface in connection with the study of control surfaces, high-lift devices, air intakes, and protuberances.

The thickness of the turbulent boundary layer can be calculated from the momentum equation (5.11) if a relation between the local skin-friction coefficient and the dependent variables θ, q, and H is known. Values of the skin-friction coefficient have been found by assuming that the relation between the boundary-layer Reynolds number R_θ and the skin-friction coefficient c_f is the same as the corresponding relation found from pipe experiments. This relation has been plotted in Fig. 47. Although this relation may not be very accurate, particularly for values of the shape parameter H differing considerably from those found in pipes, such marked differences usually occur only when dq/dx is large negatively. In this case Eq. (5.11) shows that the contribution of the skin-friction term to the rate of increase of the boundary-layer thickness $d\theta/dx$ is relatively small. It may accordingly be assumed for practical calculations that c_f is independent of the value of H.

Further simplification of the calculations is obtained by assuming a skin-friction law of the type indicated by Eq. (5.23). In this case, Eq. (5.11) may be written in the form

$$\frac{d\theta}{dx} + \left(\frac{H+2}{U}\right)\frac{dU}{dx}\,\theta = \frac{A\mu^n}{(\rho U)^n}\,\theta^{-n}$$

The value of the factor $(H+2)$ is not greatly affected by the value of H, which usually has values of about 1.4 or 1.5 with extreme variations from about 1.2 to 2.5. Detailed calculations have shown that excellent agreement can be obtained between experimental and calculated values of θ if H is assumed to be constant and to have a value[120] between 1.4 and 1.6. With H assumed constant, this equation is seen to be of the Bernoulli type, and it can be integrated to give the value of θ as follows:

$$\left(\frac{\theta}{c}\right)_{x/c} = \frac{1}{(U/V)^{H+2}}\left[\frac{(1+n)A}{2R_c{}^n}\int_{x_0/c}^{x/c}\left(\frac{U}{V}\right)^{(H+1)(n+1)+1}d\,\frac{x}{c}\right.$$
$$\left. + \left(\frac{\theta_0}{c}\right)^{n+1}\left(\frac{U_0}{V}\right)^{(H+2)(n+1)}\right]^{1/(1+n)} \quad (5.25)$$

5.11. Turbulent Separation. To determine the turbulent separation point, detailed information is needed concerning the velocity distribution in the boundary layer and the effects on the velocity distribution of such factors as the pressure distribution and the skin friction. Although the value of the shape parameter H can be found for any given velocity distribution, the value of H in itself does not define the velocity distribution. The data collected by Gruschwitz[38] and by Tetervin[135] show experimentally that the velocity distributions in turbulent boundary layers actually form a one-parameter family of curves and that the velocity distributions can be specified by the value of H. Velocity profiles for turbulent boundary layers[135] are presented for various values of H in Fig. 48. The value of H increases as the separation point is approached. It is not possible to give an exact value of H corresponding to separation, because the turbulent separation point is not very well defined. The value of H varies so rapidly near the separation point, however, that it is not necessary to fix accurately the value of H corresponding to separation. Separation has not been observed for a value of H less than 1.8 and appears definitely to have been observed for a value of H of 2.6.

An empirical expression was found[135] for the rate of change of H along the surface in terms of the ratio of the pressure gradient to the skin friction and the shape parameter itself. The following equation was derived to fit the experimental data:

$$\theta\frac{dH}{dx} = e^{4.680(H-2.975)}\left[-\frac{\theta}{q}\frac{dq}{dx}\frac{2q}{\tau} - 2.035(H-1.286)\right] \quad (5.26)$$

The skin friction corresponding to a given value of R_θ may be obtained from Fig. 47 or from Eqs. (5.22) or (5.24). Equations (5.11) and (5.26) are simultaneous first-order differential equations that can be solved by step-by-step calculation. It is usually necessary to use such a method, although,

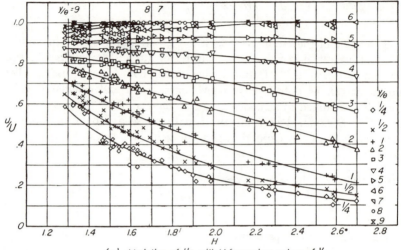

(a) - Variation of u/U with H for various values of y/θ

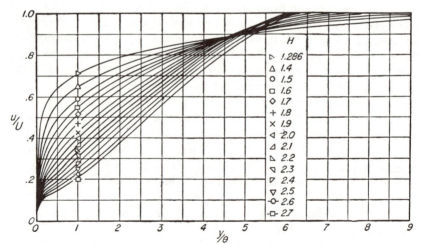

(b) - Velocity profiles for turbulent boundary layers corresponding to various values of H

Fig. 48. Velocity distributions in turbulent boundary layers.

for some particular cases, the equations may be integrated directly. The method of calculation is as follows. The values of the variables entering into the computation at the initial station are substituted in the momentum equation (5.11) and the equation for dH/dx (5.26). Values for $d\theta/dx$ and dH/dx are thus obtained at the initial station. An increment of the length along the surface of the body x is then chosen and multiplied by $d\theta/dx$ and dH/dx to give $\Delta\theta$ and ΔH, respectively. These increments of θ and H are added to the initial values and result in values of θ and H for the new value of x. The process is repeated until the separation point is reached, as shown by the value of H. For the purposes of the computations the value of H corresponding to separation may be taken as 2.6. The choice of the increments of x is a matter for the judgment of the investigator. As a general rule, the increment of x should be made small when $d\theta/dx$ or dH/dx changes rapidly from one value of x to the next.

5.12. Transition from Laminar to Turbulent Flow. Experience has shown that transition from laminar to turbulent flow takes place at a Reynolds number that depends upon the magnitude of the disturbances. For viscous flows at very low Reynolds numbers, as in oil, all disturbances are damped out by the viscosity, and the flow is laminar regardless of the magnitude of any disturbance. As the Reynolds number is increased, a condition is reached at which some particular types of disturbances are amplified and eventually cause transition. This value of the Reynolds number is called the "lower critical." Further increase of the Reynolds number causes amplification to occur for a greater variety of disturbances and increases the rate of amplification. Under these circumstances the Reynolds number at which transition occurs depends on the magnitude of the disturbances. Transition can be delayed to high values of the Reynolds number only by reducing all disturbances such as stream turbulence, unsteadiness, and surface roughness to a minimum. For flows in a pipe, the lower value of the critical Reynolds number based on the diameter and the mean velocity is about 2,300.[98] The upper value appears to depend only on the care taken in conducting the experiment, and values twenty times as great as the lower value have been obtained.

It has been found possible[99, 100, 127, 128] to compute the rate of amplification or damping of disturbances of various frequencies and wave lengths. Such computations[66a] permit curves of constant amplification coefficient $\beta i\delta^*/V$ to be plotted on coordinates of $\alpha\delta^*$ against the Reynolds number $R_{\delta*} = \rho V\delta^*/\mu$ where $\alpha = 2\pi/l$ and l is the wave length. Such a plot is given in Fig. 49. Experimental points obtained by Schubauer and Skramstad[106] are also plotted in Fig. 49. The data of Fig. 49 were obtained for a Blasius profile corresponding to flow over a flat plate. It will be noted that there is a Reynolds number below which no disturbance will be amplified. This value of the Reynolds number is the lower critical. The agreement between the experimental and theoretical results is good considering the difficulties of the experimental work and the simplifying assumptions made in the development of the theory.

Similar calculations have been made by Schlichting and Ulrich[101] for Pohlhausen's velocity profiles corresponding to various pressure gradients.

The results are presented in Fig. 50, which is a plot of the lower critical Reynolds number against the parameter λ that defines the velocity distribution through the laminar layer. According to the Pohlhausen method,

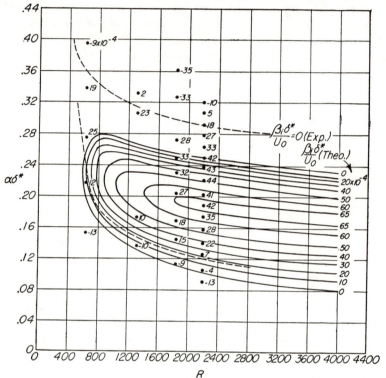

FIG. 49. Contours of equal amplification according to Schlichting. Values of $\beta_i\delta^*/U_0$ (all values to be multiplied by 10^{-4}) opposite points are amplifications determined by experiment. Faired experimental contour of zero amplification shown by broken curves.

the velocity distribution through the layer is given by the following formula

$$\frac{u}{U} = \left(\frac{12 + \lambda}{6}\right)\frac{y}{\delta} - \frac{\lambda}{2}\left(\frac{y}{\delta}\right)^2 - \left(\frac{4 - \lambda}{2}\right)\left(\frac{y}{\delta}\right)^3 + \left(\frac{6 - \lambda}{6}\right)\left(\frac{y}{\delta}\right)^4$$

and

$$\lambda = \frac{\rho\delta^2}{\mu}\frac{dU}{dx}$$

Figure 50 shows that decreasing pressures in the direction of flow are favorable to the stability of the laminar boundary layer and that increasing pressures are unfavorable.

The theory has not been developed to such a point that it can be used to predict transition, although it is useful in pointing out the fundamental reasons for the instability of the laminar layer at high Reynolds numbers. The theory does not relate the boundary-layer disturbances to either the surface irregularities or the turbulence of the air stream. The theory also does not relate the transition to the magnitude of the fluctuations of velocity in the boundary layer. Under these circumstances, it is necessary to

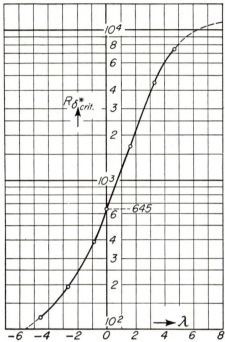

Fig. 50. Critical values of $R\delta^*$ as a function of the shape parameter λ according to Schlichting and Ulrich.[101]

rely on empirical results to predict the location of transition. The empirical data are discussed in Chap. 7.

5.13. Calculation of Profile Drag. In cases where it is known that the flow is not separated from the surface, the boundary-layer thickness at the trailing edge of a wing section and the profile drag can be calculated from a knowledge of the potential-flow pressure distribution and the location of transition from laminar to turbulent flow. The momentum thickness of the laminar boundary layer at the transition point can be calculated from Eq. (5.12). The initial momentum thickness of the turbulent boundary layer is taken to be the same as that of the laminar layer at the transition point. The thickness of the turbulent boundary layer at the trailing edge can be calculated from Eq. (5.25). If the pressure at the trailing edge were free-stream static pressure, as in the case of a flat plate, the profile drag would be

$$c_d = 2\left(\frac{\theta}{c}\right)$$

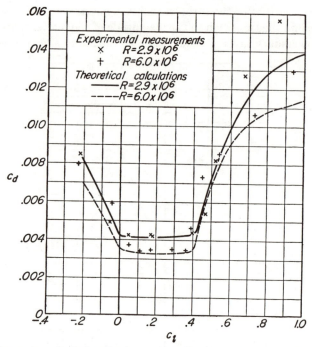

Fig. 51. Comparison of calculated and experimentally measured polars for NACA 67,1-215 airfoil.

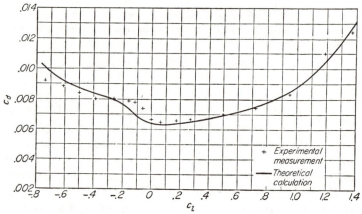

Fig. 52. Comparison of calculated and experimentally measured polars for NACA 23015 airfoil at $R = 5.9 \times 10^6$.

where θ = sum of momentum thicknesses on upper and lower surfaces

In the usual case, the pressure at the trailing edge is not the same as free-stream static pressure, and some means must be found for finding the effective momentum thickness at a point far downstream in the wake where the pressure has returned to the free-stream static value. Squire and Young[112] derived the required relation by setting the value of τ in Eq. (5.11) equal to zero in the wake and finding an empirical relation between H and q/q_0. The resulting expression for the profile drag is

$$c_d = 2\left(\frac{\theta}{c}\right)_t\left(\frac{U}{V}\right)_t^{(H+5)/2} \tag{5.27}$$

where the subscript t designates conditions at the trailing edge. The value of H is the value used in calculating the boundary-layer thickness.

The agreement to be expected between experimental and calculated profile drag coefficients is indicated by Figs. 51 and 52. The calculated drag coefficients presented in these figures were obtained by a method[76] fundamentally the same as that presented here.

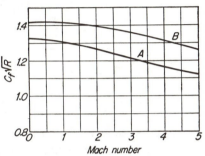

Fig. 53. Skin-friction coefficients. (*A*) No heat transferred to wall, (*B*) wall temperature one-quarter of free-stream temperature.

5.14. Effect of Mach Number on Skin Friction. Solutions for the velocity distribution through the laminar boundary layer in compressible flow and for the corresponding skin friction have been obtained by von Kármán and Tsien[142] for the case of the flat plate. The results of the skin-friction calculations are presented in Fig. 53 for the case of no heat transfer to the plate and for the case of a plate whose absolute temperature is one-quarter of that of the free stream. These results show a moderate decrease in laminar skin friction with increasing Mach number. Although these results indicate that the laminar skin friction increases with heat transfer from the fluid to the plate, this increase is small even for the extremely low plate temperatures for which these results were obtained.

For the turbulent boundary layer, Theodorsen and Regier[126] found experimentally that the skin friction is independent of the Mach number (at least up to a value of 1.69). The experimental data are presented in Fig. 54. These data were obtained by measuring the torques (moments) required to rotate smooth disks in atmospheres of air and Freon 12 (CCl_2F_2) at various pressures. Data obtained by Keenan and Neumann[55] for the skin friction in pipes appear to confirm Theodorsen's conclusion for the fully developed turbulent flow.

Analytical studies by Lees[62] on the effect of Mach number and heat transfer on the lower critical Reynolds number for the case of the flat plate indicate that increasing the Mach number has a destabilizing influence on laminar flow when there is no heat transfer to the plate. Heat transfer to the plate has a stabilizing effect, while the opposite is the case for heat transfer from the plate to the fluid. This effect becomes stronger

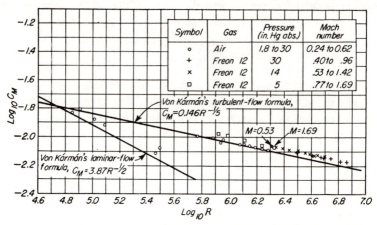

Fig. 54. Moment coefficient for disks as function of Reynolds number for several values of Mach number with air and Freon 12 as mediums. Maximum Mach number, 1.69.

as the Mach number is increased, and, at moderate supersonic speeds, comparatively small amounts of heat transfer to the plate stabilize the laminar layer to very high values of the Reynolds number.

CHAPTER 6

FAMILIES OF WING SECTIONS

6.1. Symbols.

P_R resultant pressure coefficient

R radius of curvature of surface of modified NACA four-digit series symmetrical sections at the point of maximum thickness

V velocity of the free stream

a coefficient

a mean-line designation; fraction of the chord from leading edge over which loading is uniform at the ideal angle of attack

c chord

c_{l_i} section design lift coefficient

$c_{m_{c/4}}$ section moment coefficient about the quarter-chord point

d coefficient

k_1 constant

m maximum ordinate of the mean line in fraction of the chord

p chordwise position of m

r leading-edge radius in fraction of the chord

r_t leading-edge radius corresponding to thickness ratio t

t maximum thickness of section in fraction of chord

v local velocity over the surface of a symmetrical section at zero lift

Δv increment of local velocity over the surface of a wing section associated with camber

Δv_a increment of local velocity over the surface of a wing section associated with angle of attack

x abscissa of point on the surface of a symmetrical section or a chord line

x_L abscissa of point on the lower surface of a wing section

x_U abscissa of point on the upper surface of a wing section

x_c abscissa of point on the mean line

y_L ordinate of point on the lower surface of a wing section

y_U ordinate of point on the upper surface of a wing section

y_c ordinate of point on the mean line

y_t ordinate of point on the surface of a symmetrical section

α_i design angle of attack

θ $\tan^{-1}(dy_c/dx_c)$

τ trailing-edge angle

6.2. Introduction. Until recently the development of wing sections has been almost entirely empirical. Very early tests indicated the desirability of a rounded leading edge and of a sharp trailing edge. The demand for improved wings for early airplanes and the lack of any generally accepted wing theory led to tests of large numbers of wings with shapes gradually improving as the result of experience. The Eiffel and early RAF series were outstanding examples of this approach to the problem.

111

The gradual development of wing theory tended to isolate the wing-section problem from the effects of plan form and led to a more systematic experimental approach. The tests made at Göttingen during the First World War contributed much to the development of modern types of wing sections. Up to about the Second World War, most wing sections in common use were derived from more or less direct extensions of the work at Göttingen. During this period, many families of wing sections were tested in the laboratories of various countries, but the work of the NACA was outstanding. The NACA investigations were further systematized by separation of the effects of camber and thickness distribution, and the experi-

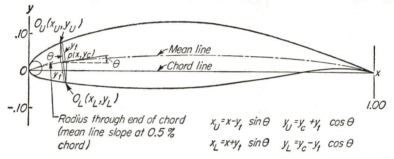

$$x_U = x - y_t \sin \theta \qquad y_U = y_c + y_t \cos \theta$$
$$x_L = x + y_t \sin \theta \qquad y_L = y_c - y_t \cos \theta$$

Sample calculations for derivation of the NACA 65,3-818 airfoil ($a = 1.0$)

x	y_t*	y_c†	$\tan \theta$	$\sin \theta$	$\cos \theta$	$y_t \sin \theta$	$y_t \cos \theta$	x_U	y_U	x_L	y_L
0	0	0	....	....	......	0	0	0	0	0	0
0.005	0.01324	0.00200	0.33696‡	0.31932	0.94765	0.00423	0.01255	0.00077	0.01455	0.00923	−0.01055
0.05	0.03831	0.01264	0.18744	0.18422	0.98288	0.00706	0.03765	0.04294	0.05029	0.05706	−0.02501
0.25	0.08093	0.03580	0.06996	0.06979	0.99756	0.00565	0.08073	0.24435	0.11653	0.25565	−0.04493
0.50	0.08593	0.04412	0	0	1.00000	0	0.08593	0.50000	0.13005	0.50000	−0.04181
0.75	0.04456	0.03580	−0.06996	−0.06979	0.99756	−0.00311	0.04445	0.75311	0.08025	0.74689	−0.00865
1.00	0	0	....	....	......	0	0	1.00000	0	1.00000	0

* Thickness distribution obtained from ordinates of the NACA 65,3-018 airfoil.

† Ordinates of the mean line, 0.8 of the ordinate for $c_{l_i} = 1.0$.

‡ Slope of radius through end of chord.

Fig. 55. Method of combining mean lines and basic-thickness forms.

mental work was performed at higher Reynolds numbers than were generally obtained elsewhere. The wing sections now in common use are either NACA sections or have been strongly influenced by the NACA investigations. For this reason, and because the NACA sections form consistent families, detailed attention will be given only to modern NACA wing sections.

6.3. Method of Combining Mean Lines and Thickness Distributions. The cambered wing sections of all NACA families of wing sections considered here are obtained by combining a mean line and a thickness distribution. The process for combining a mean line and a thickness distribution to obtain the desired cambered wing section is illustrated in Fig. 55. The leading and trailing edges are defined as the forward and rearward extremities, respectively, of the mean line. The chord line is defined as the straight line connecting the leading and trailing edges. Ordinates of the cambered wing sections are obtained by laying off the thickness distributions perpendicular to the mean lines. The abscissas, ordinates, and slopes of the mean line are designated as x_c, y_c, and $\tan \theta$, respectively. If x_U and y_U represent, respectively, the abscissa and ordinate of a typical point of the upper surface of the wing section and y_t is the ordinate of the symmetrical thickness distribution at chordwise position x, the upper-surface coordinates are given by the following relations:

$$\left. \begin{array}{l} x_U = x - y_t \sin \theta \\ y_U = y_c + y_t \cos \theta \end{array} \right\} \tag{6.1}$$

The corresponding expressions for the lower-surface coordinates are

$$\left. \begin{array}{l} x_L = x + y_t \sin \theta \\ y_L = y_c - y_t \cos \theta \end{array} \right\} \tag{6.1}$$

The center for the leading-edge radius is found by drawing a line through the end of the chord at the leading edge with a slope equal to the slope of the mean line at that point and laying off a distance from the leading edge along this line equal to the leading-edge radius. This method of construction causes the cambered wing sections to project slightly forward of the leading-edge point. Because the slope at the leading edge is theoretically infinite for the mean lines having a theoretically finite load at the leading edge, the slope of the radius through the end of the chord for such mean lines is usually taken as the slope of the mean line at x/c equals 0.005. This procedure is justified by the manner in which the slope increases to the theoretically infinite value as x/c approaches 0. The slope increases slowly until very small values of x/c are reached. Large values of the slope are thus limited to values of x/c very close to 0 and may be neglected in practical wing-section design.

The data required to construct some cambered wing sections are presented in Appendixes I and II, and ordinates for a number of cambered sections are presented in Appendix III.

6.4. NACA Four-digit Wing Sections. *a. Thickness Distributions.* When the NACA four-digit wing sections were derived,[49] it was found that the thickness distributions of efficient wing sections such as the Göttingen 398 and the Clark Y were nearly the same when their camber was removed (mean line straightened) and they were reduced to the same maximum thickness. The thickness distribution for the NACA four-digit sections was selected to correspond closely to that for these wing sections and is given by the following equation:

$$\pm y_t = \frac{t}{0.20} \left(0.29690\sqrt{x} - 0.12600x - 0.35160x^2 + 0.28430x^3 - 0.10150x^4 \right) \tag{6.2}$$

where t = maximum thickness expressed as a fraction of the chord
The leading-edge radius is

$$r_t = 1.1019t^2 \tag{6.3}$$

It will be noted from Eqs. (6.2) and (6.3) that the ordinate at any point is directly proportional to the thickness ratio and that the leading-edge radius varies as the square of the thickness ratio. Ordinates for thickness ratios of 6, 9, 12, 15, 18, 21, and 24 per cent are given in Appendix I.

b. Mean Lines. In order to study systematically the effect of variation of the amount of camber and the shape of the mean line, the shape of the mean lines was expressed analytically as two parabolic arcs tangent at the position of maximum mean-line ordinate. The equations[49] defining the mean lines were taken to be

$$\left.\begin{aligned} y_c &= \frac{m}{p^2}\,(2px - x^2) && \text{forward of maximum ordinate} \\[2mm] \text{and}& \\[1mm] y_c &= \frac{m}{(1-p)^2}\,[(1 - 2p) + 2px - x^2] && \text{aft of maximum ordinate} \end{aligned}\right\} \tag{6.4}$$

where m = maximum ordinate of mean line expressed as fraction of chord
p = chordwise position of maximum ordinate

It will be noted that the ordinates at all points on the mean line vary directly with the maximum ordinate. Data defining the geometry of mean lines with the maximum ordinate equal to 6 per cent of the chord are presented in Appendix II for chordwise positions of the maximum ordinate of 20, 30, 40, 50, 60, and 70 per cent of the chord.

c. Numbering System. The numbering system for NACA wing sections of the four-digit series is based on the section geometry. The first integer indicates the maximum value of the mean-line ordinate y_c in per cent of the chord. The second integer indicates the distance from the leading edge to the location of the maximum camber in tenths of the chord. The last two integers indicate the section thickness in per cent of the chord. Thus the NACA 2415 wing section has 2 per cent camber at 0.4 of the chord from the leading edge and is 15 per cent thick.

The first two integers taken together define the mean line, for example, the NACA 24 mean line. Symmetrical sections are designated by zeros for the first two integers, as in the case of the NACA 0015 wing section, and are the thickness distributions for the family.

d. Approximate Theoretical Characteristics. Values of $(v/V)^2$, which is equivalent to the low-speed pressure distribution, and values of v/V are presented in Appendix I for the NACA 0006, 0009, 0012, 0015, 0018, 0021, and 0024 wing sections at zero angle of attack. These values were calculated by the method of Sec. 3.6. Values of the velocity increments $\Delta v_a/V$ induced by changing angle of attack are also presented for an additional lift coefficient of approximately unity. Values of the velocity ratio v/V for intermediate thickness ratios may be obtained approximately by linear scaling of the velocity increments obtained from the tabulated values of v/V for the nearest thickness ratio; thus

$$\left(\frac{v}{V}\right)_{t_2} = \left[\left(\frac{v}{V}\right)_{t_1} - 1\right]\frac{t_2}{t_1} + 1 \tag{6.5}$$

Values of the velocity increment ratio $\Delta v_a/V$ may be obtained for intermediate thicknesses by interpolation.

The design lift coefficient c_{l_i}, and the corresponding design angle of attack α_i, the moment coefficient $c_{m_{c/4}}$, the resultant pressure coefficient P_R, and the velocity ratio $\Delta v/V$ for the NACA 62, 63, 64, 65, 66, and 67 mean lines are presented in Appendix II. These values were calculated by the method of Sec. 3.6. The tabulated values for each mean line may be assumed to vary linearly with the maximum ordinate y_c; and data for similar mean lines with different amounts of camber, within the usual range, may be obtained simply by scaling the tabulated values. Data for the NACA 22 mean line may thus be obtained simply by multiplying the data for the NACA 62 mean line by the ratio 2:6, and for the NACA 44 mean line by multiplying the data for the NACA 64 mean line by the ratio 4:6.

Approximate theoretical pressure distributions may be obtained for cambered wing sections from the tabulated data for the thickness forms and mean lines by the method presented in Sec. 4.5.

6.5. NACA Five-digit Wing Sections. *a. Thickness Distributions.* The thickness distributions for the NACA five-digit wing sections are the same as for the NACA four-digit sections (see Sec. 6.4a).

b. Mean Lines. The results of tests of the NACA four-digit series wing sections indicated that the maximum lift coefficient increased as the position of maximum camber was shifted either forward or aft of approximately the mid-chord position. The rearward positions of maximum camber were not of much interest because of large pitching-moment coefficients. Because the type of mean line used for the NACA four-digit sections was not suitable for extreme forward positions of the maximum camber, a new series of mean lines was developed, and the resulting sections are the NACA five-digit series.

The mean lines are defined[46] by two equations derived so as to produce shapes having progressively decreasing curvatures from the leading edge aft. The curvature decreases to zero at a point slightly aft of the position of maximum camber and remains zero from this point to the trailing edge. The equations for the mean line are

$$\left.\begin{array}{ll} y_c = \frac{1}{6}k_1[x^3 - 3mx^2 + m^2(3 - m)x] & \text{from } x = 0 \text{ to } x = m \\ y_c = \frac{1}{6}k_1m^3(1 - x) & \text{from } x = m \text{ to } x = c = 1 \end{array}\right\} \tag{6.6}$$

The values of m were determined to give five positions p of maximum camber, namely, $0.05c$, $0.10c$, $0.15c$. $0.20c$, and $0.25c$. Values of k_1 were initially calculated to give a design lift coefficient of 0.3. The resulting values of p, m, and k_1 are given in the following table.

Mean-line designation	Position of camber p	m	k_1
210	0.05	0.0580	361.4
220	0.10	0.1260	51.64
230	0.15	0.2025	15.957
240	0.20	0.2900	6.643
250	0.25	0.3910	3.230

This series of mean lines was later extended[48] to other design lift coefficients by scaling the ordinates of the mean lines. Data for the mean lines tabulated in the foregoing table are presented in Appendix II.

c. Numbering System. The numbering system for wing sections of the NACA five-digit series is based on a combination of theoretical aerodynamic characteristics and geometric characteristics. The first integer indicates the amount of camber in terms of the relative magnitude of the design lift coefficient; the design lift coefficient in tenths is thus three-halves of the first integer. The second and third integers together indicate the distance from the leading edge to the location of the maximum camber; this distance in per cent of the chord is one-half the number represented by these integers. The last two integers indicate the section thickness in per cent of the chord. The NACA 23012 wing section thus has a design lift coefficient of 0.3, has its maximum camber at 15 per cent of the chord, and has a thickness ratio of 12 per cent.

d. Approximate Theoretical Characteristics. The theoretical aerodynamic characteristics of the NACA five-digit series wing sections may be obtained by the same method as that previously described (Sec. 6.4d) for the NACA four-digit series sections using the data presented in Appendixes I and II.

6.6. Modified NACA Four- and Five-digit Series Wing Sections. Some early modifications of the NACA four-digit series wing section[49] included thinner nosed and blunter nosed sections which were denoted by the suffixes T and B, respectively. Some sections of this family[49] with reflexed mean lines were designated by numbers of the type $2R_112$ and $2R_212$, where the first integer indicates the maximum camber in per cent of the chord and the subscripts 1 and 2 indicate small positive and negative moments, respectively.

Another series of NACA five-digit wing sections[46] is the same as those previously described except that the mean lines are reflexed to produce theoretically zero pitching moment. These sections are distinguished by the third integer, which is always 1 instead of 0. These modified sections have been little used and will not be discussed further.

More important modifications common to both the NACA four-digit and five-digit series wing sections consisted of systematic variations of the thickness distributions.[116] These modifications are indicated by a suffix consisting of a dash and two digits as for the NACA 0012-64 or the 23012-64 sections. These modifications consist essentially of changes of the leading-edge radius from the normal value [Eq. (6.3)] and changes of the position of maximum thickness from the normal position at $0.30c$ [Eq. (6.2)]. The first integer following the dash indicates the relative magnitude of the leading-edge radius. The normal leading-edge radius is designated by 6 and a sharp leading edge by 0. The leading-edge radius varies as the square of this integer except for values larger than 8, when the variation becomes arbitrary.[116] The second integer following the dash indicates the position of maximum thickness in tenths of the chord. The suffix -63 indicates sections very nearly but not exactly the same as the sections without the suffix.

The modified thickness forms are defined by the following two equations:

$$\left. \begin{aligned} \pm\, y_t &= a_0\sqrt{x} + a_1 x + a_2 x^2 + a_3 x^3 \quad \text{ahead of maximum thickness} \\ \pm y_t &= d_0 + d_1(1-x) + d_2(1-x)^2 + d_3(1-x)^3 \quad \text{aft of maximum thickness} \end{aligned} \right\} \quad (6.7)$$

The four coefficients d_0, d_1, d_2, and d_3 are determined from the following conditions:

1. Maximum thickness, t.
2. Position of maximum thickness, m.
3. Ordinate at the trailing edge $x = 1$, $y_t = 0.01t$.
4. Trailing-edge angle, defined by the following table:

m	$\dfrac{dy}{dx}$, $(x = 1)$
0.2	$1.000t$
0.3	$1.170t$
0.4	$1.575t$
0.5	$2.325t$
0.6	$3.500t$

The four coefficients a_0, a_1, a_2, and a_3 are determined from the following conditions:

1. Maximum thickness, t.
2. Position of maximum thickness, m.
3. Leading-edge radius, $r_t = a_0{}^2/2$. $r_t = 1.1019(tI/6)^2$, where I is the first integer following the dash in the designation and the value of I does not exceed 8.
4. Radius of curvature R at the point of maximum thickness

$$R = \frac{1}{2d_2 + 6d_3(1 - m)}$$

Data[68a] for some of the modified thickness forms are presented in Appendix I. The data presented may be used together with data for the desired type of mean line to obtain approximate theoretical characteristics by the method presented in Sec. 4.5.

The family of NACA wing sections defined by these equations has been studied extensively by German aerodynamicists, who have applied designations of the following type to these sections:

$$\text{NACA} \qquad 1.8 \qquad 25 \qquad 14\text{-}1.1 \qquad 30/0.50$$

where   1.8 = maximum camber in per cent of chord

25 = location of maximum camber in per cent of the chord from the leading edge

14 = maximum thickness in per cent of the chord

1.1 = leading-edge radius parameter, r/t^2

30 = location of maximum thickness in per cent of the chord from the leading edge

0.50 = trailing-edge angle parameter, $(1/t)(\tan \tau/2)$

and t = thickness ratio

r = leading-edge radius, fraction of chord

τ = trailing-edge angle (included angle between the tangents to the upper and lower surfaces at the trailing edge)

For wing sections having this designation, the selected values of the leading-edge radius and of the trailing-edge angle are used directly with the other conditions in obtaining the coefficients of Eq. (6.7). The selected values for the maximum camber and location of maximum camber are applied to Eq. (6.4) to obtain the corresponding mean line.

6.7. NACA 1-series Wing Sections.   The NACA 1-series wing sections[45, 113] represent the first attempt to develop sections having desired types of pressure distributions and are the first family of NACA low-drag high-critical-speed wing sections. In order to meet one of the requirements for extensive laminar boundary layers, and to minimize the induced velocities, it was desired to locate the minimum-pressure point unusually far back on both surfaces and to have a small continuously favorable pressure gradient from the leading edge to the position of minimum pressure. The development of these wing sections (prior to 1939) was hampered by the lack of adequate theory, and difficulties were experienced in obtaining the desired pressure distribution over more than a very limited range of lift coefficients. As compared with later sections, the NACA 1-series wing sections are characterized by small leading-edge radii, comparatively large trailing-edge angles, and slightly higher critical speeds for a given thickness ratio. These sections have proved useful for propellers. The only commonly used sections of this series have the minimum pressure located at 60 per cent of the chord from the leading edge, and data will be presented only for these sections.

a. Thickness Distributions. Ordinates for thickness distributions with the minimum pressure located at $0.6c$ and thickness ratios of 6, 9, 12, 15, 18, and 21 per cent are presented in Appendix I. These data are similar in form to those presented for the NACA four-digit series. These sections were not developed from any analytical expression. The ordinate at any station is directly proportional to the thickness ratio, and sections of intermediate thickness may be correctly obtained by scaling the ordinates.

b. Mean Lines. The NACA 1-series wing sections, as commonly used, are cambered with a mean line of the uniform-load type [Eq. (4.26)]. Data for the mean line are tabulated in Appendix II. This type of mean line was selected because, at the design lift coefficient, it does not change the shape of the pressure distribution of the symmetrical section at zero lift. This mean line also imposes minimum induced velocities for a given design lift coefficient.

c. Numbering System. The NACA 1-series wing sections are designated by a five-digit number as, for example, the NACA 16-212 section. The first integer represents the series designation. The second integer represents the distance in tenths of the chord from the leading edge to the position of minimum pressure for the symmetrical section at zero lift. The first number following the dash indicates the amount of camber expressed in terms of the design lift coefficient in tenths, and the last two numbers together indicate the thickness in per cent of the chord. The commonly used sections of this family have minimum pressure at 0.6 of the chord from the leading edge and are usually referred to as the NACA 16-series sections.

d. Approximate Theoretical Characteristics. The theoretical aerodynamic characteristics of the NACA 1-series wing sections may be obtained by application of the method of Sec. 4.5 to the data of Appendixes I and II by the same method as that described for the NACA four-digit wing sections (Sec. 6.4*d*).

6.8. NACA 6-series Wing Sections. Successive attempts to design wing sections by approximate theoretical methods led to families of wing sections designated NACA 2- to 5-series sections.[45] Experience with these sections showed that none of the approximate methods tried was sufficiently accurate to show correctly the effect of changes in profile near the leading edge. Wind-tunnel and flight tests of these sections showed that extensive laminar boundary layers could be maintained at comparatively large values of the Reynolds number if the wing surfaces were sufficiently fair and smooth. These tests also provided qualitative information on the effects of the magnitude of the favorable pressure gradient, leading-edge radius, and other shape variables. The data also showed that separation of the turbulent boundary layer over the rear of the section, especially with rough surfaces, limited the extent of laminar layer for which the wing sections should be designed. The wing sections of these early families generally showed relatively low maximum lift coefficients and, in many cases, were designed for a greater extent of laminar flow than is practical. It was learned that, although sections designed for an excessive extent of laminar flow gave extremely low drag coefficients near the design lift coefficient when smooth, the drag of such sections became unduly large when rough, particularly at lift coefficients higher than the design value. These families of wing sections are accordingly considered obsolete.

The NACA 6-series basic thickness forms were derived by new and improved methods described in Sec. 3.8, in accordance with design criteria established with the objective of obtaining desirable drag, critical Mach number, and maximum-lift characteristics.

a. Thickness Distributions. Data for NACA 6-series thickness distributions covering a wide range of thickness ratios and positions of minimum pressure are presented in Appendix I. These data are comparable with the similar data for wing sections of the NACA four-digit series (Sec. 6.4a) except that ordinates for intermediate thickness ratios may not be correctly obtained by scaling the tabulated ordinates proportional to the thickness ratio. This method of changing the ordinates by a factor will, however, produce shapes satisfactorily approximating members of the family if the change of thickness ratio is small.

b. Mean Lines. The mean lines commonly used with the NACA 6-series wing sections produce a uniform chordwise loading from the leading edge to the point $x/c = a$, and a linearly decreasing load from this point to the trailing edge. Data for NACA mean lines with values of a equal to 0, 0.1, 0.2, 0.3, 0.4, 0.5, 0.6, 0.7, 0.8, 0.9, and 1.0 are presented in Appendix II. The ordinates were computed from Eqs. (4.26) and (4.27). The data are presented for a design lift coefficient c_{l_i} equal to unity. All tabulated values vary directly with the design lift coefficient. Corresponding data for similar mean lines with other design lift coefficients may accordingly be obtained simply by multiplying the tabulated values by the desired design lift coefficient.

In order to camber NACA 6-series wing sections, mean lines are usually used having values of a equal to or greater than the distance from the leading edge to the location of minimum pressure for the selected thickness distributions at zero lift. For special purposes, load distributions other than those corresponding to the simple mean lines may be obtained by combining two or more types of mean line having positive or negative values of the design lift coefficient.

c. Numbering System. The NACA 6-series wing sections are usually designated by a six-digit number together with a statement showing the type of mean line used. For example, in the designation NACA 65,3-218, $a = 0.5$, the 6 is the series designation. The 5 denotes the chordwise position of minimum pressure in tenths of the chord behind the leading edge for the basic symmetrical section at zero lift. The 3 following the comma gives the range of lift coefficient in tenths above and below the design lift coefficient in which favorable pressure gradients exist on both surfaces. The 2 following the dash gives the design lift coefficient in tenths. The last two digits indicate the thickness of the wing section in per cent of the chord. The designation $a = 0.5$ shows the type of mean line used. When the mean-line designation is not given, it is understood that the uniform-load mean line ($a = 1.0$) has been used.

When the mean line used is obtained by combining more than one mean line, the design lift coefficient used in the designation is the algebraic sum of the design lift coefficients of the mean lines used, and the mean lines are described in the statement following the number as in the following case:

$$\text{NACA 65,3-218} \begin{cases} a = 0.5 & c_{l_i} = 0.3 \\ a = 1.0 & c_{l_i} = -0.1 \end{cases}$$

Wing sections having a thickness distribution obtained by linearly increasing or decreasing the ordinates of one of the originally derived thickness distributions are designated as in the following example:

$$\text{NACA 65(318)-217} \qquad a = 0.5$$

The significance of all the numbers except those in the parentheses is the same as before. The first number and the last two numbers enclosed in the parentheses denote, respectively, the low-drag range and the thickness in per cent of the chord of the originally derived thickness distribution.

The more recent NACA 6-series wing sections are derived as members of thickness families having a simple relationship between the conformal transformations for wing sections of different thickness ratios but having minimum pressure at the same chordwise position. These wing sections are distinguished from the earlier individually derived wing sections by writing the number indicating the low-drag range as a subscript, for example,

$$\text{NACA 65}_3\text{-218} \qquad a = 0.5$$

Ordinates for the basic thickness distributions designated by a subscript are slightly different from those for the corresponding individually derived thickness distributions. As before, if the ordinates of the basic thickness distributions are changed by a factor, the low-drag range and thickness ratio of the original thickness distribution are enclosed in parentheses as follows·

$$\text{NACA 65}_{(318)}\text{-217} \qquad a = 0.5$$

For wing sections having a thickness ratio less than 12 per cent, the low-drag range is less than 0.1 and the subscript denoting this range is omitted from the designation, thus

$$\text{NACA 65-210}$$

or

$$\text{NACA 65}_{(10)}\text{-211}$$

The latter designation indicates that the 11 per cent thick section was obtained by linearly scaling the ordinates of the 10 per cent thick symmetrical section.

If the design lift coefficient in tenths or the thickness of the wing section in per cent of the chord are not whole integers, the numbers giving these quantities are usually enclosed in parentheses as in the following example:

$$\text{NACA } 65_{(318)}\text{-}(1.5)(16.5) \qquad a = 0.5$$

Some early experimental wing sections are designated by the insertion of the letter x immediately preceding the dash as in the designation 66,$2x$-015.

Some modifications of the NACA 6-series sections are designated by replacing the dash by a capital letter, thus

$$\text{NACA } 64_1\text{A212}$$

In this case, the letter indicates both the modified thickness distribution and the type of mean line used to camber the section. Sections designated by the letter A are substantially straight on both surfaces from about $0.8c$ to the trailing edge.

d. Approximate Theoretical Characteristics. Approximate theoretical characteristics may be obtained by applying the methods of Sec. 4.5 to the data of Appendixes I and II as described for the NACA four-digit series sections in Sec. 6.4d. If two or more of the simple mean lines are combined to camber the desired section, data for the resulting mean line may be obtained by algebraic addition of the scaled values for the component mean lines.

6.9. NACA 7-series Wing Sections. The NACA 7-series wing sections are characterized by a greater extent of possible laminar flow on the lower than on the upper surface. These sections permit low pitching-moment coefficients with moderately high design lift coefficients at the expense of some reduction of maximum lift and critical Mach number.

The NACA 7-series wing sections are designated by a number of the following type:

$$\text{NACA } 747\text{A315}$$

The first number 7 indicates the series number. The second number 4 indicates the extent over the upper surface, in tenths of the chord from the leading edge, of the region of favorable pressure gradient at the design lift coefficient. The third number 7 indicates the extent over the lower surface, in tenths of the chord from the leading edge, of the region of favorable pressure gradient at the design lift coefficient. The significance of the last group of three numbers is the same as for the NACA 6-series wing sections. The letter A which follows the first three numbers is a serial letter to distinguish different sections having parameters that would correspond to the same numerical designation. For example, a second section having the same extent of favorable pressure gradient over the upper and lower surfaces, the same design lift coefficient, and the same thickness ratio as the original wing section but having a different mean-line combination or thickness distribution would have the serial letter B. Mean lines used for the NACA 7-series sections are obtained by combining two or more of the previously described mean lines. The basic thickness distribution is given a designation similar to those of the final cambered wing sections. For example, the basic thickness distribution for the NACA 747A315 and 747A415 sections is given the designation NACA 747A015 even though minimum pressure occurs at $0.4c$ on both the upper and lower surfaces at zero lift. Data for this thickness distribution are presented in Appendix I.

The NACA 747A315 wing section is cambered with the following combination of mean lines:

$$\begin{Bmatrix} a = 0.4 & c_{l_i} = 0.763 \\ a = 0.7 & c_{l_i} = -0.463 \end{Bmatrix}$$

The NACA 747A415 wing section is cambered with the following combination of mean lines:

$$\begin{Bmatrix} a = 0.4 & c_{l_i} = 0.763 \\ a = 0.7 & c_{l_i} = -0.463 \\ a = 1.0 & c_{l_i} = 0.100 \end{Bmatrix}$$

6.10. Special Combinations of Thickness and Camber. The methods presented for combining thickness distributions and mean lines are sufficiently flexible to permit combining any thickness distribution, regardless of family, with any mean line or combination of mean lines. Approximate theoretical characteristics of such combinations may be obtained by application of the method of Sec. 4.5 to the tabulated data. In this manner, it is possible to approximate a considerable variety of pressure distributions without deriving new wing sections.

CHAPTER 7

EXPERIMENTAL CHARACTERISTICS OF WING SECTIONS

7.1. Symbols.

R Reynolds number, $\rho Vc/\mu$

U velocity just outside the boundary layer

V velocity of the free stream

a mean-line designation; fraction of the chord from leading edge over which loading is uniform at the ideal angle of attack

ac aerodynamic center

c chord

c_d section drag coefficient

$c_{d_{\min}}$ minimum section drag coefficient

c_f local skin-friction coefficient

c_l section lift coefficient

c_{l_i} design lift coefficient

$c_{l_{\max}}$ maximum section lift coefficient

$c_{m_{ac}}$ section pitching-moment coefficient about the aerodynamic center

$c_{m_{c/4}}$ section pitching-moment coefficient about the quarter-chord point

k height of roughness

t time

x abscissa measured from the leading edge

y ordinate measured from the chord line

α_0 section angle of attack

δ_f angle of flap deflection

μ viscosity of air

ρ mass density of air

7.2. Introduction.   The theories presented in Chaps. 3 to 5 permit reasonably accurate calculations to be made of certain characteristics of wing sections, but the simplifying assumptions made in the development of these theories limit their over-all applicability.   For instance, the perfect-fluid theories of Chaps. 3 and 4 permit the pressure distribution to be calculated, provided that the effects of the boundary-layer flow on the pressure distribution are small.   Similarly the viscous theories of Chap. 5 permit the calculation of some of the boundary-layer conditions, provided that the pressure distribution is known.   Obvious interactions between the viscous and potential flows are the relatively small effective changes of body shape caused by the displacement thickness of the unseparated boundary layer and the seriously large changes caused by separation.   None of the wing characteristics can be calculated with confidence if the flow is separated over an appreciable part of the surface. Under these circumstances, wing-section characteristics used for design are obtained experimentally.

Wind-tunnel investigations of wing characteristics were made before airplanes were successfully flown and still constitute an important phase of aerodynamic testing. Until recently wing characteristics were usually obtained from tests of models of finite aspect ratio. The development of wing theory led to the concept of wing-section characteristics that were derived from data obtained from tests of finite-aspect-ratio wings. These derived data were then used to predict the characteristics of wings of different plan forms. The systematic investigations in the NACA variable-density wind tunnel[46, 48, 49] were examples of this type of investigation. This method of testing is hampered by the difficulties of obtaining full-scale values of the Reynolds number and sufficiently low air-stream turbulence to duplicate flight conditions properly without excessive cost for equipment and models. Other difficulties were experienced in properly correcting the data for the support tares and interference effects and in deriving the section characteristics from tests of models necessarily having varied span-load distributions and tip effects.[45a]

In order to avoid some of these difficulties and to permit testing of models that are large relative to the size of the wind tunnel, two-dimensional testing equipment was built by the NACA. The NACA two-dimensional low-turbulence pressure tunnel[133] provides facilities for testing wing sections in two-dimensional flow at large Reynolds numbers in an air stream of very low turbulence, approaching that of the atmosphere. The wing-section data presented here were obtained from tests in this tunnel.

This tunnel[133] has a test section 3 feet wide and 7½ feet high and is capable of operation at pressures up to 10 atmospheres. The usual models are of 2-foot chord and completely span the 3-foot width of the test section. The lift is measured by integration of pressures representing the reaction on the floor and ceiling of the tunnel. The drag is obtained from wake-survey measurements, and the pitching moments are measured directly by a balance. Wing-section characteristics can be obtained from such measurements with a high degree of accuracy.

The usual tests were made over a range of Reynolds numbers from 3 to 9 million and at Mach numbers less than about 0.17.[*] This range of Reynolds numbers covers the range where large-scale effects are usually experienced between the usual low-scale test data and the large-scale flight range. Individual tests have been made to provide some data at lower Reynolds numbers applicable to small personal airplanes and at much larger Reynolds numbers to indicate trends for very large airplanes. The tunnel turbulence level is very low, of the order of a few hundredths of 1 per cent; and, although it is not definitely known that the remaining turbulence is negligible, the tunnel results appear to correspond closely to

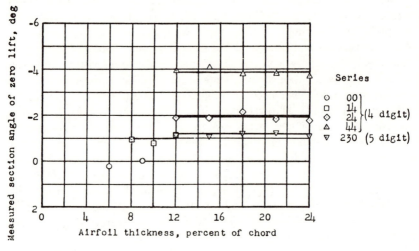

(a) NACA four- and five-digit series.

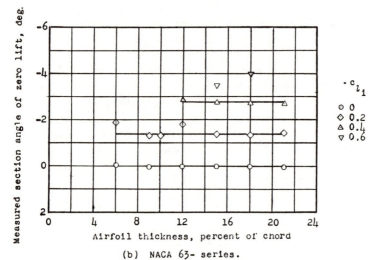

(b) NACA 63- series.

FIG. 56. Measured section angles of zero lift for a number of NACA

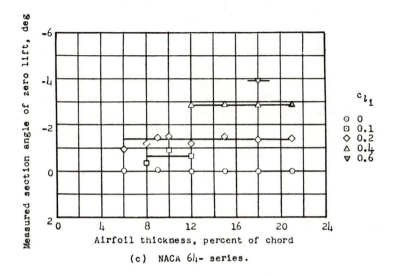

(c) NACA 64- series.

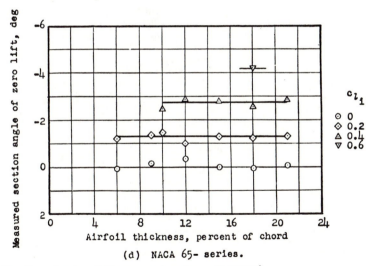

(d) NACA 65- series.

airfoil sections of various thicknesses and cambers. R, 6×10^6.

those obtained in flight. Application of these wing-section data to the prediction of the characteristics of wings of finite span depends on the adequacy of three-dimensional wing theory. These data are not applicable at high speeds where compressibility effects become important.

7.3. Standard Aerodynamic Characteristics. The resultant force on a wing section can be specified by two components of force perpendicular and parallel to the air stream (the lift and drag, respectively) and by a moment in the plane of these two forces (the pitching moment). These forces are functions of the angle of attack of the section. The standard method of presenting the characteristics of wing sections is by means of plots of the lift, drag, and moment coefficients against angle of attack or, alternately, plots of angle of attack, drag, and moment coefficients against lift coefficient.

Plots of wing-section characteristics are presented in Appendix IV for a wide range of shape parameters. On the left-hand side of each plot, the lift coefficient and the moment coefficient about the quarter-chord point are plotted against the angle of attack. On the right-hand side of each plot the drag coefficient and moment coefficient about the aerodynamic center are plotted against the lift coefficient. In most cases, the data indicated in the following table are presented.

Characteristic	Surface condition	Split flap deflection degrees	Reynolds number, millions
Left-hand side			
Lift............	Smooth	0	3, 6, 9
Lift............	Rough*	0	6
Lift............	Smooth	60	6
Lift............	Rough*	60	6
Moment........	Smooth	0	3, 6, 9
Moment........	Smooth	60	6
Right-hand side			
Drag..........	Smooth	0	3, 6, 9
Drag..........	Rough*	0	6
Moment........	Smooth	0	3, 6, 9

*0.011-inch grain carborundum spread thinly to cover 5 to 10 per cent of the area from the leading edge to $0.08c$ along both surfaces of a section with a chord of 24 inches.

7.4. Lift Characteristics. *a. Angle of Zero Lift.* As indicated in Chap. 4, the angle of zero lift of a wing section is largely determined by the camber. The theory of wing sections provides a means for computing the angle of zero lift from the mean-line data presented in Appendix II. The agreement between the calculated and the experimental angles of zero lift depends on the type of mean line used. Comparison of the theoretical data given in Appendix II with the experimental data of Appendix IV shows that the agreement is good except for the uniform-load type $(a = 1)$ of mean line. The angles of zero lift for this type of mean line are generally closer to 0 degrees than predicted.

The experimental values of the angles of zero lift for a number of NACA four- and five-digit and NACA 6-series wing sections are presented in Fig. 56. The thickness ratio of the wing section appears to have little effect on the angle of zero lift regardless of the type of thickness distribu-

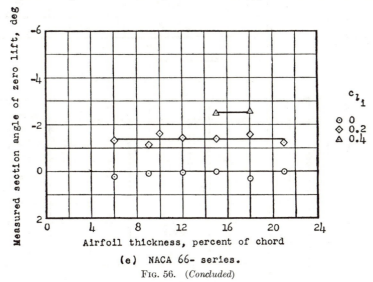

(e) NACA 66- series.

Fig. 56. (*Concluded*)

tion or camber. For the NACA four-digit series wing sections, the angles of zero lift are approximately 0.93 of the value given by the theory of thin wing sections. For the NACA 230-series wing sections, this factor is approximately 1.08; and for the NACA 6-series sections with the uniform-load type of mean line, this factor is approximately 0.74.

b. Lift-curve Slope. Lift-curve slopes for a number of NACA four- and five-digit series and NACA 6-series wing sections are plotted against thickness ratio in Fig. 57. These values of the lift-curve slope were measured for a Reynolds number of 6 million at values of the lift coefficient approximately equal to the design lift coefficient of the wing sections. This lift coefficient is approximately in the center of the low-drag range for the NACA 6-series wing sections.

In the range of thickness ratios from 6 to 10 per cent, the NACA four- and five-digit series and the NACA 64-series wing sections have values of the lift-curve slope very close to the value given by the theory of thin wing sections (2π per radian, or 0.110 per degree). Variation of Reynolds num-

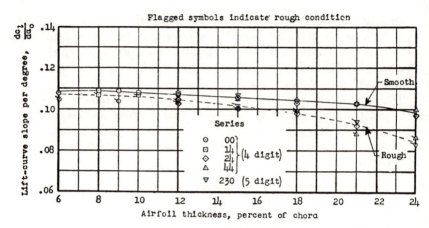

(a) NACA four- and five-digit series.

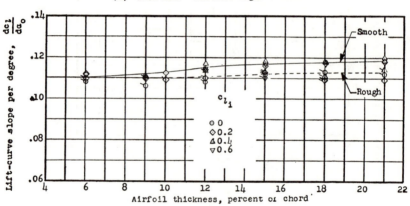

(b) NACA 63- series.

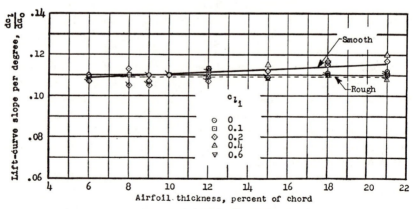

(c) NACA 64- series.

FIG. 57. Variation of lift-curve slope with airfoil thickness ratio and camber for a

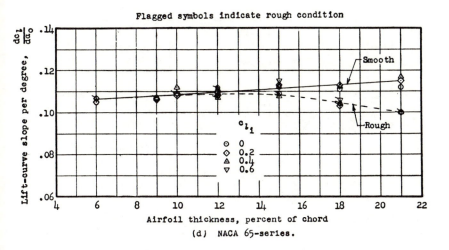

(d) NACA 65-series.

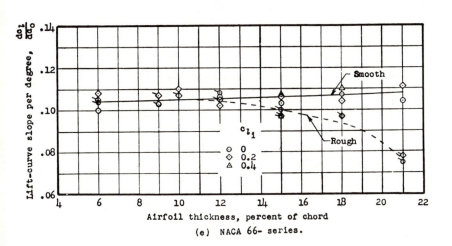

(e) NACA 66- series.

number of NACA airfoil sections in both the smooth and rough conditions. R, 6×10^6.

ber between 3 and 9 million and variation of camber up to 0.04c appear to have no systematic effect on the lift-curve slope. The thickness distribution appears to be the primary variable. For the NACA four- and five-digit series wing sections, the lift-curve slope decreases with increase of thickness ratio. For the NACA 6-series wing sections, however, the

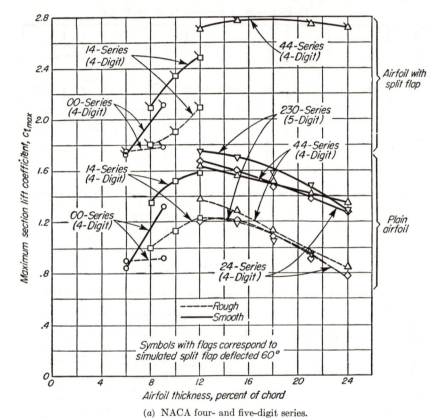

(a) NACA four- and five-digit series.

Fig. 58. Variation of maximum section lift coefficient with airfoil thickness ratio and camber for several NACA airfoil sections with and without simulated split flaps and standard roughness. R, 6×10^6.

lift-curve slope increases with increase of thickness ratio and with forward movement of the position of minimum pressure on the basic thickness form. The effect of thickness ratio is comparatively small for the NACA 66-series sections. The thick-wing-section theory (Chap. 3) shows that the slope of the lift curve should increase with increasing thickness ratio in the absence of viscous effects. For wing sections with arbitrary modifications of shape near the trailing edge, the lift-curve slope appears to decrease with increasing trailing-edge angle.

Some NACA 6-series wing sections show jogs in the lift curve at the end of the low-drag range, especially at low Reynolds numbers. This jog becomes more pronounced with increase of camber or thickness ratio and with rearward movement of the position of minimum pressure on the basic thickness form. This jog decreases rapidly in severity with increasing Reynolds number, becomes merely a change of lift-curve slope, and is practically nonexistent at a Reynolds number of 9 million for most wing

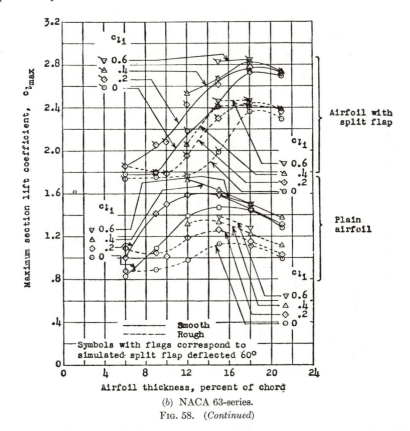

(b) NACA 63-series.

Fig. 58. (*Continued*)

sections that would be considered for practical application. This jog may be a consideration in the selection of wing sections for small low-speed airplanes. An analysis of the flow conditions leading to this jog is presented in reference 134.

The values of the lift-curve slopes presented are for steady conditions and do not necessarily correspond to the slopes obtained in transient conditions when the boundary layer has insufficient time to develop fully at each lift coefficient. Some experimental results[80a] indicate that variations of the steady value of the lift-curve slope do not result in similar variations of the gust loading.

c. Maximum Lift. The variation of maximum lift coefficient with thickness ratio at a Reynolds number of 6 million is shown in Fig. 58 for a considerable number of NACA wing sections. The sections for which data are presented in this figure have a range of thickness ratios from 6 to 24 per cent and cambers up to 4 per cent of the chord. From the data for the NACA four- and five-digit wing sections (Fig. 58a), it appears that the

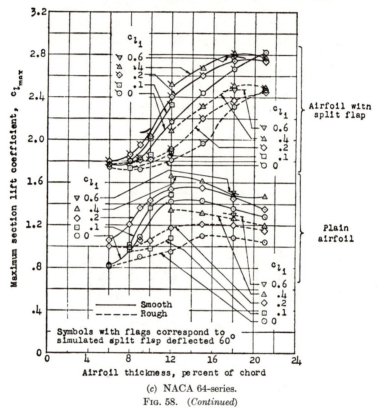

(c) NACA 64-series.

FIG. 58. (*Continued*)

maximum lift coefficients are the greatest for a thickness ratio of 12 per cent. In general, the rate of change of maximum lift coefficient with thickness ratio appears to be greater for sections having a thickness ratio less than 12 per cent than for the thicker sections. The data for the NACA 6-series sections (Figs. 58b to d) show a rapid increase of maximum lift coefficient with increasing thickness ratio for thickness ratios less than 12 per cent. The optimum thickness ratio for maximum lift coefficient increases with rearward movement of the position of minimum pressure and decreases with increase of camber.

For wing sections having thickness ratios of 6 per cent and for wing sections having thickness ratios of 18 or 21 per cent, the maximum lift

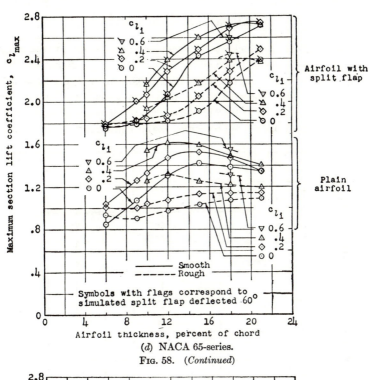

(d) NACA 65-series.

Fig. 58. (Continued)

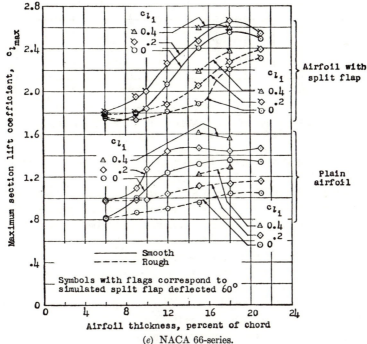

(e) NACA 66-series.

Fig. 58. (Concluded)

coefficients do not appear to be very much affected by the position of minimum pressure on the basic thickness form. The maximum lift coefficients of sections of intermediate thickness, however, decrease with rearward movement of the position of minimum pressure. The maximum lift coefficients for the NACA 64-series wing sections cambered for a design lift coefficient of 0.4 are slightly higher than those for the NACA 44-series

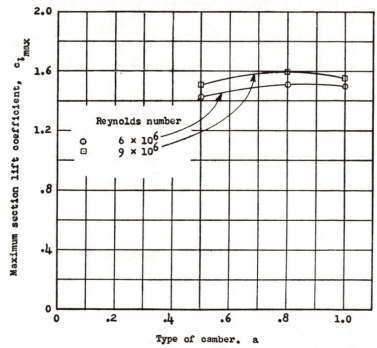

Fig. 59. Variation of maximum lift coefficient with type of camber for some NACA 65_3–418 airfoil sections from tests in the Langley two-dimensional low-turbulence pressure tunnel.

sections. The NACA 230-series sections, however, have somewhat higher maximum lift coefficients.

The maximum lift coefficients of moderately cambered sections increase with increasing camber (Fig. 58). For wing sections of about 18 per cent thickness, the rate of increase of maximum lift coefficient with camber is largest for small cambers. The effect of camber in increasing the section maximum lift coefficient becomes progressively less as the thickness ratio increases above 12 per cent. The variation of maximum lift with type of camber is shown in Fig. 59 for one condition. No systematic data are available for mean lines with values of a less than 0.5. It should be noted, however, that wing sections such as the NACA 230-series with the maximum camber far forward show large increments of maximum lift as compared with symmetrical sections. Wing sections with the maximum camber far

forward and with normal thickness ratios stall from the leading edge with
large sudden losses of lift. A more desirable gradual stall is obtained when

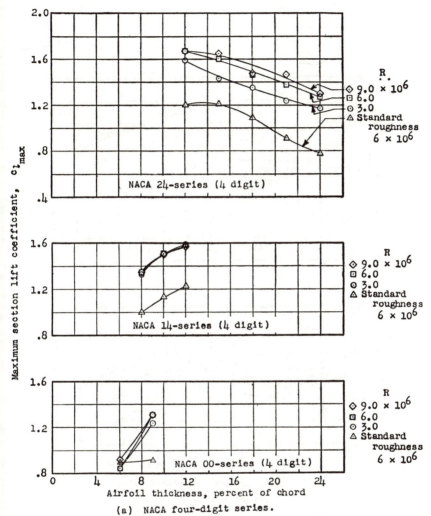

(a) NACA four-digit series.

Fig. 60. Variation of maximum section lift coefficient with airfoil thickness ratio at several
Reynolds numbers for a number of NACA airfoil sections of different cambers.

the location of maximum camber is farther back, as with the NACA 24-,
44-, and 6-series sections with normal types of camber.

The variations of maximum lift coefficient with Reynolds number (scale
effect) of a number of NACA wing sections are presented in Fig. 60 for a
range of Reynolds numbers from 3 to 9 million. The scale effect for the
NACA 24-, 44-, and 230-series wing sections (Figs. 60a and b) having

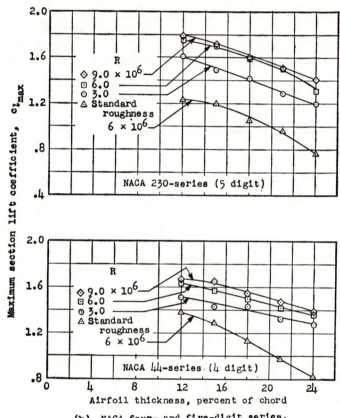

(b) NACA four- and five-digit series.

FIG. 60. (*Continued*)

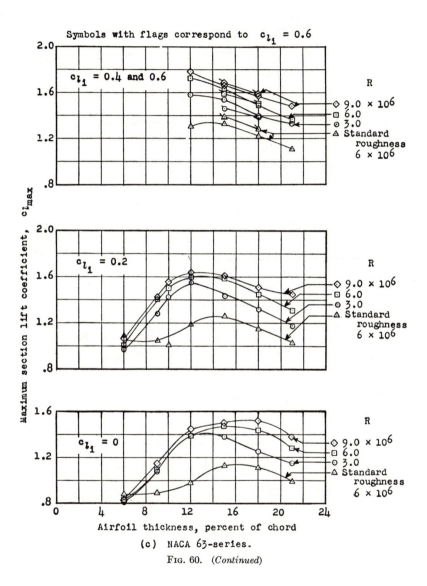

(c) NACA 63-series.

Fig. 60. (*Continued*)

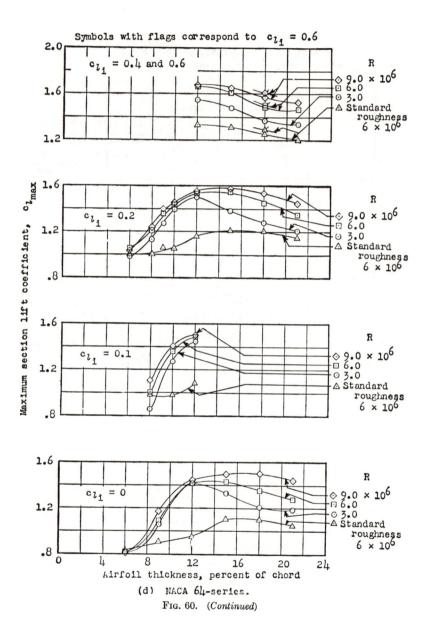

(d) NACA 64-series.

Fig. 60. (*Continued*)

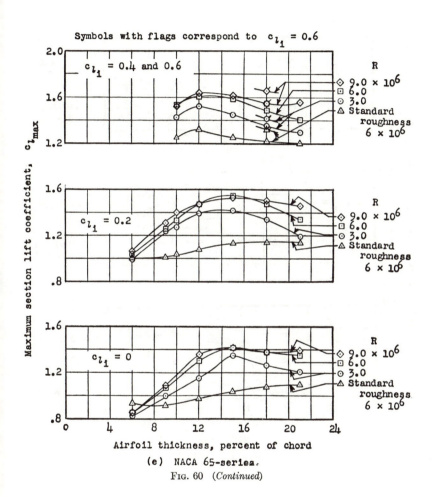

(e) NACA 65-series.

Fig. 60 (*Continued*)

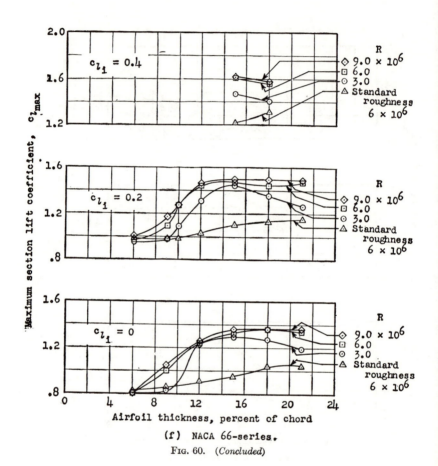

(f) NACA 66-series.

FIG. 60. (*Concluded*)

thickness ratios from 12 to 24 per cent is uniformly favorable and nearly independent of the thickness ratio. Increasing the Reynolds number from 3 to 9 million results in an increase of the maximum lift coefficient of approximately 0.15 to 0.20. The scale effect on the NACA 00- and 14- series sections having thickness ratios of 0.12c and less is very small. The scale-effect data for the NACA 6-series wing sections (Figs. 60c to e) do not show an entirely systematic variation. In general, the scale effect is favorable for these sections. For the NACA 64-series wing sections, the increase of maximum lift coefficient with increase of Reynolds number is generally small for thickness ratios less than 12 per cent, but it is some-what larger for the thicker sections. The character of the scale effect for the NACA 65- and 66-series wing sections is similar to that for the NACA 64-series, but the trends are not so well defined. The scale effect for the NACA 6-series wing sections cambered for a design lift coefficient of 0.4 or 0.6 is greater than that for these sections with less camber. The data of Fig. 61 show that the maximum lift coefficient for the NACA 63(420)-422 section continues to increase with Reynolds number up to values of at least 26 million.

The values of the maximum lift coefficient presented are for steady conditions. It is known that the maximum lift coefficient increases with the rate of change of angle of attack.[109a] The significant parameter is $(d\alpha_0/dt)(c/V)$. Even such low rates of change of angle of attack as those encountered in landing flares produce increases of the maximum lift coefficient.

d. Effect of Surface Condition on Lift Characteristics. It has long been known that surface roughness, especially near the leading edge, has large effects on the characteristics of wing sections. The maximum lift coefficient, in particular, is sensitive to leading-edge roughness. The effect on maximum lift coefficient of various degrees of roughness applied to the leading edge of the NACA 63(420)-422 wing section is shown in Fig. 62. The maximum lift coefficient decreases progressively with increasing roughness. For a given surface condition at the leading edge, the maximum lift coefficient increases slowly with increasing Reynolds number (Fig. 63). Figure 64 shows that roughness strips located more than about 0.20c from the leading edge have little effect on the maximum lift coefficient or lift-curve slope.

It is desirable to determine the relative effects of leading-edge roughness on various wing sections. In order to make a systematic investigation of this sort, it is necessary to select a standard form of roughness. The standard leading-edge roughness selected by the NACA for 24-inch chord models[3] consisted of 0.011-inch carborundum grains applied to the surface of the model at the leading edge over a surface length of 0.08c measured from the leading edge on both surfaces. The grains were thinly spread to cover 5 to 10 per cent of the area. This standard roughness is considerably

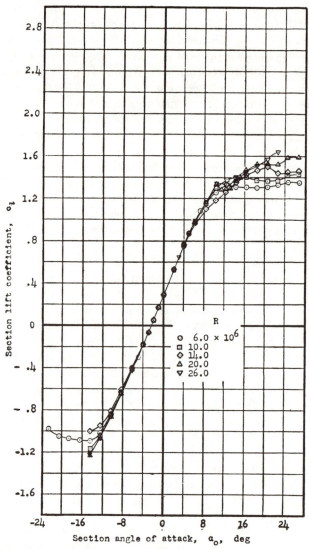

FIG. 61.   Lift and drag characteristics of the NACA 63(420)-422 airfoil at high Reynolds number; TDT tests 228 and 255.

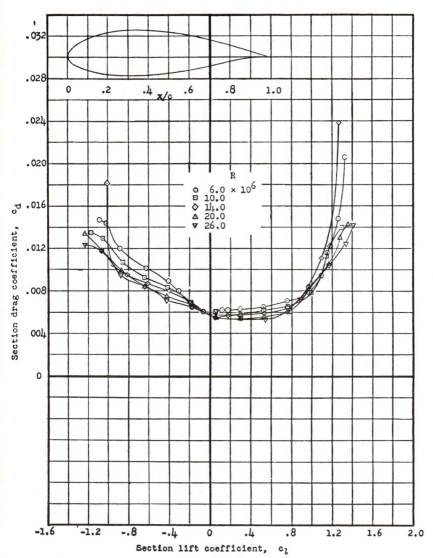

FIG. 61. (*Concluded*)

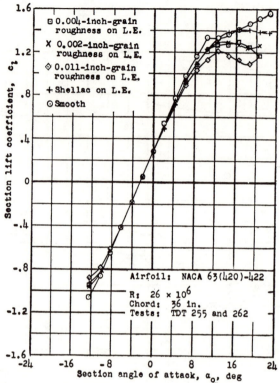

FIG. 62. Lift characteristics of a NACA 63(420)-422 airfoil with various degrees of roughness at the leading edge.

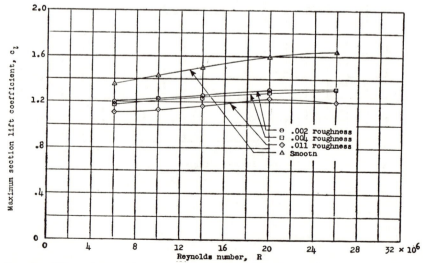

FIG. 63. Effects of Reynolds number on maximum section lift coefficient c_l of the NACA 63(420)-422 airfoil with roughened and smooth leading edge.

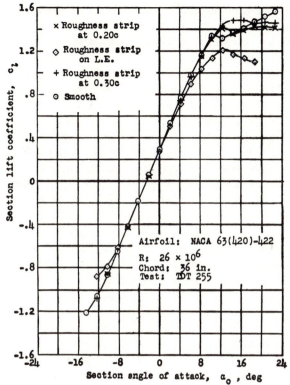

FIG. 64. Lift characteristics of a NACA 63(420)-422 airfoil with 0.011-inch-grain roughness at various chordwise locations.

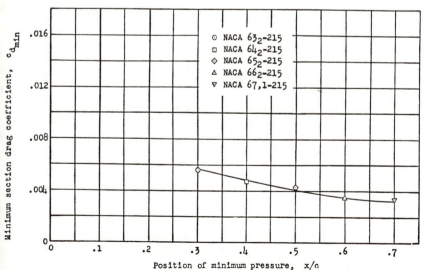

FIG. 65. Variation of minimum drag coefficient with position of minimum pressure for some NACA 6-series airfoils of the same camber and thickness. R, 6×10^6.

more severe than that caused by usual manufacturing irregularities or deterioration in service, but it is considerably less severe than that likely to be encountered in service as a result of accumulation of ice, mud, or damage in military combat.

Maximum-lift-coefficient data at a Reynolds number of 6 million for a large number of NACA wing sections are presented in Figs. 58 and 59. The variation of maximum lift coefficient with thickness for the NACA four- and five-digit series wing sections with standard roughness shows the same trends as those for the smooth sections except that the values are considerably reduced for all these sections other than the NACA 00-series of 6 per cent thickness. The values of the maximum lift coefficient for these rough sections with thickness ratios greater than 12 per cent are substantially the same for a given thickness. Much less variation of maximum lift coefficient with thickness ratio is shown by the NACA 6-series wing sections in the rough condition than in the smooth condition. The variation of maximum lift coefficient with camber, however, is about the same for the wing sections with standard roughness as for the smooth sections. The maximum lift coefficients of rough wing sections decrease with rearward movement of the position of minimum pressure; and, as in the smooth condition, the optimum thickness ratio for maximum lift increases with rearward movement of the position of minimum pressure. The NACA 64-series wing sections cambered for a design lift coefficient of 0.4 have maximum lift coefficients consistently higher than the NACA 24-, 44-, and 230-series sections of comparable thickness when rough, with the exception of the NACA 4412.

For normal wing sections, the angle of zero lift is practically unaffected by the standard leading-edge roughness. The results presented in Appendix IV show that the effect of roughness is to decrease the lift-curve slope for wing sections having thickness ratios of 18 per cent or more. The effect increases with increase of thickness ratio. For wing sections less than 18 per cent thick, the effect of roughness on the lift-curve slope is relatively small.

7.5. Drag Characteristics. *a. Minimum Drag of Smooth Wing Sections.* The value of the minimum drag coefficient for smooth wing sections is mainly a function of the Reynolds number and the relative extent of the laminar boundary layer, and it is moderately affected by thickness ratio and camber. If the extent of the laminar boundary layer is known, the minimum drag coefficient may be calculated with reasonable accuracy by the method presented in Sec. 5.13.

The effect on minimum drag of the position of minimum pressure that determines the possible extent of laminar flow is shown in Fig. 65 for some NACA 6-series wing sections. The data show a regular decrease of drag coefficient with rearward movement of minimum pressure. The variation of minimum drag coefficient with Reynolds number for several wing sections is shown in Fig. 66. The drag coefficient generally decreases with increasing Reynolds number up to Reynolds numbers of the order of 20 million. Above this Reynolds number, the drag coefficient of the NACA $65_{(421)}$-420 section remains substantially constant up to a Reynolds number of nearly 40 million. The earlier increase of drag coefficient shown by the NACA $66(2x15)$-116 section may be caused by surface irregularities because the specimen tested was a practical construction model. It may be noted that the drag coefficient for the NACA 65_3-418 wing section at low Reynolds numbers is substantially higher than that of the NACA 0012 section, whereas at high Reynolds numbers the opposite is the case. The higher drag of the NACA 65_3-418 wing section at low Reynolds numbers is caused by a relatively extensive region of laminar separation downstream from the point of minimum pressure. This region decreases in size with increasing Reynolds number. These data illustrate the inadequacy of low Reynolds number test data either to estimate the full-scale characteristics or to determine the relative merits of wing sections at flight values of the Reynolds number.

The variation of the minimum drag coefficient with camber is shown in Fig. 67 for a number of smooth 18 per cent thick NACA 6-series wing sections. These data show little change of minimum drag coefficient with increase of camber. A considerable amount of systematic data is included in Fig. 68 showing the variation of minimum drag coefficient with thickness ratio for some NACA wing sections ranging in thickness ratio from 6 to 24 per cent of the chord. The minimum drag coefficient is seen to increase with increase of thickness ratio for each series of wing sections. This increase, however, is greater for the NACA four- and five-digit series wing sections (Fig. 68a) than for the NACA 6-series sections (Figs. 68b, c, and d).

b. Variation of Profile Drag with Lift Coefficient. Most of the variation of drag with lift for wings of finite span results from the induced-drag coefficient, which varies approximately as the square of the lift coefficient for a given wing configuration. It is important to keep the induced drag in mind when considering the variation of profile drag with lift, because the variation of the wing drag coefficient with lift coefficient will be largely determined by the induced drag, which is a function of the aspect ratio.

At low and moderate lift coefficients where there is no appreciable separation of the flow, the drag is caused almost completely by skin friction. Under these circumstances, the value of the drag coefficient depends upon the relative extent of the laminar boundary layer and the induced velocities over the surfaces of the section and may be calculated by the method of Sec. 5.13. As the lift coefficient increases, the average square of the velocities over the surfaces increases, resulting in a small drag increase even though the relative extent of laminar flow was not affected. The

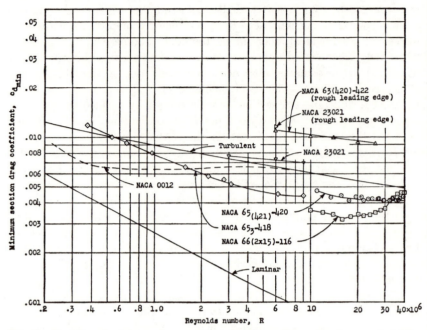

Fig. 66. Variation of minimum drag coefficient with Reynolds number for several airfoils, together with laminar and turbulent skin-friction coefficients for a flat plate.

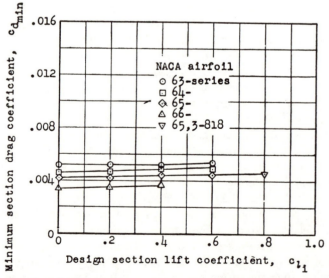

Fig. 67. Variation of section minimum drag coefficient with camber for several NACA 6-series airfoil sections of 18 per cent thickness ratio. R, 6×10^6.

more important effect of lift coefficient is to change the possible extent of laminar flow by moving the minimum-pressure points. In many of the older types of wing section, the forward movement of transition is gradual and the resulting variation of drag with lift coefficient occurs smoothly. The pressure distributions for the NACA 6-series wing sections are such as to cause transition to move forward suddenly at the end of the low-drag range of lift coefficients. A sharp increase of drag coefficient to the value corresponding to a forward location of transition on one of the surfaces results. Such sudden shifts of transition give the typical drag curve for these wing sections with a "sag" or "bucket" in the low-drag range. The same characteristic is shown to a smaller degree by some of the earlier wing sections such as the NACA 23015 when tested in a low-turbulence air stream.

The data presented in Appendix I for the NACA 6-series thickness forms show that the range of lift coefficients for low drag varies markedly with thickness ratio. It has been possible to design wing sections of 12 per cent thickness with a total theoretical low-drag range of lift coefficients of 0.2. This theoretical range increases by approximately 0.2 for each 3 per cent increase of thickness ratio. Figure 69 shows that the theoretical extent of the low-drag range is approximately realized at a Reynolds number of 9 million. Figure 69 also shows a characteristic tendency for the drag to increase to some extent toward the upper end of the low-drag range for moderately cambered wing sections, particularly for the thicker ones. All data for the NACA 6-series wing sections show a decrease of the extent of the low-drag range with increasing Reynolds number. Extrapolation of the rate of decrease observed at Reynolds numbers below 9 million would indicate a vanishingly small low-drag range at flight values of the Reynolds number. Tests of a carefully constructed model of the NACA 65$_{(421)}$-420 section showed, however, that the rate of reduction of the low-drag range with increasing Reynolds number decreased markedly at Reynolds numbers above 9 million (Fig. 70). These data indicate that the extent of the low-drag range for this wing section is reduced to about one-half the theoretical value at a Reynolds number of 35 million.

The values of the lift coefficient for which low drag is obtained are determined largely by the amount of camber. The lift coefficient at the center of the low-drag range corresponds approximately to the design lift coefficient of the mean line. The effects on the drag characteristics of various amounts of camber are shown in Fig. 71. Section data indicate that the location of the low-drag range may be shifted by even such crude camber changes as those caused by small deflections of a plain flap.[3]

The location of the low-drag range shows some variation from that predicted from the simple theory of thin wing sections. This departure appears to be a function of the type of the mean line used and the thickness

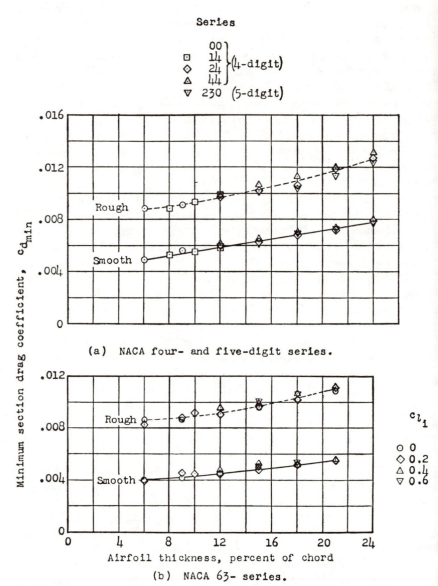

(a)   NACA four- and five-digit series.

(b)   NACA 63- series.

Fig. 68.   Variation of section minimum drag coefficient with airfoil thickness ratio for several NACA airfoil sections of different cambers in both smooth and rough conditions. R, 6×10^6.

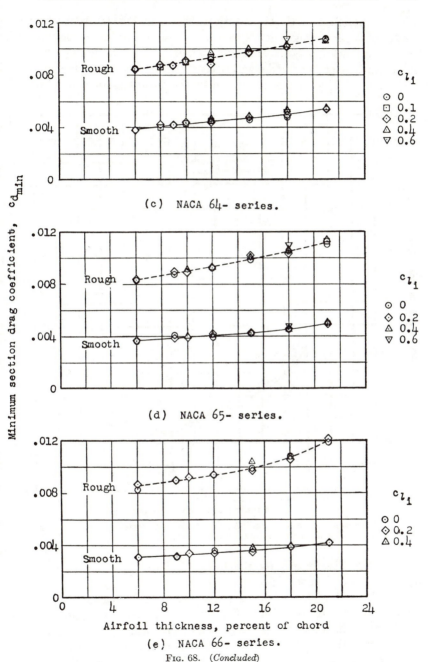

(c) NACA 64- series.

(d) NACA 65- series.

Airfoil thickness, percent of chord

(e) NACA 66- series.

FIG. 68. (*Concluded*)

ratio. The effect of thickness ratio is shown in Fig. 69 from which the center of the low-drag range is seen to shift to higher lift coefficients with increasing thickness ratio. This shift is partly explained by the increase of lift coefficient above the design lift coefficient for the mean line obtained when the velocity increments caused by the mean line are combined with

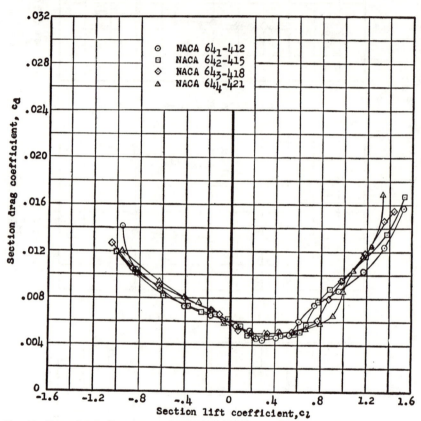

FIG. 69. Drag characteristics of some NACA 64-series airfoil sections of various thicknesses, cambered to a design lift coefficient of 0.4. R, 9×10^6.

the velocity distribution for the thickness form according to the first-approximation method of Sec. 4.5.

At the end of the low-drag range, the drag increases rapidly with increase of the lift coefficient. For symmetrical and low-cambered wing sections for which the lift coefficient at the upper end of the low-drag range is moderate, this high rate of increase does not continue (Fig. 71). For highly cambered sections for which the lift at the upper end of the low-drag range is already high, the drag coefficient shows a continued rapid increase.

Comparison of data for wing sections cambered with a uniform-load

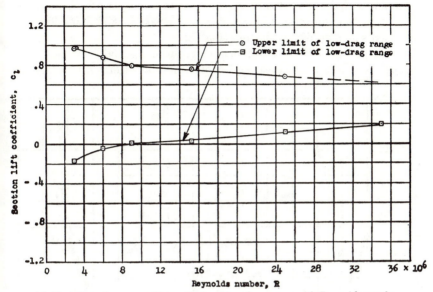

(a) Variation of upper and lower limits of low-drag range with Reynolds number.

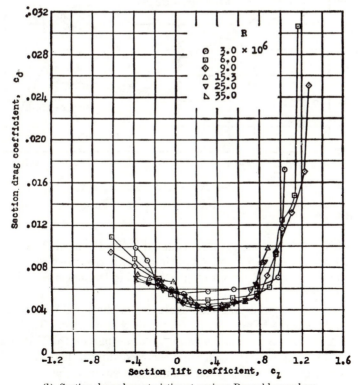

(b) Section drag characteristics at various Reynolds numbers.

Fig. 70   Variation of low-drag range with Reynolds number for the NACA $65_{(421)}$-420 airfoil.

mean line with data for sections cambered to carry the load farther forward
shows that the uniform-load mean line is favorable for obtaining low-drag
coefficients at high lift coefficients (Fig. 72).

Data for many of the wing sections given in Appendix IV show large
reductions of drag with increasing Reynolds number at high lift coefficients.

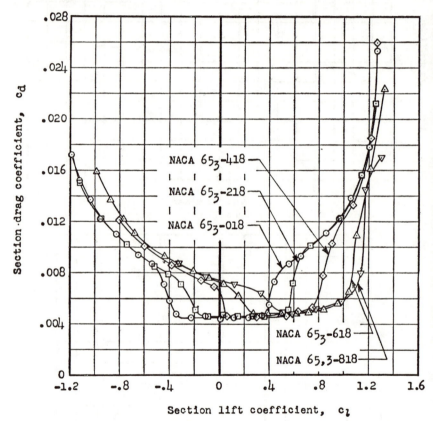

Fig. 71.   Drag characteristics of some NACA 65-series airfoil sections of 18 per cent thickness
with various amounts of camber. R, 6×10^6.

This scale effect is too large to be accounted for by the normal variation of
skin friction and appears to be associated with the effect of Reynolds
number on the onset of turbulent flow following laminar separation near
the leading edge.[134]

A comparison of the drag characteristics of the NACA 23012 and of
three NACA 6-series wing sections is presented in Fig. 73. The drag for
the NACA 6-series sections is substantially lower than for the NACA 23012
section in the range of lift coefficients corresponding to high-speed flight,
and this margin may usually be maintained through the range of lift

coefficients useful for cruising by suitable choice of camber. The NACA 6-series sections show the higher maximum values of the lift-drag ratio. At high values of the lift coefficient, however, the earlier NACA sections generally have lower drag coefficients than the NACA 6-series sections.

c. Effect of Surface Irregularities on Drag Characteristics. Numerous measurements of the effects of surface irregularities on the characteristics of wings have shown that the condition of the surface is one of the most important variables affecting the drag. Although a large part of the drag increment associated with surface roughness results from a forward movement of transition, substantial drag increments result from surface roughness in the region of turbulent flow.[42] It is accordingly important to maintain smooth surfaces even when extensive laminar flow cannot be expected. The possible gains resulting from smooth surfaces are greater, however, for wing sections such as the NACA 6-series than for sections where the extent of laminar flow is limited by a forward position of minimum pressure.

No accurate method of specifying the surface condition necessary for extensive laminar flow at high Reynolds numbers has been developed, although some general conclusions have been reached. It may be presumed that, for a given Reynolds number and chordwise location, the size of the permissible roughness will vary directly with the chord of the wing section. It is known, at one extreme, that the surfaces do not have to be polished or optically smooth. Such polishing or waxing has shown no improvement in tests[3] in the NACA two-dimensional low-turbulence tunnels when applied to satisfactorily sanded surfaces. Polishing or waxing a surface that is not aerodynamically smooth will, of course, result in improvement, and such finishes may be of considerable practical value because deterioration of the finish may be easily seen and possibly postponed. Large models having chord lengths of 5 to 8 feet tested in the NACA two-dimensional low-turbulence tunnels are usually finished by sanding in the chordwise direction with No. 320 carborundum paper when an aerodynamically smooth surface is desired.[3] Experience has shown the resulting finish to be satisfactory at flight values of the Reynolds number. Any rougher surface texture should be considered as a possible source of transition, although slightly rougher surfaces have appeared to produce satisfactory results in some cases.

Loftin[68] showed that small protuberances extending above the general surface level of an otherwise satisfactory surface are more likely to cause transition than are small depressions. Dust particles, for example, are more effective than small scratches in producing transition if the material at the edges of the scratches is not forced above the general surface level. Dust particles adhering to the oil left on wing surfaces by fingerprints may be expected to cause transition at high Reynolds numbers.

Transition spreads from an individual disturbance with an included angle of about 15 degrees.[26, 42]   A few scattered specks, especially near the

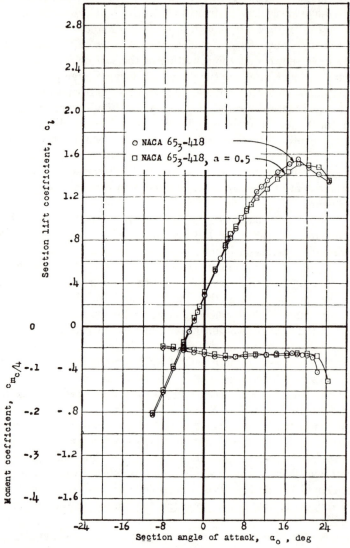

Fig. 72. Comparison of the aerodynamic characteristics of the NACA 65_3-418 and NACA 65_3-418, $a = 0.5$ airfoils. R, 9×10^6.

leading edge, will cause the flow to be largely turbulent.   This fact makes necessary an extremely thorough inspection if low drags are to be realized. Specks sufficiently large to cause premature transition can be felt by hand. The inspection procedure used in the NACA two-dimensional low-turbulence tunnels is to feel the entire surface by hand, after which the surface is thoroughly wiped with a dry cloth.

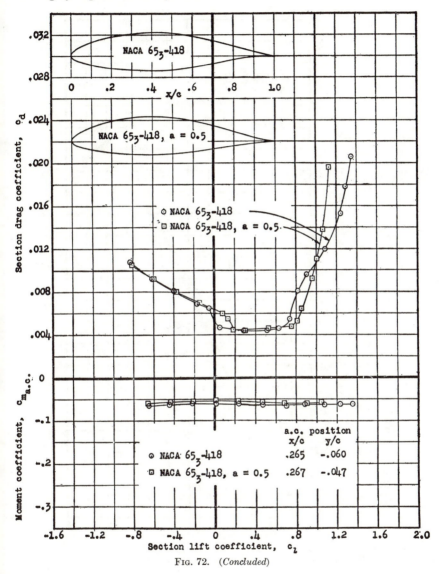

Fig. 72. (*Concluded*)

It has been noticed that transition caused by individual sharp protuberances, in contrast to waves, tends to occur at the protuberance. Transition caused by surface waviness appears to move gradually upstream toward the wave as the Reynolds number or wave size is increased. The height of a small cylindrical protuberance necessary to cause transition when located at 5 per cent of the chord with its axis normal to the surface[68] is shown in Fig. 74. These data were obtained at rather low

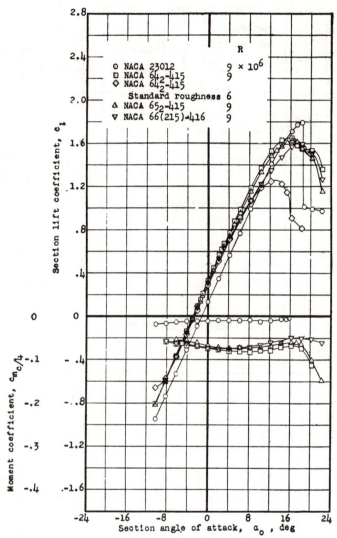

Fig. 73. Comparison of the aerodynamic characteristics of some NACA airfoils from tests in the Langley two-dimensional low-turbulence pressure tunnel.

values of the Reynolds number and show a large decrease of the allowable height with increase of Reynolds number.

Analysis[68] of these data showed that the height of the protuberance that caused transition depended on the shape of the protuberance and on the Reynolds number based on the height of the protuberance and the local velocity at the top of the protuberance. This Reynolds number is plotted against the fineness ratio of the protuberance in Fig. 75 for protuberances located at various chordwise positions on two wing sections.

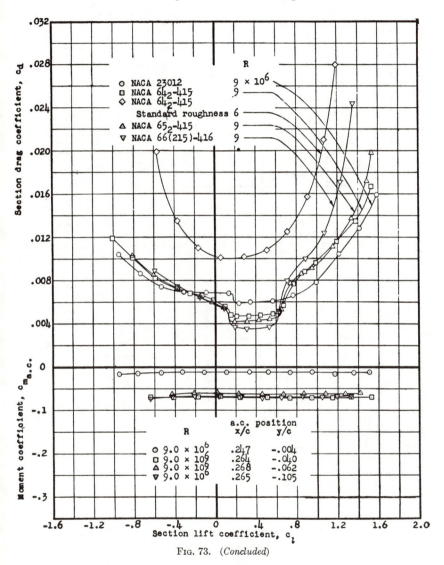

Fig. 73. (*Concluded*)

The effect of Reynolds number on permissible surface roughness[3] is also indicated in Fig. 76, in which a sharp increase of drag at a Reynolds number of approximately 20 million occurs for the model painted with camouflage lacquer. Experiments with models finished with camouflage paint[17] indicate that it is possible to obtain a matte surface without causing premature transition but that it is extremely difficult to obtain such a surface sufficiently free from specks without sanding.

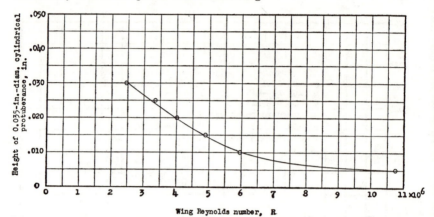

FIG. 74. Variation with wing Reynolds number of the minimum height of a cylindrical protuberance necessary to cause premature transition. Protuberance has 0.035-inch diameter with axis normal to the wing surface and is located at 5 per cent chord of a 90-inch-chord symmetrical 6-series airfoil section of 15 per cent thickness and with minimum pressure at 70 per cent chord.

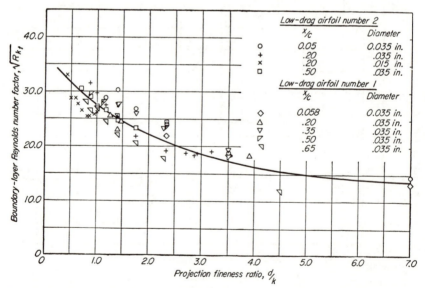

FIG. 75. Variation of boundary-layer Reynolds number factor with projection fineness ratio for two low-drag airfoils.

The magnitude of the favorable pressure gradient appears to have a small effect on the permissible surface roughness for laminar flow. Figure 77 shows that the roughness becomes more important at the extremities of the low-drag range where the favorable pressure gradient is reduced on one surface.[3] The effect of increasing the Reynolds number for a surface of

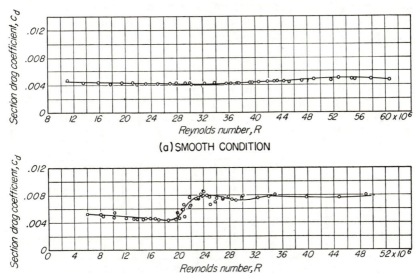

Fig. 76. Variation of drag coefficient with Reynolds number for a 60-inch-chord model of the NACA $65_{(421)}$-420 airfoil for two surface conditions.

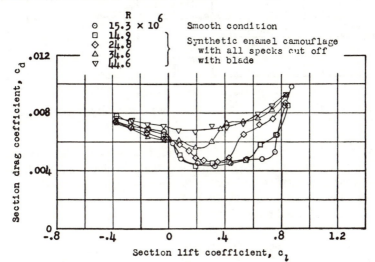

Fig. 77. Drag characteristics of NACA $65_{(421)}$-420 airfoil for two surface conditions.

marginal smoothness, which has an effect similar to increasing the surface roughness for a given Reynolds number, is to reduce rapidly the extent of the low-drag range and then to increase the minimum drag coefficient

(Fig. 77). The data of Fig. 77 were especially chosen to show this effect. In most cases, the effect of increasing the Reynolds number when the surfaces are rough is to increase the drag over the whole low-drag range. The effect of pressure gradient as shown in Fig. 77 is apparently not very powerful and tends to be evident only in cases of surfaces of marginal smoothness.[17]

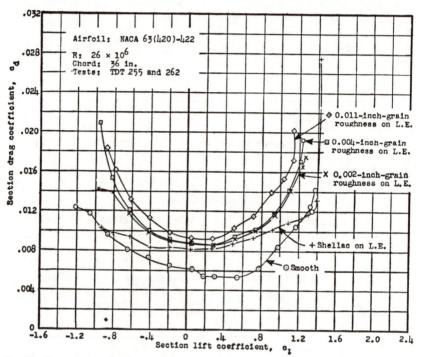

FIG. 78. Drag characteristics of a NACA 63(420)-422 airfoil with various degrees of roughness at the leading edge.

More difficulty is generally encountered in reducing the waviness to permissible values for the maintenance of laminar flow than in obtaining the required surface smoothness. In addition, the specification of the required fairness is more difficult than that of the required smoothness. The problem is not limited to finding the minimum wave size that will cause transition under given conditions because the number of waves and the shape of the waves require consideration.

If the wave is sufficiently large to affect the pressure distribution in such a manner that laminar separation is encountered, there is little doubt that such a wave will cause premature transition at all useful Reynolds numbers. Relations between the dimensions of a wave and the pressure distribution may be found by the method suggested by Allen.[9] If the pressure distribution over the wave is known, the criteria for laminar separation discussed in Sec. 5.7 may be applied. The size of the wave required to reverse the pressure gradient increases with the pressure gradient. Large negative pressure gradients would therefore appear to be favorable for

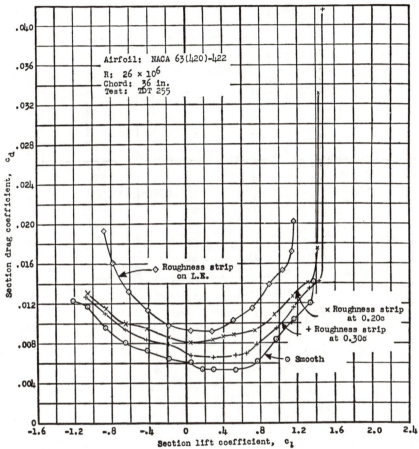

Fig. 79. Drag characteristics of a NACA 63(420)-422 airfoil with 0.011-inch-grain roughness at various chordwise locations.

wavy surfaces. Experimental results have shown this conclusion to be qualitatively correct.

For the types of waves usually found on practical-construction wings, the test of rocking a straightedge over the surface in a chordwise direction is a fairly satisfactory criterion.[3] The straightedge should rock smoothly without jarring or clicking. The straightedge test will not show the existence of waves that leave the surface convex. Tests of a large number of practical-construction models, however, have shown that those models

which passed the straightedge test were sufficiently free from small waves
to permit low drags to be obtained at flight values of the Reynolds number.

It does not appear feasible to specify construction tolerances on ordinates of wings with sufficient accuracy to ensure adequate freedom from

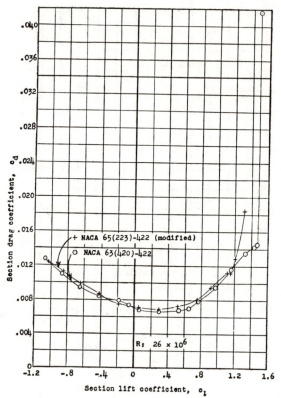

Fig. 80.   Drag characteristics of two NACA 6-series airfoils with 0.011-inch-grain roughness
at 0.30c.

waviness.   If care is taken to obtain fair surfaces, normal tolerances may
be used without causing serious alteration of the drag characteristics.

If the wing surface is sufficiently rough to cause transition near the
leading edge, large drag increases are to be expected even if the roughness
is confined to the region of the leading edge.   Figure 78 shows that, although
the degree of roughness has some effect, the increment of minimum drag
coefficient caused by the smallest roughness capable of producing transition is nearly as great as that caused by much larger grain roughness when
the roughness is confined to the leading edge.[3]   The degree of roughness has
a much larger effect on the drag at high lift coefficients.   If the roughness
is sufficiently large to cause transition at all Reynolds numbers considered,

the drag of the wing with roughness only at the leading edge decreases with increasing Reynolds number.[1]

The effect[3] of fixing transition by means of a roughness strip of carborundum of 0.011-inch grains is shown in Fig. 79. The minimum drag

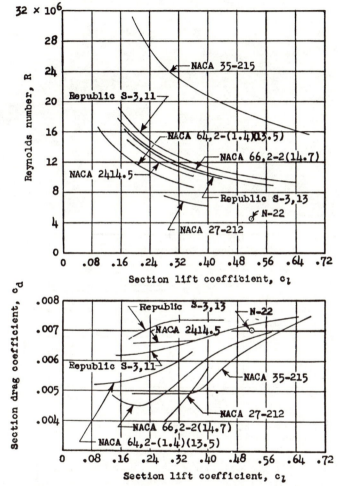

Fig. 81. Comparison of section drag coefficients obtained in flight on various airfoils. Tests of NACA 27-212 and 35-215 sections made on gloves.

increases progressively with forward movement of the roughness strip. The effect on the drag at high lift coefficients is not progressive; the drag increases rapidly when the roughness is at the leading edge. Figure 80 shows that the drag coefficients for the NACA 65(223)-422 and 63(420)-422 wing sections[3] are nearly the same throughout most of the lift range when the extent of the laminar flow is limited to 0.30c. These data indicate that, for wing sections of the same thickness ratio and camber, the drag coefficients depend almost entirely on the extent of the laminar boundary layer if no appreciable separation occurs.

The variation of minimum drag coefficient with thickness ratio for a number of NACA four-digit, five-digit, and 6-series wing sections with

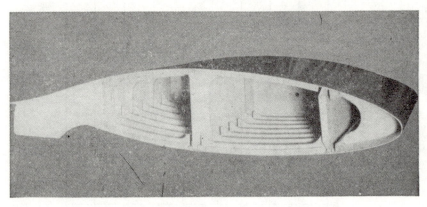

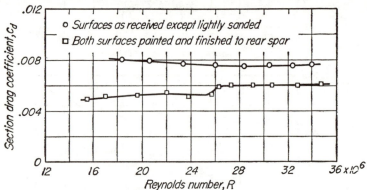

FIG. 82.   Drag scale effect on 100-inch-chord practical-construction model of the NACA 65(216)-3(16.5)(approx.) airfoil section. $c_l = 0.2$ (approx.).

leading edges rough is shown in Fig. 68. The minimum drag coefficients increase with thickness ratio but are substantially the same for all the sections of equal thickness ratio. The increments of drag coefficient caused by leading-edge roughness are correspondingly greater for the sections having the lower drag coefficients when smooth.

The section drag coefficients of several airplane wings have been measured in flight[160] by the wake-survey method, and a number of practical-construction wing sections have been tested in the NACA two-dimensional low-turbulence pressure tunnel[3] at flight values of the Reynolds number. Some of the flight data obtained by Zalovcik[160] are summarized in Fig. 81.

All wings for which data are presented in Fig. 81 were carefully finished to produce smooth surfaces. Care was taken to reduce surface waviness to a minimum for all the sections except the NACA 2415.5, N-22, Republic

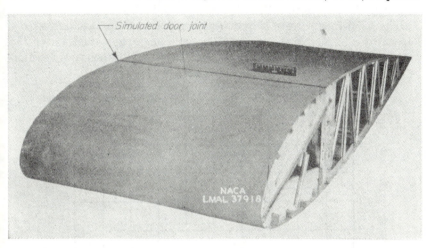

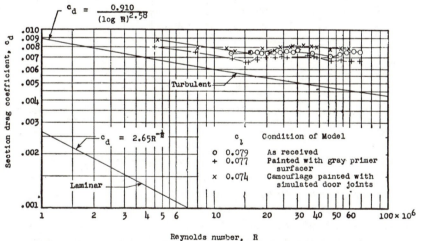

FIG. 83. Variation of the drag coefficient with Reynolds number for the NACA 23016 airfoil section together with laminar and turbulent skin-friction coefficients for a flat plate.

S-3, 13, and the NACA 27-212. Curvature gauge measurements of surface waviness for some of these sections are presented by Zalovcik.[160] These data show that the NACA 2-, 3-, and 6-series sections that permitted extensive laminar flow had substantially lower drag coefficients when smooth than did the other sections.

Data obtained in the NACA two-dimensional low-turbulence pressure tunnel[3] for typical practical-construction sections are presented in Figs. 82

to 86. Figure 87 presents a comparison of the drag coefficients obtained in this wind tunnel for a model of the NACA 0012 section and in flight for the same model mounted on an airplane.[3] For this case, the wind-tunnel and flight data agree to within the experimental error.

The wind-tunnel tests of practical-construction wing sections as delivered by the manufacturer showed minimum drag coefficients of the order of 0.0070 to 0.0080 in nearly all cases regardless of the type of section

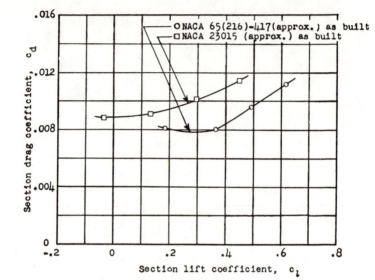

Fig. 84. Drag characteristics of the NACA 65,2-417 (approx.) and NACA 23015 (approx.) airfoil sections built by practical-construction methods by the same manufacturer. R, 10.23×10^6.

(Figs. 82 to 86). Such values may be regarded as typical for good American construction practice during the Second World War. Finishing the sections to produce smooth surfaces always resulted in substantial drag reductions, although considerable waviness usually remained. None of the sections tested had fair surfaces at the front spar. Unless special care is taken to produce fair surfaces at the front spar, the resulting wave may be expected to cause transition either at the spar location or a short distance behind it. One practical-construction specimen tested with smooth surfaces maintained relatively low drags up to Reynolds numbers of approximately 30 million [NACA 66(2x15)-116 wing section of Fig. 66]. This specimen had no spar forward of about 35 per cent chord from the leading edge and no spanwise stiffeners forward of the spar. This type of construction resulted in unusually fair surfaces.

Few data are available on the effect of propeller slipstream on transition or wing drag; the data that are available do not show consistent results.

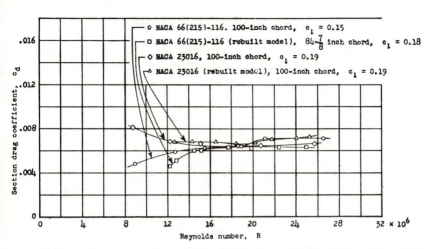

Fig. 85. Scale effect on drag of the NACA 66(215)-116 and NACA 23016 airfoil sections built by practical-construction methods by the same manufacturer and tested as received.

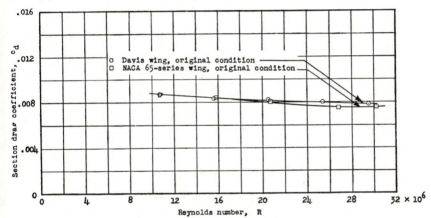

Fig. 86. Drag scale effect for a model of the NACA 65-series airfoil section 18.27 per cent thick and the Davis airfoil section 18.27 per cent thick, built by practical-construction methods by the same manufacturer. $c_l = 0.46$ (approx.).

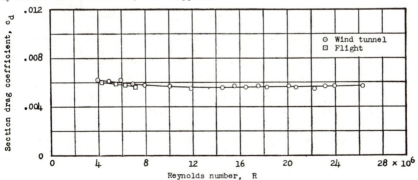

Fig. 87. Comparison of drag coefficients measured in flight and wind tunnel for the NACA 0012 airfoil section at zero lift.

This inconsistency may result from variation of lift coefficient, surface condition, air-stream turbulence, propeller advance-diameter ratio, and num-

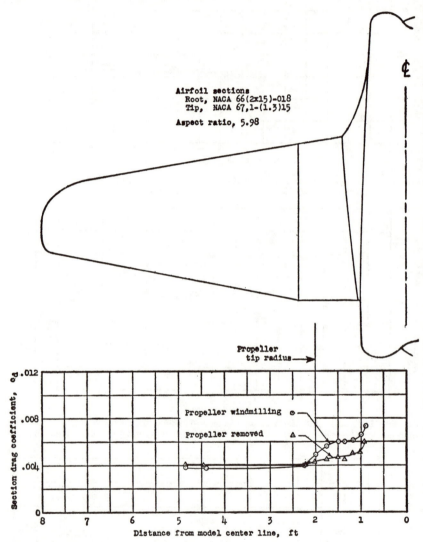

FIG. 88.   The effect of propeller operation on section drag coefficient of a fighter-type airplane, from tests of a model on the Langley 19-foot pressure tunnel. $C_L = 0.10$; R, 3.7×10^6.

ber of blades.   Some early British investigations[157, 158, 159] showed that transition occurred at 5 to 10 per cent of the chord from the leading edge in the slipstream.   Similar results were indicated by tests in the NACA 8-foot high-speed tunnel.[43]   Drag measurements in the NACA 19-foot pressure tunnel[3] (Fig. 88) indicated that only moderate drag increments resulted from a windmilling propeller. Although these data are only qualitative[3] because of the difficulty of making wake surveys in the slipstream, they seem to preclude very large drag increments such as would result from a movement of transition to a position close to the leading edge. A

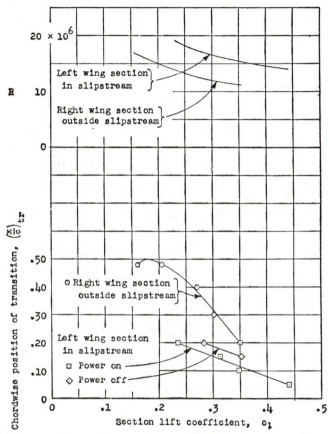

Fig. 89. Flight measurements of transition on a NACA 66-series wing within and outside the slipstream.

similar conclusion is indicated by NACA flight data[3] (Fig. 89), which show transition as far back as 20 per cent of the chord in the slipstream.

Even fewer data are available on the effects of vibration on transition. Tests in the NACA 8-foot high-speed tunnel[43] showed negligible effects, but the range of frequencies tested may not have been sufficiently wide to represent conditions encountered on airplanes. Some NACA flight data[3] showed small but consistent rearward movements of transition outside the slipstream when the propellers were feathered. This effect was noticed even when the propeller on the opposite side of the two-engined airplane was feathered, and it was accordingly attributed to vibration or noise. In some cases, increases of drag have been noted in wind tunnels when the model or its supports vibrated. A tentative conclusion may be drawn from the meager data that vibration may cause small forward movements of transition on airplane wings.

The skin friction associated with a turbulent boundary layer on a smooth surface decreases with increase of Reynolds number, as shown in Chap. 5. This favorable scale effect is not obtained at high Reynolds

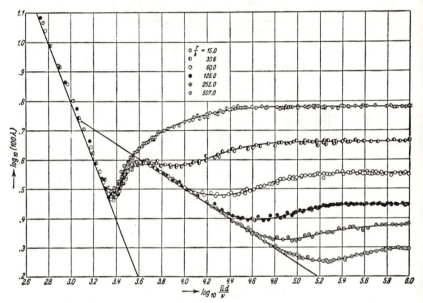

FIG. 90. Coefficient of resistance of rough and smooth tubes dependent on Reynolds number.

numbers on rough surfaces. An explanation of this effect is that the pressure drag of the individual protuberances constituting the roughness contribute to the skin friction when they project through the laminar sublayer. The shapes of the individual protuberances are usually such that the scale effect on the pressure drag should be small. The result at large Reynolds numbers for generally roughened surfaces is that the skin-friction coefficient is essentially constant.

The general nature of the scale effects for rough surfaces has been indicated by experiments in pipes.[28] These data (Fig. 90) show that at low Reynolds numbers the pipe-loss coefficient decreases with increasing Reynolds number at the rate expected for laminar flow. When transition occurs, the pipe-loss coefficient increases to the value expected for turbulent flow over smooth surfaces. For very small roughness, the pipe-loss coefficient decreases with increasing Reynolds number along the turbulent skin-friction curve for smooth surfaces until large values of the Reynolds number are reached. For any given size of roughness, however, there appears to be a Reynolds number beyond which the coefficient increases slightly to a constant value. The value of the Reynolds number decreases and the constant value of the coefficient increases as the grain size of the roughness increases. At large values of the Reynolds number, the roughness causes large increments of skin friction over the values corresponding to smooth surfaces.

On the basis of the pipe experiments and reasoning similar to the foregoing, von Kármán[139] obtained a formula for the grain size just sufficiently large to affect turbulent skin friction. This formula may be written

$$\frac{\rho U k}{\mu} = 3\sqrt{\frac{2}{c_f}}$$

where $\dfrac{\rho U k}{\mu} =$ Reynolds number based on grain size of roughness and local velocity outside boundary layer

$c_f =$ local skin-friction coefficient

For a wing of approximately 9-foot chord at a speed of 300 feet per second, the limiting grain size is approximately 0.0004 inch and varies only slightly over the surface, increasing toward the trailing edge.

The effect of roughness on the drag of wing sections with largely turbulent flow is analogous to the effect in pipes as shown[3] by Fig. 83. At the lower Reynolds numbers, the drag coefficient decreases with increasing Reynolds number at about the rate expected for turbulent flow over smooth surfaces. In this case, the drag remains essentially constant at Reynolds numbers above about 15 million. At a Reynolds number of 70 million the increment of drag caused by the moderate roughness was of the order of one-quarter of the drag that would have been obtained had the favorable scale effect continued. The sensitivity of the turbulent boundary layer to roughness is so great that airplanes should not be expected to show favorable scale effects at large Reynolds numbers unless considerable care is taken to obtain smooth surfaces.

It should be noted that, in contrast to the situation with respect to the laminar layer, it is relatively easy to obtain reductions of drag by attention to the surface conditions for turbulent flow. Even though the allowable size of the roughness is very small, each speck of roughness presumably contributes only the drag of itself and does not have any appreciable effect on the skin friction over the surface downstream. Consequently favorable effects may be expected from careful finishing of the general surface of modern high-performance airplanes even though imperfections such as rivets and seams may be present.

d. Unconservative Wing Sections. The need for low drags in order to obtain long range for large airplanes flying at speeds below the critical Mach number leads to designs having high wing loadings to reduce the wing area and profile drag together with relatively low span loadings to avoid high induced drags. These tendencies result in wings of high aspect ratio that require large spar depths for structural efficiency. The large spar depths require the use of thick root sections. The comparatively high lift coefficients corresponding to the cruise condition for such designs leads to the use of large cambers to obtain low profile drags. Such designs are encouraged by the drag characteristics of modern smooth wing sections

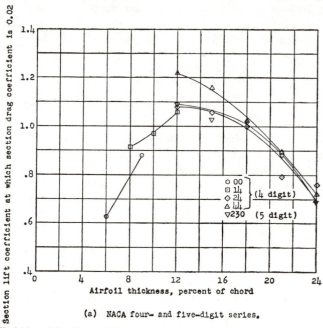

(a) NACA four- and five-digit series.

FIG. 91. Variation of the lift coefficient corresponding to a drag coefficient of 0.02 with thickness and camber for a number of NACA airfoil sections with roughened leading edges. R, 6×10^6.

which show relatively small increases of drag coefficient with increasing thickness ratio and camber (Sec. 7.5a).

Unfortunately airplane wings are not usually constructed with smooth surfaces, and, in any case, the surfaces cannot be relied upon to stay smooth under all service conditions. The effect of roughening the leading edges of thick wing sections is to cause large increases of the drag coefficient at high lift coefficients. The resulting drag coefficient may be excessive at cruising lift coefficients for heavily loaded high-altitude airplanes. Wing sections that have suitable characteristics when smooth but have excessive drag coefficients when rough at lift coefficients corresponding to cruising or climbing conditions are called "unconservative."

The decision as to whether a given wing section is conservative will depend upon the power and wing loadings of the airplane. The decision may be affected by expected service and operating conditions. For example, the ability of a multiengine airplane to fly with one or more engines inoperative in icing conditions or, in the case of military airplanes, after

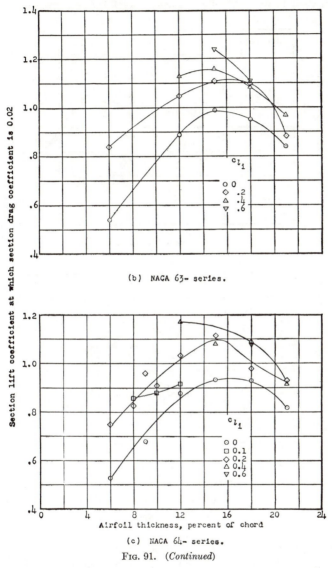

(b) NACA 63- series.

(c) NACA 64- series.

Fig. 91. (*Continued*)

suffering damage in combat may be a consideration. As an aid in judging whether the sections are conservative, the lift coefficient corresponding to a drag coefficient of 0.02 was determined from the figures of Appendix IV for a number of NACA wing sections with roughened leading edges. The variation of this lift coefficient with thickness ratio and camber[3] is shown in Fig. 91 for a Reynolds number of 6 million. These data show that, in general, the lift coefficient at which the drag coefficient is 0.02 decreases with rearward movement of the position of minimum pressure and with

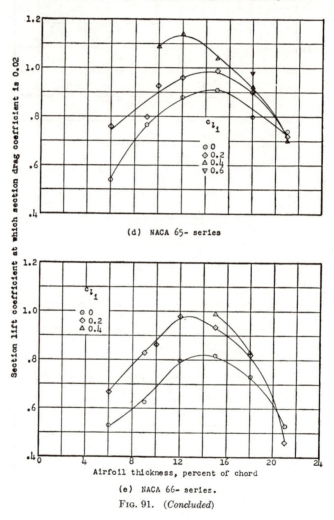

(d) NACA 65- series

(e) NACA 66- series.

Fig. 91. (*Concluded*)

increased thickness above thickness ratios of about 15 per cent. For wing sections thinner than approximately 18 per cent, the effect of camber is to increase this lift coefficient. For the thicker sections, however, increasing the camber becomes relatively ineffective and may even be harmful in extreme cases. The highest values of this lift coefficient for wing sections having thickness ratios greater than 15 per cent are obtained with the NACA 64-series sections having a design lift coefficient of 0.4.

7.6. Pitching-moment Characteristics. The variation of the quarter-chord pitching-moment coefficient at zero angle of attack with thickness

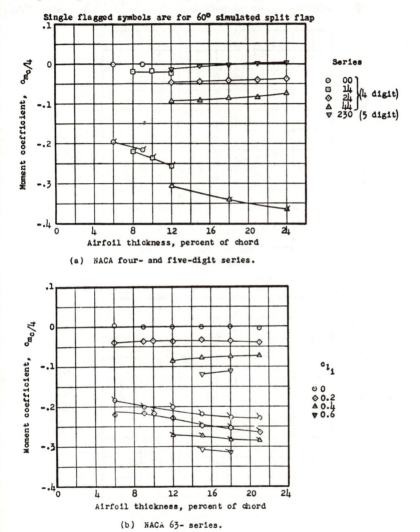

Fig. 92. Variation of section quarter-chord pitching-moment coefficient (measured at an angle of attack of zero degrees) with airfoil thickness ratio for several NACA airfoil sections of different camber. R, 6×10^6.

ratio and camber is presented in Fig. 92 for a large number of NACA wing sections.[3] The pitching-moment coefficients of the NACA four- and five-digit series sections become more negative with decreasing thickness ratio. Comparison of the experimental data in Fig. 92 with the theoretical values obtained from the theory of thin wing sections shows that the absolute

magnitudes of the pitching-moment coefficients for the NACA four- and
five-digit series sections are somewhat less than those indicated by the
theory. The pitching-moment coefficients for the NACA 6-series wing

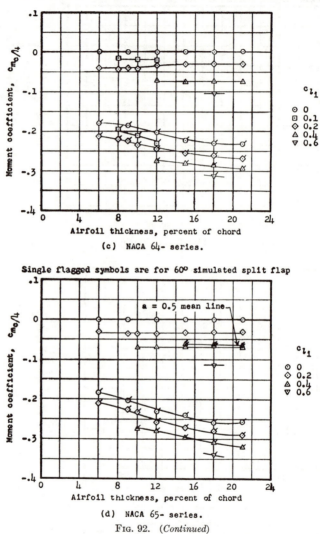

(c) NACA 64- series.

Single flagged symbols are for 60° simulated split flap

(d) NACA 65- series.

FIG. 92. (*Continued*)

sections show practically no variation with thickness ratio or position of
minimum pressure. As indicated by the theory of thin wing sections, increasing the amount of camber causes a nearly uniform negative increase
of the pitching-moment coefficient. In general, the absolute magnitude of
the quarter-chord pitching-moment coefficients for the NACA 6-series wing
sections having mean lines of the type $a = 1.0$ are approximately three-

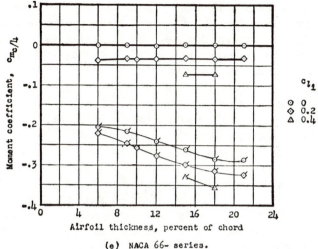

(e) NACA 66- series.

FIG. 92. (*Concluded*)

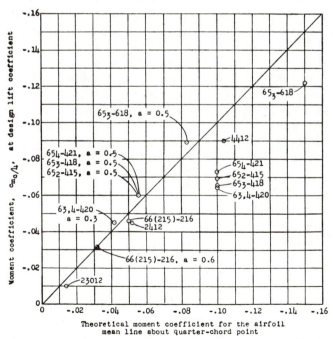

FIG. 93. Comparison of theoretical and measured pitching-moment coefficients for some NACA airfoils. R, 6×10^6.

quarters of the theoretical values. The pitching-moment coefficients for sections having mean lines of the type $a < 1.0$ are equal to or slightly more negative than the theoretical values as shown in Fig. 93. Consequently, changing the type of mean line from $a = 1.0$ to $a < 1.0$ to reduce the magnitude of the pitching-moment coefficient is relatively ineffective unless the value of a is reduced to a small value.

The variation of chordwise position of the aerodynamic center at a Reynolds number of 6 million for a large number of NACA wing sections is presented in Fig. 94. From the data presented in Appendix IV, there appears to be no systematic variation of chordwise position of the aerodynamic center with Reynolds number between 3 and 9 million. For the NACA 24-, 44-, and 230-series wing sections with thickness ratios ranging from 12 to 24 per cent, the chordwise position of the aerodynamic center is ahead of the quarter-chord point and moves forward with increases of thickness ratio. The data for the NACA 00- and 14-series sections show that the aerodynamic center is at the quarter-chord point. The aerodynamic center is aft of the quarter-chord point for the NACA 6-series wing sections and moves rearward with increase of thickness ratio. There appears to be little systematic variation of the chordwise position of the aerodynamic center with camber or position of minimum pressure for these sections. For the thick cambered sections, the chordwise position of the aerodynamic center appears to move forward as the design position of minimum pressure moves back. For wing sections with arbitrary modifications of shape near the trailing edge, the trailing-edge angle appears to be an important parameter affecting the chordwise position of the aerodynamic center. For such sections, the aerodynamic center moves forward as the trailing-edge angle is increased.[92]

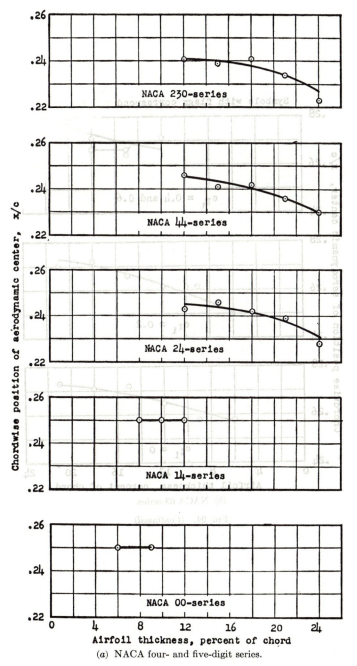

(a) NACA four- and five-digit series.

FIG. 94. Variation of section chordwise position of the aerodynamic center with airfoil thickness ratio for several NACA airfoil sections of different cambers. R, 6×10^6.

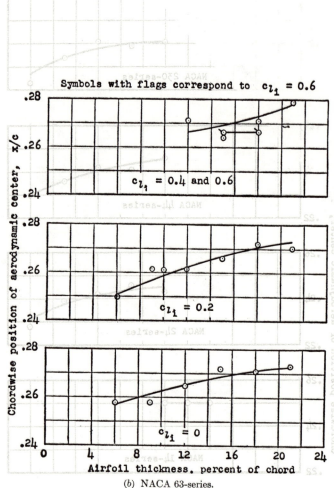

(b) NACA 63-series.

FIG. 94. (Continued)

FIG. 94. Variation of section chordwise position of the aerodynamic center with airfoil thickness ratio for several NACA airfoil sections of different cambers. $R = 6 \times 10^6$.

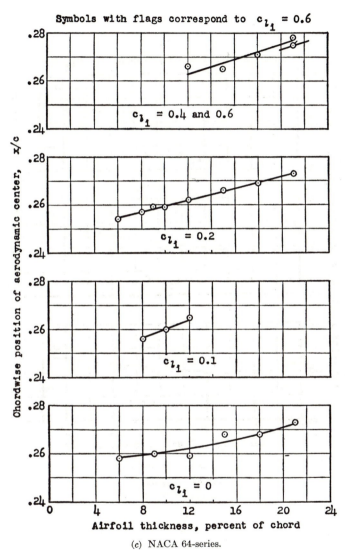

(c) NACA 64-series.

Fig. 94. (*Continued*)

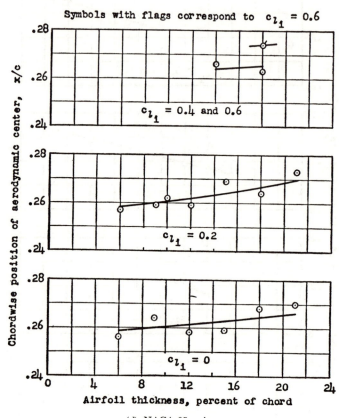

(d) NACA 65-series.

FIG. 94. (Continued)

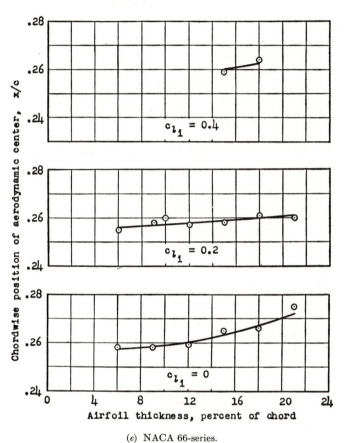

(e) NACA 66-series.

FIG. 94. (*Concluded*)

CHAPTER 8

HIGH-LIFT DEVICES

8.1. Symbols.

C_D drag coefficient

$C_{D_{\min}}$ minimum lift coefficient

$C_{L_{\max}}$ maximum lift coefficient

$\Delta C_{L_{\max}}$ increment of maximum lift coefficient

C_N normal-force coefficient

C_{N_f} flap normal-force coefficient based on the area of the flap

C_Q volume-flow coefficient, Q/VS

E flap-chord ratio, c_f/c

G moment arm, in terms of the chord, of the basic normal force about the quarter-chord point

M Mach number

P_{a_δ} incremental additional load distribution associated with flap deflection

P_{b_δ} incremental basic load distribution associated with flap deflection

P_δ incremental load distribution associated with flap deflection

Q volume of flow through slot

R Reynolds number, $\rho Vc/\mu$

S wing area

V velocity of the free stream

c chord

c_d section drag coefficient

c_f flap chord

c_h section flap hinge-moment coefficient, $h/\frac{1}{2}\rho V^2 c_f^2$

c_l section lift coefficient

c_{l_f} section flap lift coefficient, $l_f/\frac{1}{2}\rho V^2 c_f$

$c_{l_{\max}}$ maximum section lift coefficient

$\Delta c_{l_{\max}}$ increment of maximum section lift coefficient

c_m section pitching-moment coefficient, $m/\frac{1}{2}\rho V^2 c^2$

$c_{m_{c/4}}$ section pitching-moment coefficient about the quarter-chord point

c_{m_1} section pitching-moment coefficient about the quarter-chord point with flap neutral

c_{m_2} section pitching-moment coefficient about the quarter-chord point with flap deflected

Δc_m incremental section pitching-moment coefficient about the quarter-chord point

c_{n_1} section normal-force coefficient with flap neutral

c_{n_2} section normal-force coefficient with flap deflected

$c_{n_{a_\delta}}$ incremental additional normal-force coefficient associated with flap deflection

$c_{n_{b_\delta}}$ incremental basic normal-force coefficient associated with flap deflection

$c_{n_{f_\delta}}$ incremental flap normal-force coefficient

Δc_n incremental section normal-force coefficient

$(cp)_f$ center of pressure of the load on a flap measured from the leading edge of the flap

h section flap hinge-moment coefficient (positive in direction of positive flap deflection)

l_f lift force acting on the flap per unit span

m section pitching moment

n_{f_δ} incremental flap normal force per unit span

q dynamic pressure, $\frac{1}{2}\rho V^2$

r leading-edge radius

t thickness of wing section

v local velocity over the surface of a symmetrical section at zero lift

Δv_a increment of local velocity over the surface of a wing section associated with angle of attack

x distance parallel to chord

y distance normal to chord

α angle of attack

α_0 section angle of attack

$\Delta\alpha_0$ increment of section angle of zero lift

$\gamma_{a_\delta}, \gamma_{b_\delta}$ ratios of the flap normal force to the section normal force for the incremental additional and the incremental basic normal forces

δ flap deflection

δ_{f_1} deflection of fore flap or vane

δ_{f_2} deflection of main flap of a double-slotted flap configuration

$\Delta\delta$ increment of flap deflection

ρ mass density of air

τ turbulence factor, effective Reynolds number/test Reynolds number

τ_n, τ_m see Eq. (8.8)

8.2. Introduction. The auxiliary devices discussed in this chapter are essentially movable elements that permit the pilot to change the geometry and aerodynamic characteristics of the wing sections to control the motion of the airplane or to improve the performance in some desired manner. The desire to imp ove performance by increasing the wing loading while maintaining acceptable landing and take-off speeds led to the development of retrac able devices to improve the maximum lift coefficients of wings without changing the characteristics for the cruising and high-speed flight conditions. Some typical high-lift devices are illustrated in Fig. 95. The aerodynamic characteristics of some typical high-lift devices are presented and discussed in the following sections. Primary emphasis is given to the capabilities and relative merits

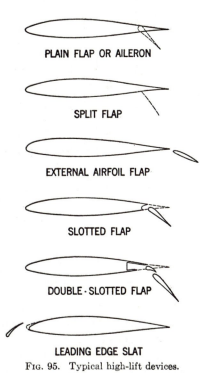

PLAIN FLAP OR AILERON

SPLIT FLAP

EXTERNAL AIRFOIL FLAP

SLOTTED FLAP

DOUBLE · SLOTTED FLAP

LEADING EDGE SLAT

FIG. 95. Typical high-lift devices.

of the various devices. Numerous references to the literature are given to provide design information.

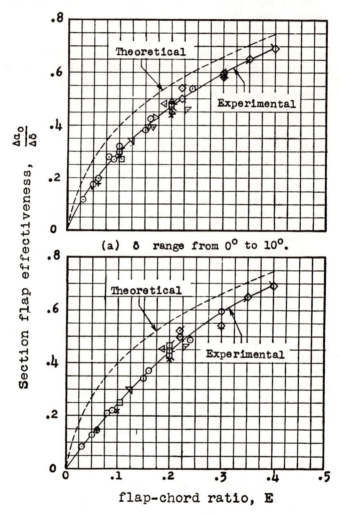

(a) δ range from $0°$ to $10°$.

(b) δ range from $0°$ to $20°$.

Fig. 96. Variation of section flap effectiveness with flap chord ratio for true-airfoil-contour flaps without exposed overhang balance on a number of airfoil sections; gaps sealed; $c_l = 0$.

8.3. Plain Flaps. Plain trailing-edge flaps are formed by hinging the rearmost part of the wing section about a point within the contour. Downward deflections of the trailing edge are called "positive-flap deflections." Deflection of a plain flap with no gap effectively changes the camber of the wing section, and some of the resulting changes of the aerodynamic charac-

Sym-bol	Basic airfoil	Type of flap	Air-flow characteristics		
			τ	M	R
○	NACA 0009	Plain	1.93	0.08	....
+	NACA 0015	Plain	1.93	0.10	1.4×10^6
×	NACA 23012	Plain	1.60	0.11	2.2×10^6
□	NACA 66(2x15)-009	Plain, straight contour	1.93	0.10	1.4×10^6
◇	NACA 66-009	Plain	1.93	0.11	1.4×10^6
▽	NACA 63,4-4(17.8) approx.	Internally balanced	Approaching 1.00	0.17	2.5×10^6
△	NACA 66(2x15)-216, $a = 0.6$	Internally balanced	Approaching 1.00	0.18	5.3×10^6
◁	NACA 66(2x15)-116, $a = 0.6$	Internally balanced	Approaching 1.00	0.14	6.0×10^6
▷	NACA 64,2-(1.4)(13.5)	Plain	Approaching 1.00	....	13.0×10^6
◢	NACA 65,2-318 approx.	Internally balanced	Approaching 1.00	0.14	6.0×10^6
◺	NACA 63(420)-521 approx.	Internally balanced	Approaching 1.00	....	8.0×10^6
◹	NACA 66(215)-216 $a = 0.6$	Internally balanced	Approaching 1.00	0.20 to 0.48	2.8×10^6 to 6.8×10^6
▽	NACA 66(215)-014	Plain	1.93	0.09	1.2×10^6
◯	NACA 66(215)-216 $a = 0.6$	Plain	Approaching 1.00	....	6.0×10^6
◻	NACA 65_2-415	Plain	Approaching 1.00	0.13	6.0×10^6
◰	NACA 65_3-418	Plain	Approaching 1.00	0.13	6.0×10^6
◱	NACA 65_4-421	Plain	Approaching 1.00	0.13	6.0×10^6
◇	NACA $65_{(112)}$-213	Internally balanced	Approaching 1.00	0.14	8.0×10^6
◇	NACA 745A317 approx.	Internally balanced	Approaching 1.00	0.13	6.0×10^6
◇	NACA 64,3-013 approx.	Internally balanced	Approaching 1.00	0.13	6.0×10^6
◇	NACA 64,3-1(15.5) approx.	Internally balanced	Approaching 1.00	0.13	6.0×10^6

(c) Supplementary information.

Fig. 96. (*Concluded*)

191

teristics may be calculated from the theory of thin wing sections (Chap. 4) if the flow does not separate from the surface. For most commonly used wing sections, this condition is satisfied reasonably well for flap deflections of not over 10 to 15 degrees. The theory permits calculation of the angle of zero lift, the pitching-moment coefficient, and the chordwise load distribution with reasonable accuracy. The flap loads and hinge moments may also be obtained, but the accuracy of these quantities is relatively poor because the effects of viscosity are particularly pronounced over the aft portion of the section. The effectiveness of the flap in increasing the maximum lift coefficient cannot be calculated.

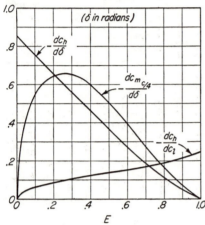

FIG. 97. Theoretical hinge moment and pitching moment characteristics of plain trailing-edge flaps.

Glauert[35] calculated the effect of plain flaps in changing the angle of zero lift, the pitching moment, and the flap hinge moments. The calculated effect of flap deflection on the angle of zero lift is shown in Fig. 96, where the flap effectiveness[3] $\Delta\alpha_0/\Delta\delta$ is plotted against the ratio of the flap chord to the section chord c_f/c. Numerous experimental points are also shown for flap deflections of 0 to 10 degrees and 0 to 20 degrees. In general, the flap effectiveness is less than that indicated by the theory, and the discrepancy increases with increased flap deflection for small chord flaps. The calculated effectiveness of a plain flap in changing the section pitching-moment coefficient about the quarter-chord point $dc_m/d\delta$ is shown in Fig. 97. The calculated values of the rate of change of hinge moment with lift coefficient dc_h/dc_l and the rate of change of hinge moment with flap deflection $dc_h/d\delta$ are also shown in Fig. 97. For symmetrical wing sections, the theoretical pitching-moment and hinge-moment coefficients are

$$c_m = \delta\left(\frac{dc_m}{d\delta}\right)$$

$$c_h = c_l\left(\frac{dc_h}{dc_l}\right) + \delta\left(\frac{dc_h}{d\delta}\right)$$

The linearity of the theory of thin wing sections permits the application of the values obtained from Figs. 96 and 97 to cambered as well as symmetrical sections. In the case of cambered sections, these values are applied as increments to the corresponding values for the unflapped section.

Pinkerton[81] calculated the flap loads that are presented in Fig. 98. The rate of change of the flap lift coefficient with the section lift coefficient dc_{l_f}/dc_l and with flap deflection $dc_{l_f}/d\delta$ are plotted against the ratio of flap chord to section chord.

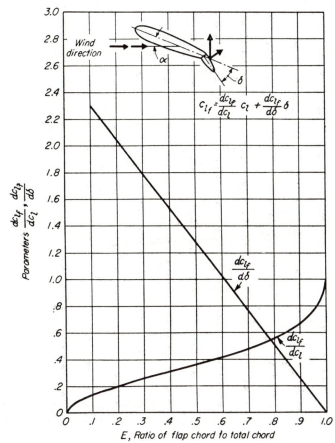

$$c_{l_f} = \frac{dc_{l_f}}{dc_l} c_l + \frac{dc_{l_f}}{d\delta} \delta$$

Fig. 98. Theoretical variation of flap lift coefficient with section lift coefficient and flap deflection.

The pressure distribution over a wing section with a deflected plain flap may be calculated by the wing-section theory of Chap. 3, but such calculations for a number of flap deflections would be unduly laborious. Some knowledge of the chordwise load distribution is given by the theory of thin wing sections. The change of load distribution caused by flap deflection can be resolved into two components, one of which is simply the additional load distribution associated with angle of attack. The other component is the difference between the load distribution on the original and deflected mean lines at their respective ideal angles of attack. These load distributions may be calculated by the theory of thin wing sections (Chap. 4) as explained by Allen,[6] but the results are not very accurate. For example, the simple theory predicts infinite load concentrations at the leading edge and hinge location. The infinite load concentration at the leading edge may be avoided by use of the additional types of load distribution associated with angle of attack as obtained from thick-wing-section theory and presented in Appendix I. Allen[6] obtained empirical load distributions analogous to those associated with changes of the camber that permit calculation of reasonable approximations to the load distributions of flapped

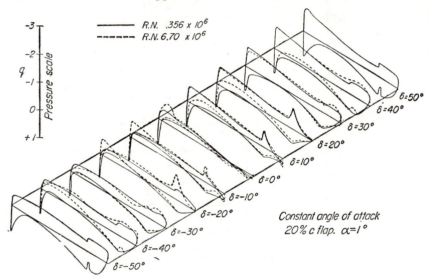

Fig. 99.   Distribution of pressure over a wing section with a plain flap.

wing sections. Allen's method is presented in Sec. 8.8. Once the load distribution is obtained, the pressure distribution may be calculated by the method of Sec. 4.5. Some typical pressure distributions for sections with plain flaps[47] are presented in Fig. 99.

The lift, drag, and moment characteristics of a typical NACA 6-series wing section with a 0.20c plain flap[3] are presented in Fig. 100 for several flap deflections up to 65 degrees. These and other data[2] show that the angle of maximum lift coefficient with the flap deflected is generally somewhat less than for the plain wing section. Curves of the maximum lift coefficient plotted against flap deflection for two sections[3] are presented in Fig. 101. Plain flaps of 0.20c appear to be capable of producing increments of the section maximum lift coefficient ranging up to about 1.0 and are more effective when applied to sections with small amounts of camber. Additional data on plain flaps are presented by Jacobs,[47] Abbott,[2] and Wenzinger.[148, 149, 156]

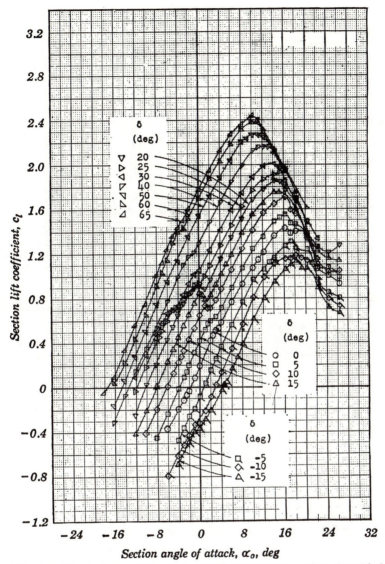

Fig. 100. Aerodynamic characteristics of the NACA 66(215)-216 airfoil section with 0.20c sealed plain flap.

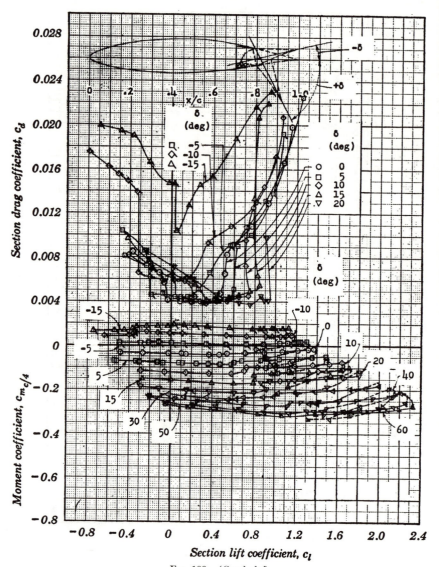

FIG. 100. (*Concluded*)

When the turbulent boundary layer over the aft portion of the wing section is thin and resistant to separation (as when extensive laminar flow is obtained), small deflections of a plain flap do not cause separation but shift the range of lift coefficients for which low drag is obtained.[3] If the extent of laminar flow is not large or if the flap deflection is sufficiently

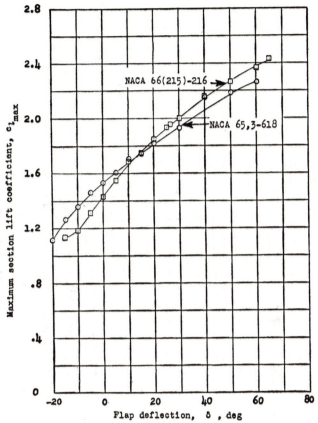

FIG. 101. Maximum lift coefficients for the NACA 65,3-618 and NACA 66(215)-216 airfoils fitted with 0.20-airfoil-chord plain flaps. R, 6×10^6.

large to be of interest as a high-lift device, the flow over the flap separates and large drag increments result. A typical variation of drag with lift coefficient for the NACA 23012 wing section with a 0.20c plain flap deflected the optimum amount at each lift coefficient[2] is shown in Fig. 102.

8.4. Split Flaps. The split flap is one of the simplest of the high-lift devices. The usual split flap is formed by deflecting the aft portion of the lower surface about a hinge point on the surface at the forward edge of the deflected portion (see Fig. 95). One variation of the simple split flap has the hinge point located forward of the deflected portion of the surface in such a manner as to leave a gap between the deflected flap and the wing surface. In another variation the leading edge of the flap is moved aft as the flap is deflected, either with or without a gap between the flap and the wing surface. Split flaps derive their effectiveness from the large increase of camber produced and, in the case of some of the variations, from the effective increase of wing area.

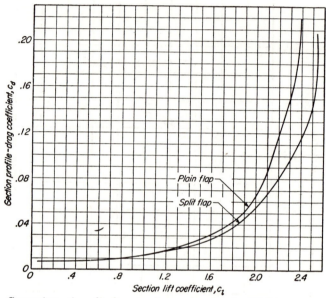

FIG. 102.   Comparison of profile-drag envelope polars for the NACA 23012 airfoil with 0.20c plain and split flaps.

The lift and moment characteristics[3] for typical NACA 6-series wing sections with 0.20c split flaps deflected 0, 40, 50, 60, and 70 degrees are presented in Figs. 103 and 104.   Similar data for a flap deflection of 60 degrees are presented for most of the wing sections of Appendix IV.   The lift-curve slope with split flaps is higher and the angle of maximum lift is somewhat lower than for the plain section.   There appears to be a tendency for the lift-curve slope for large flap deflections to be less than that for moderate deflections.[2, 154]

Inspection of Figs. 103 and 104 shows that the split flap is more effective in increasing the maximum lift coefficient when applied to the thick than when applied to the thin wing sections.   Figure 58 shows that the maximum lift coefficients of NACA 6-series wing sections with 0.20c split flaps deflected 60 degrees increase with thickness ratio up to ratios of at least 18 per cent.   Figure 58 shows comparatively little variation of maximum lift coefficient for NACA four-digit wing sections with 0.20c split flaps for thickness ratios greater than 12 per cent, but the maximum lift coefficient

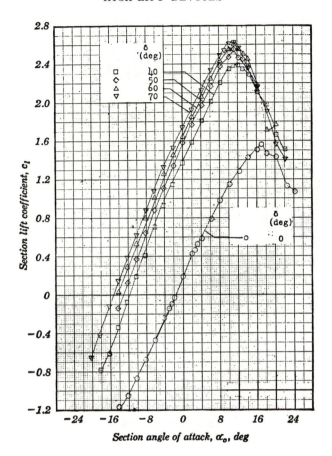

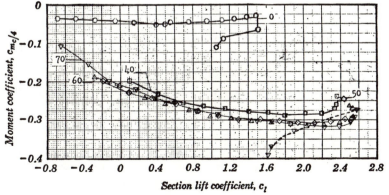

Fig. 103. Lift and moment characteristics of the NACA 66(215)-216 airfoil section with 0.20c split flap. R, 6×10^6.

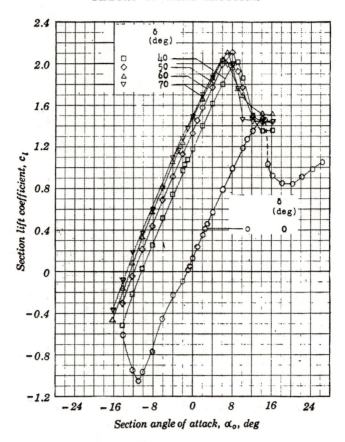

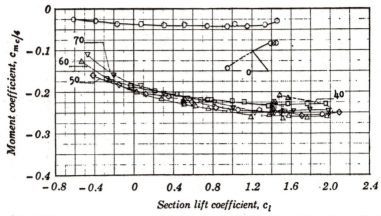

Fig. 104. Lift and moment characteristics of the NACA 65$_1$-212 airfoil section with 0.20c split flap. R, 6 × 10^6.

decreases for smaller thickness ratios. These results are in essential agreement with those[154] of Fig. 105, which shows that, for the NACA 230-series sections, the maximum lift coefficient generally increases with thickness ratio for large-chord split flaps but decreases for small-chord flaps.

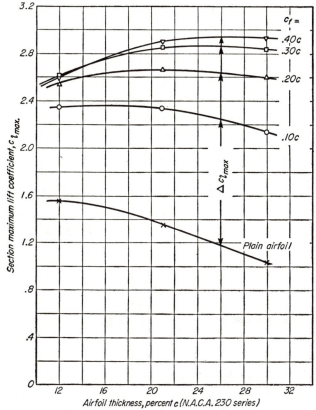

FIG. 105. Effect of airfoil thickness on maximum lift coefficient of NACA 230-series airfoils with and without split flaps.

The variation of the increment of maximum lift coefficient resulting from deflection of split flaps of 10, 20, 30, and 40 per cent chord [154] is shown in Fig. 106. For wing sections of normal thickness ratios, substantially full benefit is obtained for flap deflections of 60 or 70 degrees. The effect of the chord of split flaps on the increment of maximum lift coefficient[154] is shown in Fig. 107. For wings of normal thickness ratio, comparatively little benefit is obtained by increasing the flap chord to more than 20 or 25 per cent of the section chord, although large flap chords are more effective on thick than on thin sections.

Deflection of a split flap produces a bluff body, and large drag increments are to be expected. Figure 102 shows that the drag of the NACA 23012 section with a 0.20c split flap deflected the optimum amount at each lift coefficient is about the same or less than that for a plain flap.[2]

The pressure distributions over wing sections with highly deflected split flaps are similar to those for plain flaps (Fig. 99). The pressure distributions with deflected split flaps may be predicted by the semiempirical method of Allen.[6]

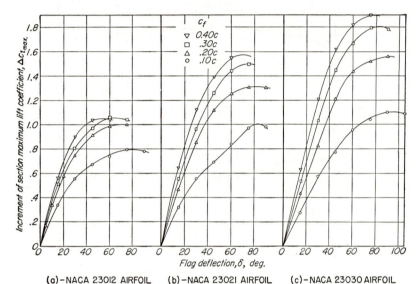

(a)—NACA 23012 AIRFOIL   (b)—NACA 23021 AIRFOIL   (c)—NACA 23030 AIRFOIL

FIG. 106. Effect of split-flap deflection on increment of maximum lift coefficient for various airfoils and flaps.

Data required for the structural design of split flaps are given by Wenzinger and Rogallo.[156] The normal force coefficients and centers of pressure for a 0.20c split flap at various deflections are shown in Fig. 108. The data of Wenzinger and Rogallo show that the chordwise position of the forward edge of the flap has only a small effect on the flap loads.

The effect of a gap between the wing surface and a split flap with a nominal chord of 0.20c hinged at 0.80c from the leading edge of the section[150] is shown in Fig. 109. The data show that the loss of maximum lift coefficient associated with the gap is considerably greater than that caused by removing the same area from the trailing edge of the flap.

Figure 110 summarizes the effect[144] on maximum lift coefficient of moving a 0.20c split flap toward the trailing edge from its normal position. These data show that the percentage increment of maximum lift coefficient obtained by moving the split flap to the rear is about the same as the percentage increase of wing area measured as projected onto the original chord line. Similar results have been obtained[144] for 0.30c and 0.40c split flaps.

8.5. Slotted Flaps. *a. Description of Slotted Flaps.* Slotted flaps provide one or more slots between the main portion of the wing section and the

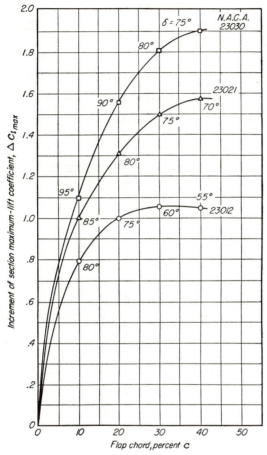

Fig. 107. Effect of chord of split flap on increment of maximum lift coefficient for three airfoil thicknesses.

deflected flap; and they derive their effectiveness from increasing the camber and, in some cases, from increasing the effective chord of the section. The slots duct high-energy air from the lower surface to the upper surface and direct this air in such a manner as to delay separation of the flow over the flap by providing boundary-layer control.

The numerous types of slotted flap are classified by their geometry. Several types are shown in Fig. 111. The primary classification is the number of slots. The single-slotted flap is the simplest and most generally used type. Double-slotted flaps have been used to some extent, and multiple-slotted or venetian-blind flaps have been investigated. An important consideration in the design of slotted flaps, especially of the single-

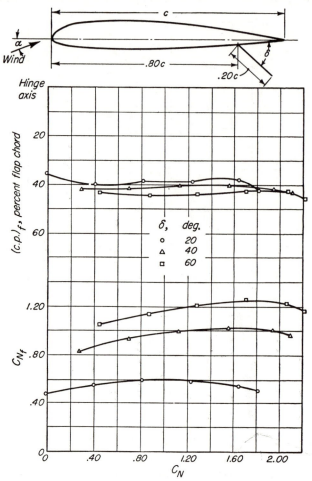

Fig. 108. Normal force coefficients and centers of pressure of a 0.20c split flap at 0.80c on a NACA 2212 wing.

slotted type, is the extent to which the flap moves aft as it is deflected. The movement of the flap may vary from a simple rotation about a fixed point to a combined rotation and translation that moves the leading edge of the flap to the vicinity of the normal trailing-edge position. Rearward movement of the flap requires an extension of the upper surface over some or all of the flap in the retracted position. This extension of the upper surface serves to direct the flow of air through the slot in the proper direction and is called the "lip." The external-airfoil flap (Fig. 111) may be considered as a special case of the single-slotted flap with the distinguishing feature that it does not retract within the section.

The flow about a wing section with a deflected slotted flap is very complicated, and no adequate theory has been developed to predict the aerodynamic characteristics. Consequently the information required for

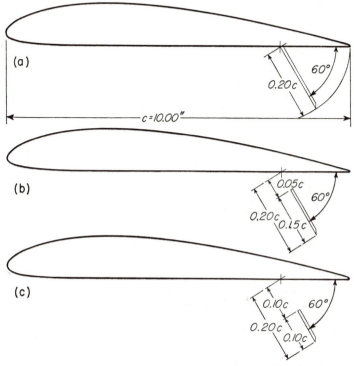

Fig. 109. (a) Details of split flaps with gaps tested on Clark Y wing.

design is obtained entirely by empirical methods. Although many experimental data have been accumulated, the large number of configurations possible and the sensitivity of the chacteristics to small changes in the slot configuration make the design problem a difficult one.

b. Single-slotted Flaps. One important parameter in the design of single-slotted flaps is the chordwise position of the lip. Although completely comparable data are not available for configurations with varied positions of the lip, the maximum lift coefficients appear to increase as the lip position approaches the trailing edge for wing sections of moderate thickness. This effect[69, 153] is shown in Fig. 112, where increments of maximum lift coefficient are presented for the NACA 23012 section with single-slotted flaps having lips located at $0.827c$, $0.900c$, and $1.000c$. The configuration with the lip located at $0.827c$ is not exactly comparable with the others because the chord of this flap is $0.2566c$ as compared with $0.30c$ for the others. It is apparent, nevertheless, that the increment of the maximum lift coefficient is considerably higher for the configuration having the lip at the normal trailing-edge position than for those with a more forward lip location. It is uncertain whether any of the difference between the

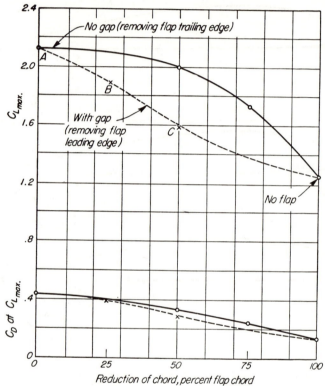

FIG. 109. (*Concluded*) (*b*) Effect on $C_{L_{max}}$ and on C_D at $C_{L_{max}}$ of reducing the chord of a $0.20c$ split flap. $\delta = 60$ degrees.

other two configurations should be attributed to the difference in lip position. Envelope polars for these flaps are presented[69] in Fig. 113.

When the lip is located at or near the normal trailing-edge position, the thickness of the flap is necessarily less than that for a more forward location of the lip[20] (Fig. 114a). In the case of thin wing sections, especially of the NACA 6-series type, the flap thickness may become too small with a rearward location of the lip to permit favorable slot configurations. Under such conditions, the favorable effect on the maximum lift coefficient of moving the lip toward the normal trailing-edge position may not be realized. Cahill[20] shows (Fig. 114b) that the maximum lift coefficients for

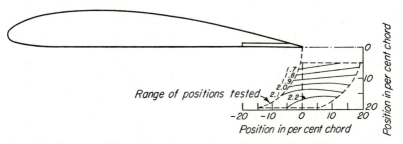

Fig. 110. Contours of $C_{L_{\max}}$ for various positions of trailing edge of 20 per cent flap.

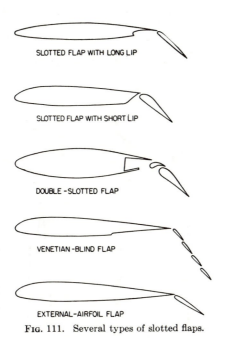

SLOTTED FLAP WITH LONG LIP

SLOTTED FLAP WITH SHORT LIP

DOUBLE-SLOTTED FLAP

VENETIAN-BLIND FLAP

EXTERNAL-AIRFOIL FLAP

Fig. 111. Several types of slotted flaps.

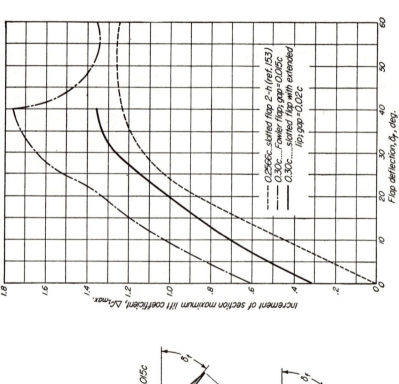

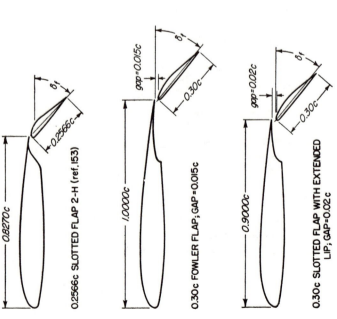

Fig. 112. Comparison of increments of section maximum lift coefficient for three flaps on a NACA 23012 airfoil.

the NACA 65-210 section are essentially the same for lip positions of $0.84c$, $0.90c$, and $0.975c$. The structural difficulties presented by a long thin lip extension and the mechanism necessary for the corresponding large rearward movement of the flap are such as to discourage the use of rearward locations of the lip unless such configurations result in substantial improvement of the maximum lift coefficient.

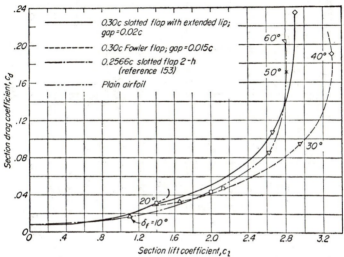

Fig. 113. Envelope polar curves for three slotted flaps on a NACA 23012 wing section.

The effect of flap chord on the increment of maximum lift[39, 153] is indicated by Fig. 115. The data presented for the $0.2566c$ and $0.40c$ flaps are reasonably comparable in that the shapes of the slots are generally similar. Figure 115 shows that larger increments of maximum lift coefficient are obtained with the larger chord flap, but the increased effectiveness is small compared with the increase of flap chord. The slightly higher maximum lift coefficients obtainable with large chord flaps do not appear to justify the structural difficulties encountered with such flaps, and flap chords in excess of $0.25c$ to $0.30c$ are seldom used.

The maximum lift coefficients obtained[39, 96, 153] with various arrangements of slotted flap on NACA 23012, 23021, and 23030 wing sections are plotted in Fig. 116. The flap chord was 25.66 per cent of the section chord in all cases. These data indicate little variation of the maximum lift coefficient with thickness ratio from 12 to 30 per cent for this type of wing section. A few data[3, 20] are also shown in Fig. 116 for comparable slotted flaps on NACA 6-series wing sections. In this case, the flap chords are 25 or 30 per cent of the section chord. These limited data indicate that, for the NACA 6-series sections, the maximum lift coefficients obtainable with

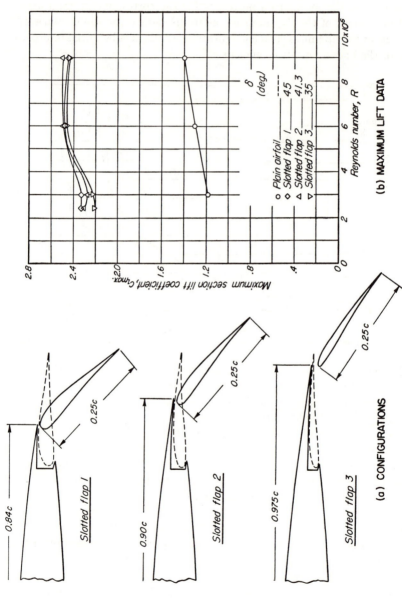

Fig. 114. Variation of maximum section lift coefficient with Reynolds number for several slotted flaps on the NACA 65-210 wing section.

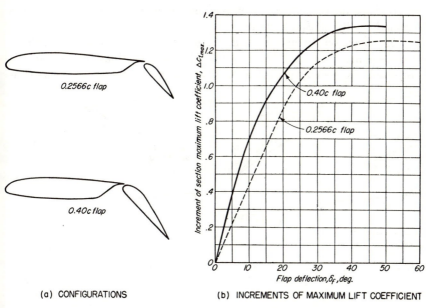

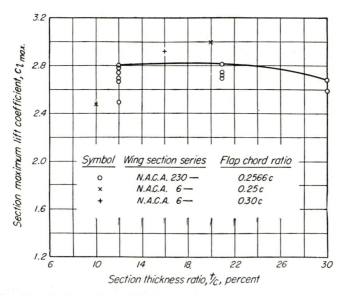

Fig. 115. Effect of flap chord on increments of section maximum lift coefficient for the NACA 23012 wing section.

Fig. 116. Maximum lift coefficients for various arrangements of slotted flaps.

slotted flaps on 10 per cent thick sections are appreciably less than those obtainable with thicker sections.

The effects on the maximum lift coefficient of some variations of shape of the slot are illustrated[153] in Fig. 117. In configurations 1-*a* and 1-*e*, the slot is only slowly converging, if at all, at the end of the lip, and these con-

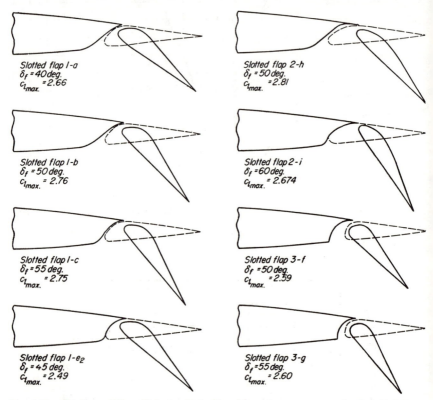

Slotted flap 1-*a*
δ_f = 40 deg.
$c_{l_{max.}}$ = 2.66

Slotted flap 1-*b*
δ_f = 50 deg.
$c_{l_{max.}}$ = 2.76

Slotted flap 1-*c*
δ_f = 55 deg.
$c_{l_{max.}}$ = 2.75

Slotted flap 1-*e₂*
δ_f = 45 deg.
$c_{l_{max.}}$ = 2.49

Slotted flap 2-*h*
δ_f = 50 deg.
$c_{l_{max.}}$ = 2.81

Slotted flap 2-*i*
δ_f = 60 deg.
$c_{l_{max.}}$ = 2.674

Slotted flap 3-*f*
δ_f = 50 deg.
$c_{l_{max.}}$ = 2.59

Slotted flap 3-*g*
δ_f = 55 deg.
$c_{l_{max.}}$ = 2.60

Fig. 117. Maximum lift coefficients attainable with various arrangements of slotted flaps on the NACA 23012 wing section.

figurations have the lowest maximum lift coefficients of the 1-series configurations. A short extension of the lip as in configuration 1-*b*, which makes the slot definitely convergent and directs the air downward toward the flap surface, is effective in increasing the maximum lift coefficient. It may be concluded from these and other data that the slot should be definitely convergent in the vicinity of the lip and shaped to direct the air downward toward the flap. The effects of changing the radius of curvature at the entry to the slot from the lower surface are shown by configurations 1-*b* and 1-*c* of Fig. 117 for a flap having a comparatively small rearward displacement when deflected. Decreasing the radius of curvature from about 0.08*c* to 0.04*c* did not produce a significant difference in the maximum lift coefficient. Other data indicate that this radius of curvature is of little importance when the flap is displaced rearward enough to produce a large area for the entry of air into the slot.

It is difficult to draw general conclusions about the proper shape of a slotted flap. Figure 117 shows that the highest maximum lift coefficients were obtained with flap 2, which is shaped more like a good wing section than flaps 1 or 3. The difference between the maximum lift coefficients produced by flaps 1 and 2 is small and may be caused by the difference in slot shape and lip extension rather than by the difference in flap shape.

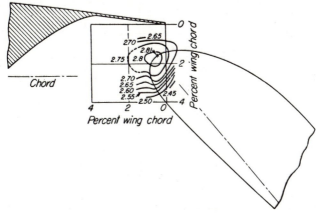

Fig. 118. Contours of flap location for $c_{l_{max}}$. Slotted flap 2-h, $\delta_f = 50$ degrees on NACA 23012 wing section.

Flap 3, however, appears to be too blunt with a too small radius of curvature on the upper surface aft of the lip in the deflected position.

Typical contours[153] of the maximum lift obtainable with various flap positions at one flap deflection are shown in Fig. 118. In general, the optimum flap position for good flaps at large deflections appears to be that which produces a slot opening of the order of 0.01c or slightly more and which locates the foremost point of the flap about 0.01c forward of the lip. The maximum lift coefficient, however, is frequently sensitive to the flap position, and the optimum position is best determined by test.

A complete set of section characteristics[153] for a typical single-slotted flap configuration is shown in Fig. 119. This figure illustrates the characteristic ability of slotted flaps to produce high lift coefficients with comparatively small profile drag coefficients.

The increment of moment coefficient associated with the use of single-slotted flaps[23] is illustrated by Fig. 120. This figure shows the ratio of the increment of the section pitching-moment coefficient to the increment of the section lift coefficient at an angle of attack of 0 degree for three wing sections and several flaps. The moment coefficients used in this case are

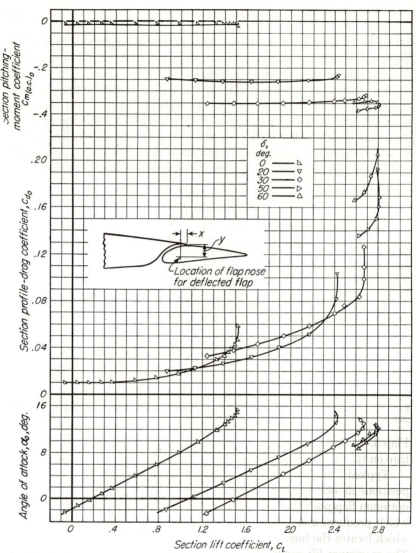

Path of flap nose for various flap deflections. Distances measured from lower edge of lip in per cent airfoil chord c.

δ_f, degrees	x	y	δ_f, degrees	x	y
0	8.36	3.91	40	1.35	2.43
10	5.41	3.63	50	0.50	1.63
20	3.83	3.45	60	0.12	1.48
30	2.63	3.37			

Fig. 119. Section aerodynamic characteristics of NACA 23012 airfoil with slotted flap 2-h.

based on the total chord with the flap deflected. The ratio plotted appears to be fairly constant at values between about -0.17 and -0.21, instead of varying with flap-chord ratio as in the case of plain flaps at small deflections.

The normal-force coefficient, pitching-moment coefficient, and center of pressure for the flap of the configuration of Fig. 119 (configuration 2-h of reference 153) are shown[152] in Fig. 121. All the coefficients are based on the flap chord. The pitching moments are taken about the quarter-chord point of the flap, and the center of pressure is given in per cent of the flap

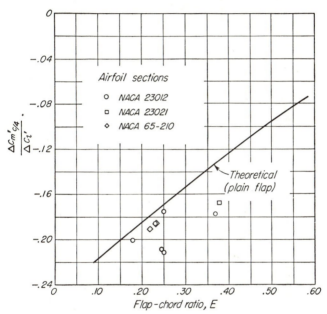

Fig. 120. Variation of ratio of increment of section pitching-moment coefficient to increment of section lift coefficient with flap-chord ratio for several sections with slotted flaps. α_0 = 0 degrees.

chord from the leading edge of the flap. These data are useful in determining the loads on the flap and the flap linkages, and they show that the normal force coefficients on the flap are less than those for the wing section.

c. External-airfoil Flaps. The external-airfoil flap investigated by Wragg,[156a] Platt,[83 84 85] and Wenzinger[151] may be considered as a special case of the single-slotted flap in which the flap does not retract within the wing section. The maximum lift coefficients[84] obtained at an effective Reynolds number of 8 million for a NACA 23012 wing section with a 0.20c external-airfoil flap of the same section are shown plotted against flap deflection in Fig. 122. These maximum lift coefficients are based on the combined areas of the wing and flap. The values presented were obtained from tests of a

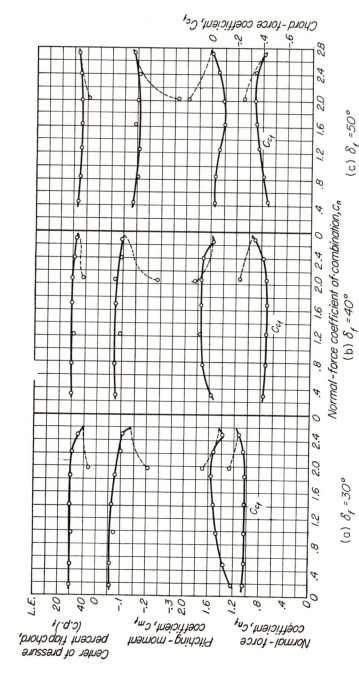

FIG. 121. Section characteristics of the flap alone of the NACA 23012 airfoil with a 0.2566c slotted flap.

finite span model and were only partly corrected to section data. The corresponding section maximum lift coefficients are judged to be about 4 per cent higher than those presented. If these maximum lift coefficients were based on the chord of the wing section, as is customary for other types of flap, the resulting values would be about the same as those for single-slotted flaps. Although the external-airfoil flap appeared[83, 85] to offer some

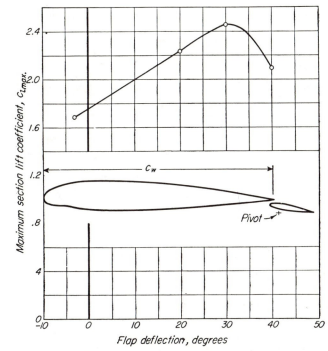

Fig. 122. Variation of maximum section lift coefficient, based on chord of wing section plus flap, with flap deflection for the NACA 23012 wing section with 0.20 c_w external-airfoil flap.

advantages as a full-span flap, it has not been used extensively because of its failure to show definite advantages over retractable flaps and because its use would probably aggravate the icing hazard.

d. Double-slotted flaps. Double-slotted flaps[3] (Fig. 123) produce substantial increments of maximum lift coefficient over that obtainable with single-slotted flaps. The fore flap or vane of the double-slotted flap assists in turning the air downward over the main flap, thus delaying the stall of the flap to higher deflections. It has frequently been found possible to develop flap-vane combinations that can be retracted into the wing section without relative motion between the vane and the flap. Such an arrangement is desirable because the linkage system is much less complicated than when relative motion is required between the flap and vane.

Investigations by Harris,[40] Purser,[91] and Fischel[31] of approximately 0.30c
and 0.40c double-slotted flaps on the NACA 23021 and 23012 wing sections
indicated maximum lift coefficients of the order of 3.3 and 3.5 for the two
sizes of flap. Typical results for the NACA 23012 section are shown in
Figs. 124 and 125. The fore flap and the main flap did not deflect together
as a unit for these configurations. Aerodynamic characteristics[3] are presented in Fig. 126 for the NACA 65$_3$-118 section with the 0.309c double-
slotted flap shown in Fig. 123. For this configuration, the vane and flap

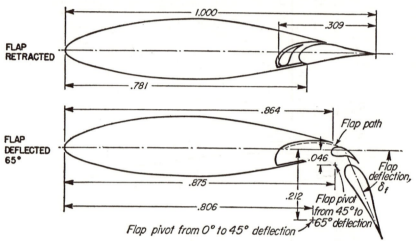

FIG. 123.   NACA 65$_3$-118 airfoil section with 0.309c double-slotted flap.

moved together as a unit up to deflections of 45 degrees. At higher angles,
the flap rotated about a pivot and the vane remained fixed.

The variation of maximum lift obtained for thin NACA 64-series
sections[21] with approximately 0.30c double-slotted flaps is shown in Fig.
127. The type of flap used for these tests is illustrated in Fig. 128. The
flap and vane deflected as a unit. These data (Fig. 127) show that the
maximum lift coefficient decreases rapidly as the thickness ratio of the wing
section is decreased to values below 10 or 12 per cent. The maximum lift
coefficient obtained for the NACA 1410 wing section[21] is also plotted to
indicate the effect of type of section. The maximum lift coefficient obtained for the NACA 65$_3$-118 wing section with a flap deflection of 45 degrees (Fig. 126) is also plotted to indicate the effect of larger thickness
ratios. The NACA 65$_3$-118 data are not exactly comparable, but the
indicated gradual rise of the maximum lift coefficient as the thickness ratio
is increased to 18 per cent is believed to be representative.

The effect of the design position of minimum pressure on the maximum
lift coefficients obtained with double-slotted flaps on 10 per cent thick
NACA 6-series wing sections[21] is shown in Fig. 129. The type of flap used

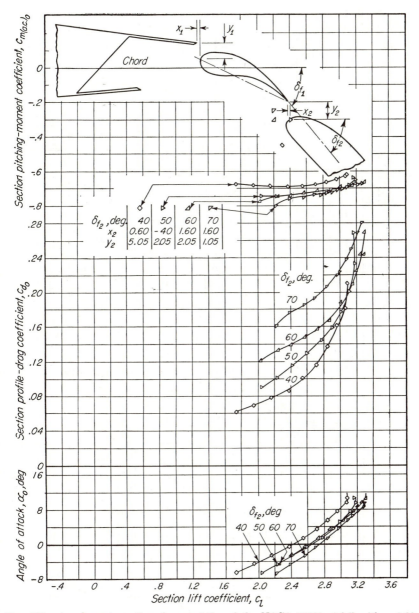

Fɪɢ. 124. Aerodynamic section characteristics of the NACA 23012 airfoil with a 0.30c double-slotted flap. $\delta_{f_1} = 25$ degrees; $x_1 = 0.41$; $y_1 = 1.72$. (Values of x_1, y_1, x_2, and y_2 are given in per cent of airfoil chord.)

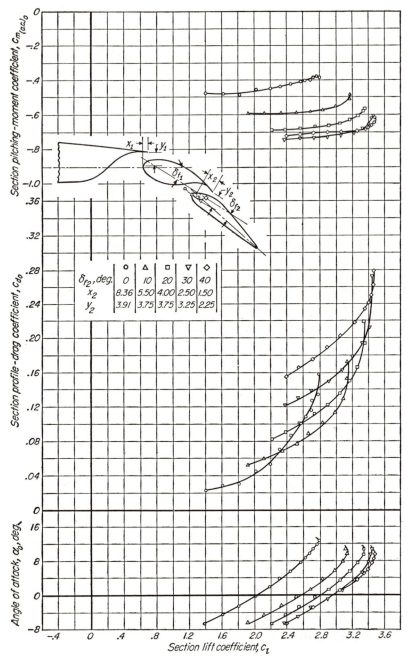

FIG. 125.   Aerodynamic characteristics of the NACA 23012 airfoil section with 40-per cent-chord double-slotted flap. $\delta_{f_1} = 30$ degrees; $x_1 = 1.50$; $y_1 = 3.50$.   (Values of x_1, y_1, x_2, and y_2 are given in per cent of airfoil chord.)

was similar to that shown in Fig. 128. These data show that the highest maximum lift coefficient was obtained for the NACA 64-series sections and that the maximum lift coefficient decreases rapidly as the minimum-pressure

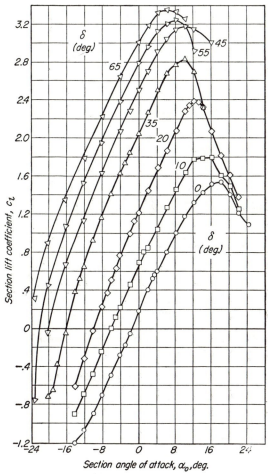

Fig. 126. NACA 65_3-118 airfoil section with 0.309c double-slotted flap.

position moves farther aft. It is thought that the rather large variations shown in Fig. 129 are associated with the fact that the thickness ratio of the sections was in a critical range, as indicated by Fig. 127.

Load data for typical double-slotted flaps are given by Cahill.[22] These data show that a disproportionately large part of the load is carried by the vane. Normal-force coefficients for the vane based on the vane chord and the dynamic pressure of the free stream reached a value in excess of 5.0.

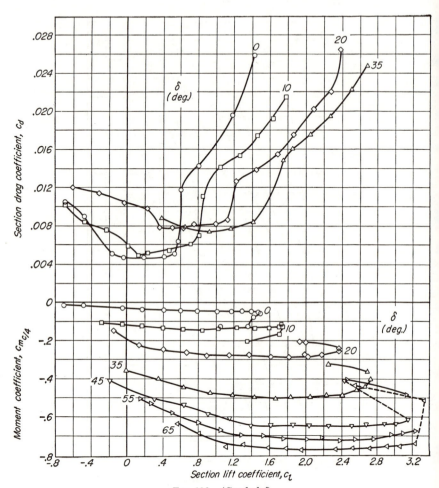

FIG. 126. (Concluded)

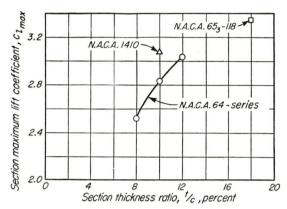

F<small>IG</small>. 127. Effect of thickness ratio and type of wing section on maximum lift with double-slotted flaps.

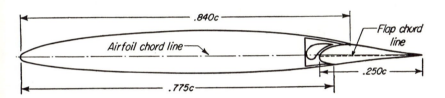

(a) AIRFOIL WITH FLAP

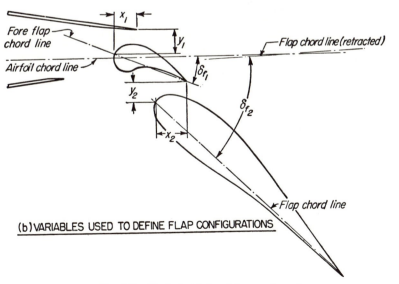

(b) VARIABLES USED TO DEFINE FLAP CONFIGURATIONS

F<small>IG</small>. 128. Typical airfoil and flap configuration.

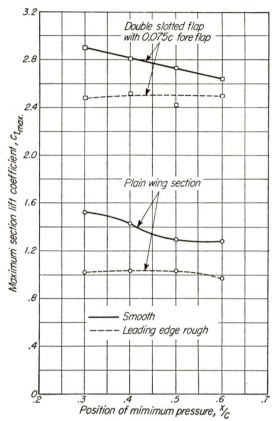

Fig. 129. Variation of maximum section lift coefficient with position of minimum pressure for some NACA 6-series wing sections of 10 per cent thickness and a design lift coefficient of 0.2. R, 6×10^6.

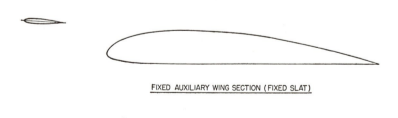

FIXED AUXILIARY WING SECTION (FIXED SLAT)

LEADING EDGE RETRACTABLE SLAT

Fig. 130.   Examples of fixed and retractable slats.

8.6. Leading-edge High-lift Devices. *a. Slats.* Leading-edge slats are airfoils mounted ahead of the leading edge of the wing in such an attitude as to assist in turning the air around the leading edge at high angles of attack and thus delay leading-edge stalling. They may be either fixed in

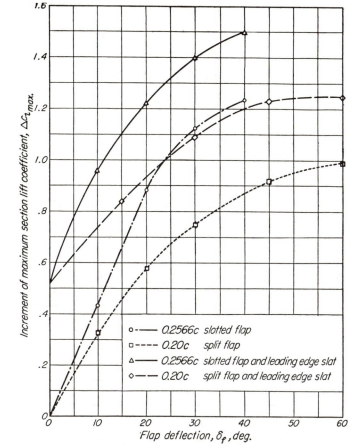

FIG. 131. Effect of flaps and leading-edge slat on increment of maximum section lift coefficient.

position or retractable (Fig. 130). The fixed leading-edge slat consisting of an auxiliary airfoil mounted ahead of the wing leading edge has been investigated in detail by Weick and Sanders.[146] This investigation showed that leading-edge slats of this type with chords varying from 7.5 to 25 per cent of the wing chord and with various sections all produced substantially the same maximum lift coefficient when located in the optimum position for the ratio

$$\frac{C_{L_{\max}}^{2}}{C_{D_{\min}}}$$

The value of this maximum lift coefficient was about 1.64 for the rather low Reynolds numbers of the tests. It is doubtful whether such configurations would experience much beneficial scale effect.

The effectiveness of the retractable leading-edge slat[107] shown in Fig. 130 in increasing the maximum lift coefficient and the angle of attack for

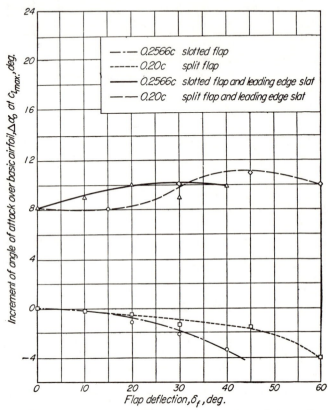

FIG. 132. Effect of flaps and leading-edge slat on angle of attack for maximum lift.

maximum lift is shown in Figs. 131 and 132. These data, which were obtained on a NACA 23012 wing section, indicate an increment of about 0.5 for the maximum lift coefficient and about 8 degrees for the angle of attack for maximum lift. The increment of maximum lift decreased to about one-half of its value for the unflapped section when either a split or slotted flap was deflected to optimum deflections despite readjustment of the slat to optimum positions with flap deflected. Weick and Platt[145] obtained considerably larger increments of maximum lift coefficient with a special retractable slat (Fig. 133) having a shape providing a rounded entrance into the slot. The increments of maximum lift coefficient with this configuration were 0.81 for the unflapped section and 0.45 with a deflected slotted flap.

b. Slots. Slots to permit the passage of high-energy air from the lower surface to control the boundary layer on the upper surface are common features of many high-lift devices. The most common application is the slotted flap. When the slot is located near the leading edge, the configuration differs only in detail from the leading-edge slat. Additional slots may be introduced at various chordwise stations. Weick and Shortal[147] made a systematic study of slots on a Clark Y airfoil. The results of this investigation are summarized in Fig. 134. For the unflapped section, the most effective position for a single slot is near the leading edge, and the effectiveness decreases as the slot is moved aft. Multiple slots are relatively ineffective on the plain airfoil unless they include a slot near the

FIG. 133. Special retractable slat on Clark Y wing section.

leading edge, in which case a total of three slots, all located well forward, is optimum. For the flapped section, the slot located near the leading edge was effective. A single or double slot at the flap changed the type of flap from plain to slotted with a corresponding increase of the maximum lift coefficient. Load data for the leading-edge slot are given by Harris and Lowry.[41] These data show resultant-force coefficients as large as about 7.5 for that portion of the wing section ahead of a slot near the leading edge.

If slots are considered as a fixed high-lift device, the profile drag in the high-speed flight attitude is an important characteristic. Figure 134 shows that any of the slots investigated cause large increments in the minimum profile drag. This increment increases with the number of slots and decreases with rearward movement of the slots. Attempts have been made to maintain low drags with slots open by locating the slots so that there would be no flow through them in the high-speed condition.[108] Such configurations have failed to improve the ratio of maximum lift to minimum drag over that for the plain wing section.

c. Leading-edge Flaps. A leading-edge flap may be formed by bending down the forward portion of the wing section in a manner similar to that in which the trailing edge is deflected in the case of plain flaps. Other types of leading-edge flap are formed by extending a surface downward and forward from the vicinity of the leading edge. As shown in Fig. 135, such flaps may extend smoothly from the upper surface near the leading edge, may be hinged at the center of the leading-edge radius, or may be hinged on the lower surface somewhat aft of the leading edge. Although

Slot combination	$C_{L_{max}}$	$C_{D_{min}}$	$\dfrac{C_{L_{max}}}{C_{D_{min}}}$	$\alpha_{C_{L_{max}}}$ degrees
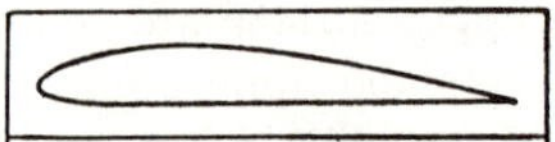	1.291	0.0152	85.0	15
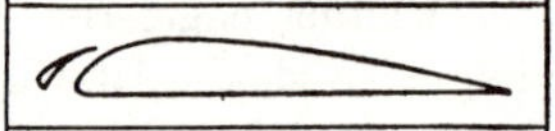	1.772	0.0240	73.8	24
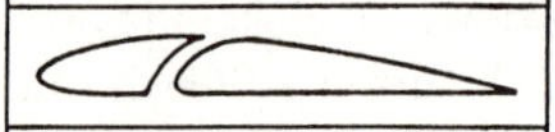	1.596	0.0199	80.3	21
	1.548	0.0188	82.3	19
	1.440	0.0164	87.8	17
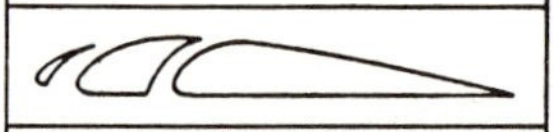	1.902	0.0278	68.3	24
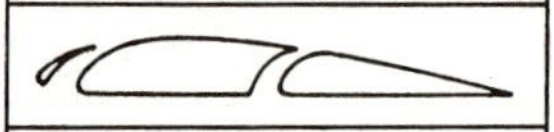	1.881	0.0270	69.7	24
	1.813	0.0243	74.6	23
	1.930	0.0340	56.8	25
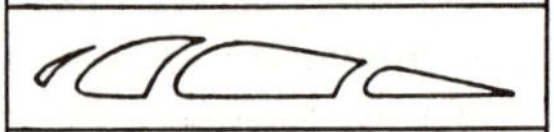	1.885	0.0319	59.2	24
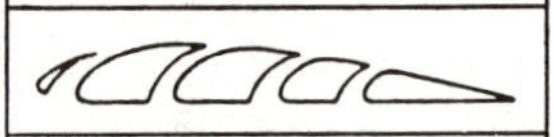	1.885	0.0363	51.9	25
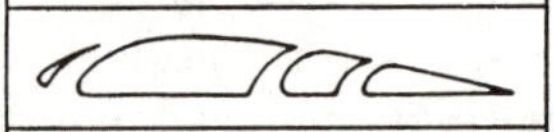	1.850	0.0298	62.1	24
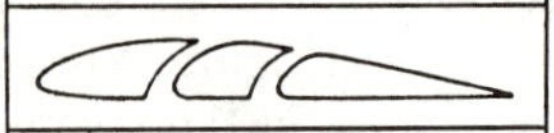	1.692	0.0228	74.2	22
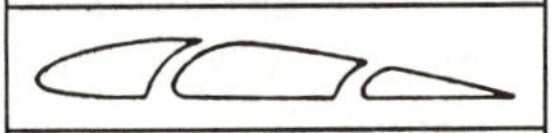	1.672	0.0214	78.2	22
	1.510	0.0208	72.6	19
	1.662	0.0258	64.4	22

(a) Multiple fixed slots.

FIG. 134. Aerodynamic characteristics of a Clark Y wing with slots and flaps.

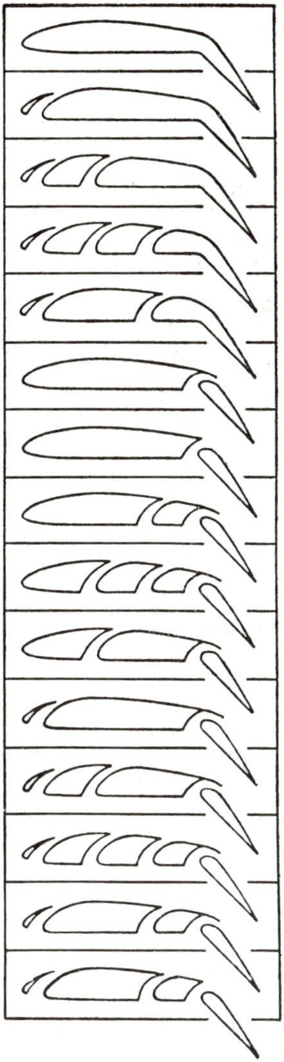

Slot combination	$C_{L_{max}}$	$C_{D_{min}}$ *	$\dfrac{C_{L_{max}}}{C_{D_{min}}}$	$\alpha_{C_{L_{max}}}$ degrees
	1.950	0.0152	128.2	12
	2.182	0.0240	91.0	19
	2.235	0.0278	80.3	20
	2.200	0.0340	64.7	21
	2.210	0.0270	81.8	20
	1.980	0.0164	120.5	12
	1.770	0.0164	108.0	14
	2.442	0.0208	117.5	16
	2.500	0.0258	96.8	18
	2.185	0.0214	102.0	18
	2.261	0.0243	93.2	19
	2.320	0.0319	72.7	20
	2.535	0.0363	69.8	20
	2.600	0.0298	87.3	20
	2.035	0.0298	68.3	21

* $C_{D_{min}}$ with flap neutral.

(b) Multiple fixed slots and a slotted flap deflected 45 degrees.

FIG. 134. (*Concluded*)

none of these devices is as powerful as trailing-edge flaps, they may be used full span without mechanical interference with lateral-control devices and they are effective when combined with trailing-edge high-lift devices. Leading-edge flaps reduce the severity of the pressure peak ordinarily associated with high angles of attack and thereby delay separa-

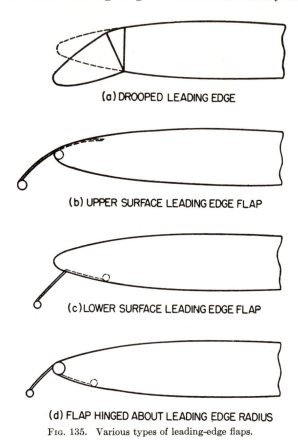

(a) DROOPED LEADING EDGE

(b) UPPER SURFACE LEADING EDGE FLAP

(c) LOWER SURFACE LEADING EDGE FLAP

(d) FLAP HINGED ABOUT LEADING EDGE RADIUS

FIG. 135. Various types of leading-edge flaps.

tion. Leading-edge flaps received little consideration until German investigators became interested in them during the Second World War.

Krueger,[59, 60] Lemme,[63, 64, 65] and Koster[58] showed increments of the maximum lift coefficient of as much as 0.7. These increments were, however, applied to maximum lift coefficients for the plain wings of the order of 0.72. These low maximum lift coefficients resulted from the small leading-edge radii of the wing sections usually used for these investigations and the very low Reynolds numbers of the tests. These investigations indicated that the effectiveness of leading-edge flaps increased with decreasing leading-edge radius. Typical results[59] are shown in Fig. 136. The increment of maximum lift coefficient, $\Delta C_{L_{max}}$, is plotted against the leading-edge-radius parameter $(r/c)/(t/c)^2$ where r is the leading-edge radius. The value of this parameter for NACA four- and five-digit wing sections is 1.1.

Fullmer[32] investigated two types of leading-edge flap on the NACA 64_1-012 section at a Reynolds number of 6 million. The chord of the flap was 10 per cent of the section chord, and the configurations corresponded to

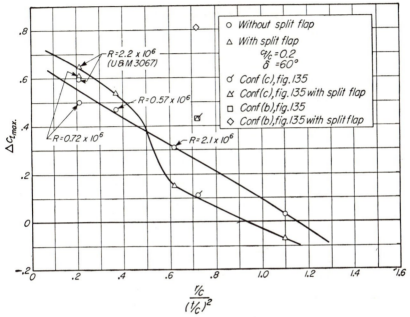

Fig. 136. Effectiveness of leading-edge flaps for various wing sections. Increment of lift resulting from leading-edge flap, $c_{l_{max}}$ as a function of the leading-edge radius coefficient, $(r/c)/(t/c)^2$.

b and c of Fig. 135. The increments of maximum lift coefficient and of the angle of attack for maximum lift are shown in Fig. 137. The leading-edge radius parameter for the NACA 64_1-012 section is 0.72. The maximum increments of Fig. 137 are shown plotted on Fig. 136 for comparison with the German results. The maximum lift coefficient of the plain wing section for these tests was 1.42.

8.7. Boundary-layer Control. The idea of removing the low-energy air of the boundary layer, or of adding kinetic energy to the boundary layer, as a means for increasing the maximum lift has been obvious since the basic mechanism for separation was first understood.[87] The kinetic energy of the layers of air close to the surface may be increased by removing low-energy air through suction slots or a porous surface. Another common method is to blow high-energy air through backward-directed slots. The air handled through either the suction or blowing slots may be carried through the interior of the wing and the necessary energy supplied by a blower. Alternately, the pressure difference required may be obtained from the variation of pressure about the wing section itself. This method

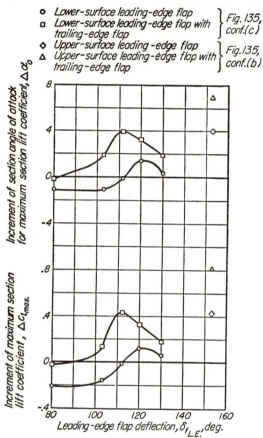

FIG. 137. Variation of the increment of maximum section lift coefficient and the increment of section angle of attack for maximum section lift coefficient with leading-edge flap deflection. NACA 64_1-012 airfoil section with leading- and trailing-edge split flaps. R, 6.0×10^6.

has been used successfully only in the case of blowing slots on the upper surface using air taken from the lower surface. Examples of such arrangements are the previously discussed slotted flaps, slots, and leading-edge slats.

It is obvious that, if boundary-layer control is applied at sufficiently close intervals along the upper surface of a wing section, separation can be avoided up to very high values of the lift coefficient. The increments of maximum lift coefficient are obtained as an extension of the lift curve to higher angles of attack as contrasted with the displacement of the lift curve resulting from deflection of trailing-edge flaps. The problem confronting the designer is to apply sufficient boundary-layer control to obtain the desired values of the maximum lift coefficient without increasing the weight and complexity of the airplane to such an extent as to nullify the gains in performance expected from the higher lift coefficients. These considerations of weight and complexity have prevented extensive use of

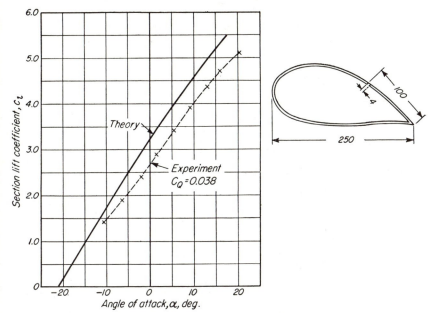

FIG. 138. Variation of lift coefficient with angle of attack for a thick wing section with boundary-layer control.

any except the simple types of boundary-layer control such as slotted flaps and leading-edge slots and slats.

Successful application of boundary-layer control must delay both turbulent separation over the aft portion of the wing and permanent laminar separation over the forward portion (see Sec. 5.3). The prevention of permanent laminar separation for thin or even moderately thick sections by means of suction slots is a difficult problem involving critically located slots in a region of greatly reduced pressure that requires high-pressure blowers. Nearly all the early investigations avoided this problem by using very thick sections with large leading-edge radii.[102, 103, 104] Using such sections, Schrenk[103] obtained lift coefficients of over 5.0 using a single suction slot (Fig. 138) with a volume-flow coefficient C_Q of 0.038.

The high drag coefficients of such thick wing sections in the high-speed flight condition led to investigations of boundary-layer control on thinner sections.[13, 57, 97] Knight and Bamber[57] obtained maximum lift coefficients of about 3.0 (Fig. 139) for the NACA 84-M wing section with a single backward-opening blowing slot.

NACA 84-M wing section. Slot width=0.667%c
Slot at 53.9 % c from leading edge. Internal
pressure 12 q higher than free-stream static
pressure

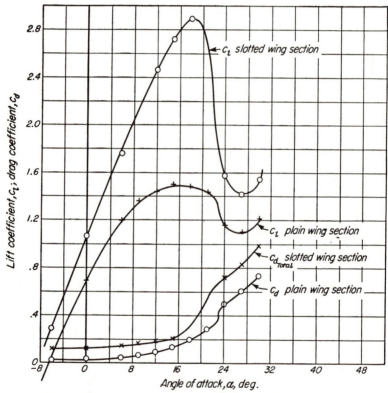

FIG. 139. Effect of boundary-layer control by means of a backward-opening blowing slot.

Much of the more recent work[94, 95] on boundary-layer control has combined suction slots with flaps and leading-edge slots. Quinn[94] obtained maximum lift coefficients exceeding 4.0 for the NACA 65_3-418 section with a double-slotted flap and a single suction slot (Fig. 140). The maximum lift coefficient for the unflapped section increased rapidly with small

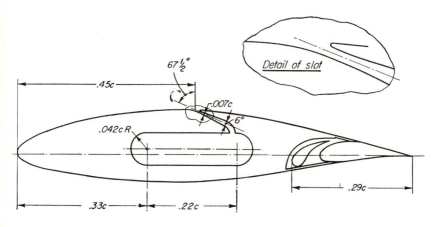

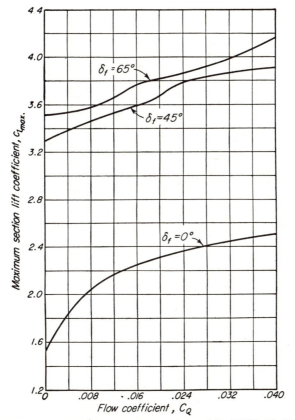

Fig. 140. Configuration and maximum lift characteristics of the NACA 65_3-418 wing section with double-slotted flap and boundary-layer control by suction. R, 6.0×10^6.

amounts of flow through the slot, but the variation of maximum lift coefficient for the flapped section was nearly linear with the flow coefficient up to a value of 0.040. Quinn[93] has concluded that the maximum effectiveness of suction slots is nearly reached when the quantity of air removed is equal to that which would pass through the displacement thickness of the boundary layer at the slot location with the local velocity outside the boundary layer. Quinn found that the maximum lift was limited by leading-edge stalling for the 18 per cent thick sections. He used a leading-

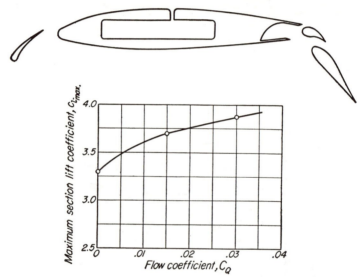

Fig. 141. Configuration and maximum lift characteristics for the NACA 64_1A212 wing section with leading-edge slat, double-slotted flap and boundary-layer control.

edge slat[95] to control the leading-edge stalling of the NACA 64_1A212 section with a suction slot and a double-slotted flap. This combination of high-lift devices gave a maximum lift coefficient of nearly 4.0 for this thin section[95] (Fig. 141).

8.8. The Chordwise Load Distribution over Flapped Wing Sections. Allen[6] obtained empirical load distributions that permit calculation of reasonable approximations to the load distributions of wing sections with deflected plain or split flaps. These load distributions are analogous to those associated with changes of camber in that the ratio $P_{b\delta}/c_{n_{b\delta}}$ is independent of the angle of attack where $P_{b\delta}$ is the incremental basic load distribution associated with flap deflection and $c_{n_{b\delta}}$ is the basic normal-force coefficient associated with flap deflection. The total load distribution is the sum of the incremental basic distribution $P_{b\delta}$, the incremental additional load distribution $P_{a\delta}$, and the load distribution of the unflapped section at the same angle of attack. The incremental additional load distribution $P_{a\delta}$ is taken to be the same as the additional load distribution associated with angle of attack. That is, for a normal additional force coefficient of unity,

$$P_{a\delta} \doteq 4 \frac{\Delta v_a}{V} \frac{v}{V}$$

where the values of $\Delta v_a/V$ and v/V are tabulated in Appendix I for various wing sections.

Allen also found empirically that the values of $P_{b\delta}/c_{n_{b\delta}}$ at any point $(x/c)/(1 - E)$ (points ahead of the flap hinge) and $[1 - (x/c)]/E$ (points aft of the hinge) were independent of the flap-chord ratio E for any given angular deflection of the flap. Values of $P_{b\delta}/c_{n_{b\delta}}$ are given in Table 6 for both plain and split flaps. The incremental load distribution associated with flap deflection is thus defined in terms of two known types of distributions. The remaining problem is to determine the magnitudes of these two types of distributions associated with a given flap deflection.

The magnitudes of the load distributions are determined from force test data obtained at the desired angle of attack with the flap neutral and deflected. The additional load distribution contributes no moment about the quarter-chord point. The incremental basic distribution is therefore selected to produce the measured increment of moment coefficient between the flap-neutral and flap-deflected conditions. The remainder of the difference between the normal-force coefficients with the flap neutral and deflected is equal to the incremental additional load distribution.

The normal-force distribution is assumed to act as shown in Fig. 142. From force tests of the wing section with flap neutral, c_{m_1} (quarter-chord pitching-moment coefficient) and c_{n_1} (normal-force coefficient) corresponding to the normal-force distribution shown in Fig. 142a are obtained. From force tests of the wing section at the same angle of attack with the flap deflected, c_{m_2} and c_{n_2} corresponding to the normal-force distribution shown in Fig. 142b are obtained. Let

$$\left.\begin{aligned} \Delta c_m &= c_{m_2} - c_{m_1'} \\ \Delta c_n &= c_{n_2} - c_{n_1'} \end{aligned}\right\} \tag{8.1}$$

where $c_{m_1'}$ and $c_{n_1'}$ are the pitching-moment and normal-force coefficients corresponding to the normal-force distribution for the section with flap neutral when plotted normal to the chord of the section with flap deflected as shown in Fig. 142c. Then Δc_m and Δc_n are the pitching-moment and normal-force coefficients of the incremental normal-force distribution when the incremental distribution is plotted normal to the chord of the section with the flap deflected as shown in Fig. 142d.

In many cases, the approximation

$$\left.\begin{aligned} c_{n_1} &= c_{n_1'} \\ c_{m_1} &= c_{m_1'} \end{aligned}\right\} \tag{8.2}$$

is sufficiently exact except when the flap-chord ratio E and the flap deflection δ are simultaneously large or when the shape of the mean line is such that the load is large near the trailing edge. In such cases, values of $c_{m_1'}$ and $c_{n_1'}$ may be obtained graphically from the theoretical load distribution for the unflapped section adjusted to agree with the experimental data.

Let $\Delta c_{m'}$ and $\Delta c_{n'}$ be the pitching-moment and the normal-force coefficients of the incremental normal-force distribution plotted normal to the flap-neutral chord as shown in Fig. 142e. Because the incremental basic

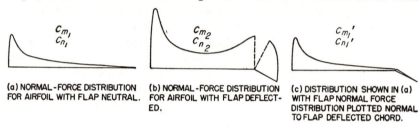

(a) NORMAL-FORCE DISTRIBUTION FOR AIRFOIL WITH FLAP NEUTRAL.

(b) NORMAL-FORCE DISTRIBUTION FOR AIRFOIL WITH FLAP DEFLECTED.

(c) DISTRIBUTION SHOWN IN (a) WITH FLAP NORMAL FORCE DISTRIBUTION PLOTTED NORMAL TO FLAP DEFLECTED CHORD.

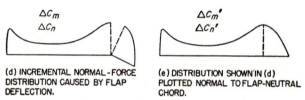

(d) INCREMENTAL NORMAL-FORCE DISTRIBUTION CAUSED BY FLAP DEFLECTION.

(e) DISTRIBUTION SHOWN IN (d) PLOTTED NORMAL TO FLAP-NEUTRAL CHORD.

FIG. 142. Normal-force distribution and incremental normal-force distribution for flaps neutral and deflected.

normal-force distribution is responsible for the increment of the quarter-chord pitching moment, then, if G is the moment arm in terms of the chord of the basic normal force about the quarter-chord point,

$$\Delta c_{m'} = G c_{n_{b\delta}}$$
$$\Delta c_{n'} = c_{n_{a\delta}} + c_{n_{b\delta}}$$

where $c_{n_{a\delta}}$ = additional normal-force coefficient associated with flap deflection

$$\left. \begin{aligned} c_{n_{b\delta}} &= \frac{\Delta c_{m'}}{G} \\ c_{n_{a\delta}} &\doteq \Delta c_{n'} - \frac{\Delta c_{m'}}{G} \end{aligned} \right\} \tag{8.3}$$

The value of G is a function of E and δ. Values of G are given in Table 7.

The correlation between the fictitious values of $\Delta c_{m'}$ and $\Delta c_{n'}$ and the measured values of Δc_m and Δc_n must be established in order to determine $c_{n_{b\delta}}$ and $c_{n_{a\delta}}$ from force tests. The incremental flap normal-force coefficient is

$$c_{n_{f\delta}} = \frac{n_{f\delta}}{qEc} \tag{8.4}$$

where $n_{f\delta}$ is the incremental flap normal force per unit span.

Then

$$\left.\begin{aligned} \Delta c_{n'} &= \Delta c_n + E c_{n_{f\delta}}(1 - \cos \delta) \\[2mm] \Delta c_{m'} &= \Delta c_m - E c_{n_{f\delta}}(1 - \cos \delta)(\tfrac{3}{4} - E) \end{aligned}\right\} \tag{8.5}$$

and

The incremental flap normal force may be considered as a combination of two components due to the incremental additional and the incremental basic normal-force distributions. Let γ_{a_δ} and γ_{b_δ} be the ratios of the flap normal force to the section normal force for the incremental additional and the incremental basic normal forces, respectively. Then

$$c_{n_{f\delta}} = \gamma_{a_\delta} c_{n_{a_\delta}} + \gamma_{b_\delta} c_{n_{b_\delta}} \tag{8.6}$$

or

$$c_{n_{f\delta}} = \gamma_{a_\delta}\left(\Delta c_{n'} - \frac{\Delta c_{m'}}{G}\right) + \gamma_{b_\delta}\frac{\Delta c_{m'}}{G} \tag{8.7}$$

The contribution to the flap load of the additional normal-force distribution is small compared with the basic contribution; and, for the purpose of determining $\Delta c_{m'}$ and $\Delta c_{n'}$, the following approximation may therefore be used:

$$c_{n_{f\delta}} = \gamma_{b_\delta}\frac{\Delta c_{m'}}{G}$$

and Eqs. (8.5) become

$$\Delta c_{n'} = \Delta c_n + E\gamma_{b_\delta}\frac{\Delta c_{m'}}{G}(1 - \cos \delta)$$

$$\Delta c_{m'} = \Delta c_m - E\gamma_{b_\delta}\frac{\Delta c_{m'}}{G}(1 - \cos \delta)\left(\frac{3}{4} - E\right)$$

so that

$$\left.\begin{aligned} \Delta c_{n'} &= \Delta c_n + \tau_n \Delta c_m \\[2mm] \Delta c_{m'} &= \tau_m \Delta c_m \end{aligned}\right\} \tag{8.8}$$

where

$$\tau_n = \frac{E(1 - \cos \delta)(\gamma_{b_\delta}/G)}{1 + E(1 - \cos \delta)(\tfrac{3}{4} - E)(\gamma_{b_\delta}/G)}$$

$$\tau_m = \frac{1}{1 + E(1 - \cos \delta)(\tfrac{3}{4} - E)(\gamma_{b_\delta}/G)}$$

The values of τ_n and τ_m as determined by Allen are given in Tables 8 and 9.

In using this method, the values of Δc_m and Δc_n are obtained from force tests of the flapped and unflapped section at the same angle of attack. From Eq. (8.8), values of $\Delta c_{n'}$ and $\Delta c_{m'}$ are obtained using Tables 8 and 9.

TABLE 6.—$P_{b_\delta}/c_{n_{b_\delta}}$ DISTRIBUTION

a. Plain flap at δ = 5, 10, and 15 degrees

E	0.05	0.10	0.15	0.20	0.25	0.30	0.35	0.40	0.45	0.50	0.55	0.60	0.65	0.70
0	0	0	0	0	0	0	0	0	0	0	0	0	0	0
0.05	0.15	0.15	0.16	0.16	0.17	0.17	0.18	0.19	0.20	0.21	0.22	0.23	0.25	0.27
0.10	0.22	0.23	0.23	0.24	0.25	0.25	0.26	0.27	0.29	0.30	0.32	0.34	0.36	0.39
0.20	0.33	0.33	0.34	0.35	0.36	0.38	0.39	0.40	0.42	0.44	0.46	0.49	0.52	0.56
0.30	0.43	0.44	0.45	0.46	0.47	0.49	0.50	0.52	0.54	0.57	0.60	0.63	0.67	0.72
0.40	0.54	0.55	0.56	0.57	0.58	0.60	0.62	0.64	0.67	0.70	0.73	0.77	0.82	0.88
0.50	0.66	0.66	0.67	0.69	0.71	0.73	0.75	0.77	0.80	0.84	0.88	0.92	0.98	1.05
0.60	0.79	0.80	0.82	0.83	0.85	0.88	0.90	0.93	0.96	1.00	1.05	1.10	1.17	1.25
0.70	0.97	0.98	1.00	1.01	1.03	1.05	1.08	1.11	1.15	1.19	1.24	1.30	1.39	1.48
0.80	1.23	1.24	1.26	1.27	1.29	1.31	1.34	1.38	1.42	1.46	1.52	1.59	1.67	1.78
0.90	1.73	1.74	1.75	1.73	1.77	1.79	1.81	1.84	1.88	1.92	1.98	2.06	2.16	2.29
1.00	8.74	6.04	4.89	4.40	4.01	3.71	3.50	3.35	3.23	3.15	3.11	3.06	3.04	3.02
0.90	6.45	4.48	3.78	3.32	2.99	2.77	2.60	2.48	2.39	2.32	2.26	2.22	2.19	2.16
0.80	4.96	3.51	2.92	2.57	2.31	2.12	2.00	1.90	1.81	1.74	1.69	1.64	1.60	1.57
0.70	3.85	2.72	2.23	1.97	1.77	1.64	1.53	1.44	1.37	1.32	1.27	1.22	1.19	1.16
0.60	3.09	2.18	1.78	1.57	1.41	1.30	1.21	1.14	1.08	1.04	1.00	0.96	0.93	0.91
0.50	2.46	1.72	1.42	1.24	1.11	1.02	0.95	0.89	0.85	0.81	0.77	0.75	0.72	0.70
0.40	1.90	1.34	1.10	0.96	0.86	0.78	0.73	0.69	0.65	0.62	0.59	0.57	0.55	0.53
0.30	1.40	0.99	0.81	0.70	0.63	0.57	0.53	0.50	0.47	0.45	0.43	0.42	0.40	0.39
0.20	0.92	0.65	0.53	0.46	0.41	0.38	0.35	0.33	0.31	0.29	0.28	0.27	0.26	0.25
0.10	0.48	0.34	0.27	0.24	0.21	0.20	0.18	0.17	0.16	0.15	0.15	0.14	0.13	0.13
0.05	0.32	0.19	0.16	0.14	0.12	0.11	0.10	0.10	0.09	0.08	0.08	0.08	0.07	0.07
0	0	0	0	0	0	0	0	0	0	0	0	0	0	0

Rows 0 to 0.90 (first group): $\dfrac{x/c}{1-E}$ (ahead of hinge). Rows 1.00 down to 0 (second group): $\dfrac{1-x/c}{E}$ (back of hinge).

b. Plain flap at δ = 20 degrees

E	0.05	0.10	0.15	0.20	0.25	0.30	0.35	0.40	0.45	0.50	0.55	0.60
0	0	0	0	0	0	0	0	0	0	0	0	0
0.05	0.15	0.15	0.16	0.16	0.17	0.17	0.18	0.19	0.20	0.21	0.22	0.23
0.10	0.22	0.23	0.23	0.24	0.25	0.25	0.26	0.27	0.29	0.30	0.32	0.34
0.20	0.33	0.33	0.34	0.35	0.36	0.38	0.39	0.40	0.42	0.44	0.46	0.49
0.30	0.43	0.44	0.45	0.46	0.47	0.49	0.50	0.52	0.54	0.57	0.60	0.63
0.40	0.54	0.55	0.56	0.57	0.58	0.60	0.62	0.64	0.67	0.70	0.73	0.77
0.50	0.66	0.66	0.67	0.69	0.71	0.73	0.75	0.77	0.80	0.81	0.88	0.92
0.60	0.79	0.80	0.82	0.83	0.85	0.88	0.90	0.93	0.96	1.00	1.05	1.10
0.70	0.97	0.98	1.00	1.01	1.03	1.05	1.08	1.11	1.15	1.19	1.24	1.30
0.80	1.23	1.24	1.26	1.27	1.29	1.31	1.34	1.38	1.42	1.46	1.52	1.59
0.90	1.73	1.74	1.75	1.76	1.77	1.79	1.81	1.84	1.88	1.92	1.98	2.06
1.00	5.83	4.05	3.38	3.02	2.83	2.70	2.63	2.58	2.56	2.56	2.58	2.62
0.90	5.08	3.53	2.98	2.61	2.36	2.18	2.05	1.96	1.88	1.83	1.78	1.75
0.80	4.23	2.99	2.50	2.19	1.97	1.81	1.70	1.62	1.55	1.49	1.44	1.40
0.70	3.71	2.63	2.16	1.90	1.71	1.58	1.48	1.39	1.32	1.27	1.22	1.18
0.60	3.33	2.35	1.92	1.69	1.52	1.40	1.30	1.22	1.17	1.12	1.08	1.04
0.50	2.98	2.09	1.73	1.50	1.35	1.24	1.15	1.08	1.03	0.98	0.94	0.91
0.40	2.65	1.87	1.54	1.34	1.19	1.09	1.02	0.96	0.91	0.86	0.83	0.80
0.30	2.32	1.63	1.34	1.16	1.04	0.95	0.88	0.82	0.78	0.75	0.72	0.69
0.20	1.93	1.35	1.11	0.96	0.86	0.79	0.73	0.68	0.65	0.62	0.59	0.56
0.10	1.39	0.98	0.79	0.69	0.62	0.57	0.53	0.50	0.47	0.45	0.43	0.41
0.05	1.16	0.70	0.58	0.50	0.45	0.40	0.38	0.35	0.33	0.31	0.30	0.28
0	0	0	0	0	0	0	0	0	0	0	0	0

	c. Plain flap at δ = 30 degrees										*d.* Plain or split flap at δ = 40 degrees							
E	0.05	0.10	0.15	0.20	0.25	0.30	0.35	0.40	0.45	0.50	0.05	0.10	0.15	0.20	0.25	0.30	0.35	0.40
$\frac{x/c}{1-E}$ (ahead of hinge)																		
0	0	0	0	0	0	0	0	0	0	0	0	0	0	0	0	0	0	0
0.05	0.15	0.15	0.16	0.16	0.17	0.17	0.18	0.19	0.20	0.21	0.15	0.15	0.16	0.16	0.17	0.17	0.18	0.19
0.10	0.22	0.23	0.23	0.24	0.25	0.25	0.26	0.27	0.29	0.30	0.22	0.23	0.23	0.24	0.25	0.25	0.26	0.27
0.20	0.33	0.34	0.34	0.35	0.36	0.38	0.39	0.40	0.42	0.44	0.33	0.34	0.35	0.36	0.38	0.39	0.40	0.42
0.30	0.43	0.44	0.45	0.46	0.47	0.49	0.50	0.52	0.54	0.57	0.43	0.44	0.45	0.46	0.47	0.49	0.51	0.52
0.40	0.54	0.55	0.56	0.57	0.58	0.60	0.62	0.64	0.67	0.70	0.54	0.55	0.56	0.57	0.58	0.60	0.62	0.64
0.50	0.66	0.66	0.67	0.69	0.71	0.73	0.75	0.77	0.80	0.84	0.66	0.66	0.68	0.69	0.71	0.73	0.75	0.77
0.60	0.79	0.80	0.82	0.83	0.85	0.88	0.90	0.93	0.96	1.00	0.79	0.80	0.82	0.83	0.85	0.88	0.90	0.93
0.70	0.97	0.98	1.00	1.01	1.03	1.05	1.08	1.11	1.15	1.19	0.97	0.98	1.00	1.01	1.03	1.05	1.08	1.11
0.80	1.23	1.24	1.26	1.27	1.29	1.31	1.34	1.38	1.42	1.46	1.22	1.23	1.25	1.26	1.28	1.30	1.33	1.36
0.90	1.67	1.68	1.69	1.70	1.71	1.72	1.73	1.74	1.77	1.80	1.58	1.59	1.60	1.61	1.61	1.62	1.62	1.63
$\frac{1-x/c}{E}$ (back of hinge)																		
1.00	4.50	3.12	2.66	2.36	2.22	2.13	2.08	2.05	2.01	1.99	4.10	2.90	2.42	2.12	1.99	1.88	1.81	1.76
0.90	4.61	3.20	2.70	2.37	2.15	2.01	1.90	1.83	1.77	1.73	4.24	3.01	2.52	2.23	2.02	1.88	1.78	1.72
0.80	4.34	3.07	2.56	2.24	2.02	1.86	1.75	1.66	1.59	1.52	4.29	3.04	2.53	2.22	2.00	1.84	1.73	1.66
0.70	4.08	2.89	2.37	2.09	1.88	1.73	1.62	1.52	1.45	1.40	4.27	3.02	2.48	2.18	1.96	1.81	1.70	1.59
0.60	3.82	2.69	2.20	1.94	1.74	1.60	1.49	1.40	1.34	1.28	4.14	2.91	2.39	2.10	1.89	1.74	1.62	1.52
0.50	3.55	2.49	2.05	1.79	1.61	1.47	1.37	1.29	1.22	1.17	3.92	2.75	2.27	1.98	1.77	1.62	1.51	1.42
0.40	3.25	2.29	1.88	1.64	1.46	1.34	1.25	1.17	1.11	1.06	3.64	2.56	2.11	1.83	1.64	1.50	1.40	1.31
0.30	2.89	2.01	1.68	1.45	1.29	1.18	1.10	1.03	0.98	0.93	3.28	2.31	1.90	1.64	1.47	1.34	1.24	1.17
0.20	2.44	1.72	1.41	1.22	1.09	1.00	0.93	0.87	0.82	0.78	2.82	1.98	1.62	1.40	1.26	1.15	1.07	1.00
0.10	1.81	1.27	1.03	0.90	0.81	0.74	0.69	0.65	0.61	0.58	2.11	1.48	1.20	1.05	0.95	0.87	0.81	0.76
0.05	1.53	0.92	0.76	0.66	0.59	0.53	0.49	0.46	0.44	0.41	1.79	1.08	0.89	0.77	0.69	0.63	0.58	0.54
0	0	0	0	0	0	0	0	0	0	0	0	0	0	0	0	0	0	0

TABLE 6.—$P_{b_\delta}/c_{n_{b_\delta}}$ DISTRIBUTION.—(Continued)

e. Plain or split flap at δ = 50 degrees

E	0.05	0.10	0.15	0.20	0.25	0.30	0.35	0.40
$\frac{x/c}{1-E}$ (ahead of hinge)								
0	0	0	0	0	0	0	0	0
0.05	0.15	0.15	0.16	0.16	0.17	0.17	0.18	0.19
0.10	0.22	0.23	0.23	0.24	0.25	0.25	0.26	0.27
0.20	0.33	0.34	0.35	0.35	0.36	0.38	0.39	0.40
0.30	0.43	0.44	0.45	0.46	0.47	0.49	0.51	0.52
0.40	0.54	0.55	0.56	0.57	0.58	0.60	0.62	0.64
0.50	0.66	0.66	0.68	0.69	0.71	0.73	0.75	0.77
0.60	0.79	0.80	0.82	0.83	0.85	0.88	0.90	0.93
0.70	0.97	0.98	1.00	1.03	1.05	1.08	1.11	1.14
0.80	1.21	1.22	1.23	1.25	1.26	1.29	1.32	1.35
0.90	1.54	1.55	1.56	1.57	1.57	1.57	1.57	1.57
$\frac{1-x/c}{E}$ (back of hinge)								
1.00	3.81	2.68	2.27	2.00	1.86	1.77	1.71	1.67
0.90	4.18	2.90	2.45	2.15	1.95	1.81	1.72	1.66
0.80	4.25	3.00	2.50	2.20	1.98	1.82	1.72	1.64
0.70	4.27	3.02	2.48	2.18	1.96	1.81	1.70	1.59
0.60	4.21	2.97	2.43	2.14	1.92	1.77	1.65	1.55
0.50	4.07	2.86	2.36	2.06	1.84	1.69	1.57	1.48
0.40	3.86	2.71	2.24	1.95	1.73	1.59	1.48	1.39
0.30	3.52	2.49	2.04	1.76	1.58	1.44	1.34	1.25
0.20	3.06	2.15	1.76	1.53	1.37	1.25	1.16	1.09
0.10	2.34	1.64	1.33	1.16	1.05	0.96	0.90	0.84
0.05	2.02	1.22	1.00	0.87	0.78	0.70	0.65	0.61
0	0	0	0	0	0	0	0	0

f. Plain or split flap at δ = 60 degrees

E	0.05	0.10	0.15	0.20	0.25	0.30	0.35	0.40
$\frac{x/c}{1-E}$ (ahead of hinge)								
0	0	0	0	0	0	0	0	0
0.05	0.15	0.15	0.16	0.16	0.17	0.17	0.18	0.19
0.10	0.22	0.23	0.23	0.24	0.25	0.25	0.26	0.27
0.20	0.33	0.34	0.35	0.35	0.36	0.38	0.39	0.40
0.30	0.43	0.44	0.45	0.46	0.47	0.49	0.51	0.52
0.40	0.54	0.55	0.56	0.57	0.58	0.60	0.62	0.64
0.50	0.66	0.67	0.68	0.69	0.71	0.73	0.75	0.77
0.60	0.79	0.80	0.82	0.83	0.85	0.88	0.90	0.93
0.70	0.97	0.98	0.99	1.01	1.02	1.05	1.07	1.10
0.80	1.20	1.21	1.22	1.24	1.25	1.28	1.29	1.32
0.90	1.52	1.53	1.54	1.55	1.54	1.53	1.51	1.50
$\frac{1-x/c}{E}$ (back of hinge)								
1.00	3.84	2.59	2.22	1.97	1.81	1.71	1.63	1.59
0.90	4.04	2.80	2.36	2.08	1.88	1.76	1.68	1.62
0.80	4.16	2.94	2.45	2.15	1.94	1.78	1.68	1.59
0.70	4.27	3.02	2.48	2.18	1.96	1.81	1.70	1.59
0.60	4.27	3.01	2.46	2.17	1.95	1.79	1.67	1.57
0.50	4.18	2.93	2.42	2.11	1.89	1.73	1.62	1.52
0.40	4.01	2.82	2.33	2.02	1.80	1.65	1.54	1.45
0.30	3.71	2.62	2.15	1.86	1.66	1.52	1.41	1.32
0.20	3.27	2.29	1.88	1.63	1.46	1.33	1.24	1.16
0.10	2.51	1.76	1.43	1.25	1.13	1.03	0.96	0.90
0.05	2.18	1.31	1.08	0.93	0.84	0.76	0.71	0.66
0	0	0	0	0	0	0	0	0

a. Split flap at δ = 20 degrees

E	0.05	0.10	0.15	0.20	0.25	0.30	0.35	0.40	0.45	0.50
0.05	0.15	0.15	0.16	0.16	0.17	0.17	0.18	0.19	0.20	0.21
0.10	0.22	0.23	0.23	0.24	0.25	0.25	0.26	0.27	0.29	0.30
0.20	0.33	0.34	0.35	0.35	0.36	0.38	0.39	0.40	0.42	0.44
0.30	0.43	0.44	0.45	0.46	0.47	0.49	0.51	0.52	0.55	0.57
0.40	0.54	0.55	0.56	0.57	0.58	0.60	0.62	0.64	0.67	0.70
0.50	0.66	0.66	0.68	0.69	0.71	0.73	0.75	0.77	0.80	0.84
0.60	0.79	0.80	0.82	0.83	0.85	0.88	0.90	0.93	0.96	1.00
0.70	0.97	0.98	1.00	1.01	1.03	1.05	1.08	1.11	1.15	1.19
0.80	1.23	1.24	1.26	1.27	1.29	1.31	1.34	1.38	1.42	1.46
0.90	1.69	1.70	1.71	1.72	1.73	1.73	1.73	1.73	1.74	1.74
1.00	4.79	3.33	2.79	2.44	2.22	2.08	1.97	1.90	1.86	1.84
0.90	4.84	3.36	2.83	2.49	2.26	2.09	1.98	1.90	1.83	1.79
0.80	4.65	3.29	2.74	2.40	2.16	1.99	1.87	1.78	1.70	1.63
0.70	4.32	3.06	2.51	2.21	1.99	1.84	1.72	1.61	1.54	1.48
0.60	3.94	2.77	2.27	2.00	1.80	1.65	1.54	1.45	1.38	1.32
0.50	3.51	2.46	2.03	1.77	1.59	1.46	1.36	1.28	1.21	1.16
0.40	3.05	2.15	1.77	1.54	1.37	1.26	1.17	1.10	1.04	0.99
0.30	2.60	1.83	1.51	1.30	1.16	1.06	0.98	0.92	0.88	0.84
0.20	2.13	1.50	1.23	1.06	0.95	0.87	0.81	0.76	0.72	0.68
0.10	1.54	1.08	0.88	0.77	0.69	0.63	0.59	0.55	0.52	0.49
0.05	1.29	0.78	0.64	0.55	0.50	0.45	0.42	0.39	0.37	0.35
0	0	0	0	0	0	0	0	0	0	0

Row labels: upper block $\dfrac{x/c}{1-E}$ (ahead of hinge); lower block $\dfrac{1-x/c}{E}$ (back of hinge).

h. Split flap at δ = 30 degrees

E	0.05	0.10	0.15	0.20	0.25	0.30	0.35	0.40	0.45	0.50
0.05	0.15	0.15	0.16	0.16	0.17	0.17	0.18	0.19	0.20	0.21
0.10	0.22	0.23	0.23	0.24	0.25	0.25	0.26	0.27	0.29	0.30
0.20	0.33	0.34	0.35	0.35	0.36	0.38	0.39	0.40	0.42	0.44
0.30	0.43	0.44	0.45	0.46	0.47	0.49	0.51	0.52	0.55	0.57
0.40	0.54	0.55	0.56	0.57	0.58	0.60	0.62	0.64	0.67	0.70
0.50	0.66	0.66	0.68	0.69	0.71	0.73	0.75	0.77	0.80	0.84
0.60	0.79	0.80	0.82	0.83	0.85	0.88	0.90	0.93	0.96	1.00
0.70	0.97	0.98	1.00	1.01	1.03	1.05	1.08	1.11	1.15	1.19
0.80	1.23	1.24	1.26	1.27	1.29	1.31	1.34	1.38	1.42	1.46
0.90	1.64	1.65	1.66	1.67	1.68	1.68	1.68	1.69	1.70	1.70
1.00	4.55	3.15	2.63	2.32	2.14	1.98	1.89	1.83	1.78	1.73
0.90	4.60	3.19	2.69	2.36	2.14	1.99	1.88	1.80	1.73	1.68
0.80	4.49	3.18	2.65	2.32	2.09	1.92	1.81	1.72	1.64	1.58
0.70	4.27	3.02	2.48	2.19	1.97	1.82	1.70	1.59	1.52	1.46
0.60	4.02	2.83	2.32	2.04	1.83	1.69	1.57	1.48	1.41	1.35
0.50	3.71	2.60	2.14	1.87	1.68	1.54	1.43	1.35	1.28	1.22
0.40	3.35	2.36	1.94	1.69	1.51	1.38	1.29	1.21	1.15	1.09
0.30	2.95	2.08	1.71	1.48	1.32	1.21	1.12	1.05	1.00	0.95
0.20	2.68	1.88	1.54	1.33	1.19	1.09	1.01	0.95	0.90	0.85
0.10	1.91	1.34	1.08	0.95	0.86	0.79	0.73	0.68	0.64	0.61
0.05	1.63	0.98	0.81	0.70	0.63	0.57	0.53	0.49	0.47	0.44
0	0	0	0	0	0	0	0	0	0	0

Row labels: upper block $\dfrac{x/c}{1-E}$ (ahead of hinge); lower block $\dfrac{1-x/c}{E}$ (back of hinge).

The incremental basic $c_{n_{b\delta}}$ and additional $c_{n_{a\delta}}$ normal-force coefficients are obtained from Eqs. (8.3) and Table 7. The appropriate values of the

TABLE 7.—VALUES OF G

δ, deg / E	5, 10, 15	20	30	40	50	60
a. Plain flaps						
0.05	−0.474	−0.476	−0.477	−0.478	−0.479	−0.479
0.10	−0.448	−0.451	−0.453	−0.455	−0.456	−0.456
0.15	−0.423	−0.428	−0.431	−0.434	−0.435	−0.435
0.20	−0.397	−0.404	−0.408	−0.411	−0.414	−0.415
0.25	−0.372	−0.380	−0.387	−0.392	−0.395	−0.396
0.30	−0.347	−0.357	−0.366	−0.372	−0.375	−0.377
0.35	−0.320	−0.334	−0.346	−0.352	−0.357	−0.360
0.40	−0.294	−0.311	−0.325	−0.332	−0.339	−0.342
0.45	−0.268	−0.288	−0.304			
0.50	−0.242	−0.265	−0.283			
0.55	−0.215	−0.242				
0.60	−0.189	−0.220				
0.65	−0.163					
0.70	−0.136					
b. Split flaps						
0.05	. . . . . .	−0.476	−0.477	−0.478	−0.479	−0.479
0.10	. . . . . .	−0.452	−0.454	−0.455	−0.456	−0.456
0.15	. . . . . .	−0.430	−0.432	−0.434	−0.435	−0.436
0.20	. . . . . .	−0.407	−0.409	−0.411	−0.414	−0.415
0.25	. . . . . .	−0.385	−0.388	−0.392	−0.395	−0.396
0.30	. . . . . .	−0.363	−0.367	−0.372	−0.375	−0.377
0.35	. . . . . .	−0.342	−0.347	−0.352	−0.357	−0.360
0.40	. . . . . .	−0.320	−0.327	−0.333	−0.339	−0.342
0.45	. . . . . .	−0.299	−0.307			
0.50	. . . . . .	−0.278	−0.287			

incremental basic $P_{b\delta}$ normal-force distribution are obtained by use of Table 6 as follows:

$$P_{b\delta} = \frac{P_{b\delta}}{c_{n_{b\delta}}} c_{n_{b\delta}} \qquad (8.9)$$

The appropriate value of the incremental additional $P_{a\delta}$ normal-force distribution is obtained from the data of Appendix I as follows:

$$P_{a\delta} = 4 \frac{\Delta v_a}{V} \frac{v}{V} c_{n_{a\delta}} \qquad (8.10)$$

TABLE 8.—VALUES OF τ_n

δ, deg / E	10	20	30	40	50	60
a. Plain flaps						
0	0	0	0	0	0	0
0.05	−0.00	−0.02	−0.05	−0.09	−0.14	−0.21
0.10	−0.01	−0.03	−0.07	−0.14	−0.23	−0.34
0.15	−0.01	−0.04	−0.09	−0.18	−0.30	−0.47
0.20	−0.01	−0.05	−0.11	−0.23	−0.38	−0.59
0.25	−0.01	−0.06	−0.14	−0.27	−0.46	−0.72
0.30	−0.02	−0.07	−0.16	−0.32	−0.54	−0.85
0.35	−0.02	−0.08	−0.19	−0.37	−0.63	−0.98
0.40	−0.02	−0.09	−0.21	−0.42	−0.71	−1.11
0.45	−0.03	−0.11	−0.25			
0.50	−0.03	−0.12	−0.28			
0.55	−0.04	−0.14				
0.60	−0.05	−0.17				
0.65	−0.06					
0.70	−0.07					
b. Split flaps						
0	. . . . .	0	0	0	0	0
0.05	. . . . .	−0.02	−0.05	−0.09	−0.14	−0.21
0.10	. . . . .	−0.03	−0.07	−0.14	−0.23	−0.34
0.15	. . . . .	−0.04	−0.10	−0.18	−0.30	−0.47
0.20	. . . . .	−0.05	−0.12	−0.23	−0.38	−0.59
0.25	. . . . .	−0.06	−0.15	−0.27	−0.46	−0.72
0.30	. . . . .	−0.07	−0.17	−0.32	−0.54	−0.85
0.35	. . . . .	−0.08	−0.20	−0.37	−0.63	−0.98
0.40	. . . . .	−0.10	−0.23	−0.42	−0.71	−1.11
0.45	. . . . .	−0.11	−0.26			
0.50	. . . . .	−0.13	−0.29			

The entire incremental normal-force distribution is obtained by adding the incremental basic and additional distributions

$$P_\delta = P_{a_\delta} + P_{b_\delta} \tag{8.11}$$

The final load distribution is obtained by adding this incremental distribution to the load distribution for the unflapped wing section at the same angle of attack (see Sec. 4.5).

Approximate hinge-moment coefficients may be obtained by an extension of this analysis.[6] Such hinge-moment coefficients are not considered to be reliable because the discrepancies between the predicted and actual

TABLE 9.—VALUES OF τ_m

δ, deg E	10	20	30	40	50	60
a. Plain flaps						
0	1.00	1.00	1.00	1.00	1.00	1.00
0.05	1.00	1.01	1.03	1.06	1.10	1.15
0.10	1.00	1.02	1.05	1.09	1.15	1.22
0.15	1.01	1.02	1.06	1.11	1.18	1.28
0.20	1.01	1.03	1.06	1.12	1.21	1.32
0.25	1.01	1.03	1.07	1.14	1.23	1.36
0.30	1.01	1.03	1.07	1.14	1.24	1.38
0.35	1.01	1.03	1.07	1.15	1.25	1.39
0.40	1.01	1.03	1.08	1.15	1.25	1.39
0.45	1.01	1.03	1.07			
0.50	1.01	1.03	1.07			
0.55	1.01	1.03				
0.60	1.01	1.03				
0.65	1.01					
0.70	1.00					
b. Split flaps						
0	....	1.00	1.00	1.00	1.00	1.00
0.05	....	1.01	1.03	1.06	1.10	1.15
0.10	....	1.02	1.05	1.09	1.15	1.22
0.15	....	1.02	1.06	1.11	1.18	1.28
0.20	....	1.03	1.07	1.12	1.21	1.32
0.25	....	1.03	1.07	1.14	1.23	1.36
0.30	....	1.03	1.08	1.14	1.24	1.38
0.35	....	1.03	1.08	1.15	1.25	1.39
0.40	....	1.03	1.08	1.15	1.25	1.39
0.45	....	1.03	1.08			
0.50	....	1.03	1.07			

load distributions tend to be appreciable at the trailing edge where the contribution to the hinge-moment coefficient is the largest.

In applying this method, it should be remembered that, although the tabulated characteristics are differentiated by flap deflection, the important variable is the degree to which the flap is stalled. The tabulated characteristics for plain flap deflections up to 15 degrees are for unstalled conditions, whereas those for larger flap deflections represent progressively increased separation. Consideration should be given to the selection of the characteristics to be used in order to represent the actual flow conditions properly.

CHAPTER 9

EFFECTS OF COMPRESSIBILITY AT SUBSONIC SPEEDS

9.1. Symbols.

C_p pressure coefficient, $(p - p_\infty)/\frac{1}{2}\rho V^2$

C_{p_c} critical pressure coefficient corresponding to a local Mach number of unity

E total energy per unit mass

K constant

M Mach number, V/a

R universal gas constant, $c_p - c_v$

S cross-sectional area of a stream tube

T absolute temperature

V velocity of the free stream

V_s component of velocity along the span

V_n component of velocity normal to span

a speed of sound

a lift-curve slope

b span

c_l section lift coefficient

c_p specific heat at constant pressure

c_v specific heat at constant volume

e base of Naperian logarithms, 2.71828

$\ln$ logarithm to the base e

m mass flow per unit area

p local static pressure

p_0 total pressure

p_∞ static pressure of the free stream

t time

t thickness ratio

u component of velocity parallel to the x axis

v volume

v component of velocity parallel to the y axis

w component of velocity parallel to the z axis

x, y, z Cartesian coordinates

α angle of attack

β angle of sweep

γ ratio of the specific heats, c_p/c_v

μ $1/\sqrt{1 - M^2}$

π ratio of the circumference of a circle to its diameter

ρ mass density of the fluid

9.2. Introduction. The wing-section theory and experimental data presented in the preceding chapters are applicable to conditions where the variation of pressure is small compared with the absolute pressure. This condition is well satisfied when the speed is low compared with the speed of sound. At a flight speed of 100 knots at sea level, the impact pressure is only about 34 pounds per square foot as compared with an ambient static pressure of 2,116 pounds per square foot. At 300 knots, the impact pressure increases to approximately 321 pounds per square foot; and, at 600 knots, the impact pressure is 1,494 pounds per square foot. At the higher speeds, the pressure and corresponding volume changes are obviously not negligible. It is to be expected, therefore, that wing-section characteristics at high speeds will not agree with those predicted by incompressible flow theory or with experimental data obtained at low speeds.

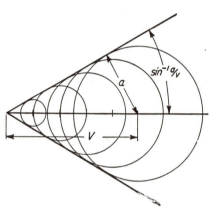

Examination of the complete equations of motion[89] shows that the parameter determining the effect of speed is the ratio of the speed to the speed of sound. This ratio is called the "Mach number." The physical significance of the speed of sound becomes evident in considering the difference between subsonic and supersonic flows. The speed of sound is the speed at which pressure impulses are transmitted through the air. At slow flight speeds, the pressure impulses caused by motion of the wing are transmitted at relatively high speed in all directions and cause the air approaching the wing to change its pressure and velocity gradually. Slow speed flows are accordingly free of discontinuities of pressure and velocity. At supersonic speeds, no pressure impulses can be transmitted ahead of the wing, and the pressure and velocity of the air remain unaltered until it reaches the immediate vicinity of the wing. Supersonic flow is thus characterized by discontinuities of pressure and velocity. As shown in Fig. 143, the pressure impulses are propagated in all directions at the speed of sound a while the wing moves through the air at the velocity V. The envelope of the pressure impulses is a straight line (for small impulses) which makes a slope $\sin^{-1} a/V$ or $\sin^{-1} M$ with the direction of motion, where M is the Mach number. This line is called the "Mach line." The air ahead of the Mach line is unaffected by the approaching wing.

FIG. 143. Significance of Mach angle.

The present discussion of the characteristics of wing sections in compressible flow is limited to subsonic speeds. The pressure-velocity relations for a stream tube will be developed, and a brief summary will be presented of compressible flow theory as applied to wing sections. The theory of wing sections in subsonic flow generally consists of the determination of the first-order compressibility effects on the incompressible flow solutions previously developed. The theory is generally limited to speeds at which the velocity of sound is not exceeded in the field of flow. A brief discussion of experimental data will, however, include data in the lower transonic region where the speed of sound is exceeded locally.

9.3. Steady Flow through a Stream Tube. *a. Adiabatic Law.* The relation among pressure, density, and temperature for a perfect gas is

$$p = \rho R T \tag{9.1}$$

where (in consistent units)

p = pressure
ρ = density, moles per unit volume
T = absolute temperature
R = universal gas constant = $c_p - c_v$
c_p = specific heat per mole at constant pressure
c_v = specific heat per mole at constant volume

This equation may be written

$$\frac{p_1}{\rho_1 T_1} = \frac{p_2}{\rho_2 T_2}$$

or

$$\frac{p_1 v_1}{T_1} = \frac{p_2 v_2}{T_2}$$

where the subscripts 1 and 2 refer to any two conditions of the gas. When the temperature is constant (isothermal changes), the relation between the pressure and the density is obviously

$$\frac{p_1}{p_2} = \frac{\rho_1}{\rho_2}$$

Except for this special case, pressure and volume changes are accompanied by temperature changes. In the high-speed flows of primary interest, these changes occur so quickly that individual elements of the gas generally do not lose or gain any appreciable heat. Changes in state of the gas are thus assumed to be adiabatic. In this case,

$$\frac{p_1}{p_2} = \left(\frac{\rho_1}{\rho_2}\right)^{\gamma} \tag{9.2}$$

where γ = ratio of specific heats, c_p/c_v

From Eqs. (9.1) and (9.2), the corresponding temperature relation is

$$\frac{p_1}{p_2} = \left(\frac{T_1}{T_2}\right)^{\gamma/(\gamma-1)} \tag{9.3}$$

b. Velocity of Sound. Expressions for the velocity of sound may be derived from consideration of the equations of motion and of continuity. The equation of continuity for a compressible fluid may be developed in a manner parallel to that for Eq. (2.1) and is

$$\frac{\partial(\rho u)}{\partial x} + \frac{\partial(\rho v)}{\partial y} + \frac{\partial(\rho w)}{\partial z} = -\frac{\partial\rho}{\partial t} \tag{9.4}$$

For motion in one dimension, this equation reduces to

$$\frac{\partial(\rho u)}{\partial x} = -\frac{\partial\rho}{\partial t}$$

or

$$\frac{\partial\rho}{\partial t} + \rho\frac{\partial u}{\partial x} + u\frac{\partial\rho}{\partial x} = 0$$

dividing by ρ

$$\frac{1}{\rho}\frac{\partial\rho}{\partial t} + \frac{\partial u}{\partial x} + \frac{u}{\rho}\frac{\partial\rho}{\partial x} = 0 \tag{9.5}$$

In the concept of a velocity of sound, the disturbances are considered to be small so that the changes in density are small compared with the density and the corresponding velocities are small. Retaining only the first powers of small quantities, Eq. (9.5) becomes

$$\frac{1}{\rho}\frac{\partial\rho}{\partial t} + \frac{\partial u}{\partial x} = 0 = \frac{\partial\ln\rho}{\partial t} + \frac{\partial u}{\partial x} \tag{9.6}$$

For one-dimensional flow, Eq. (2.3) becomes

$$-\frac{\partial p}{\partial x} = \rho\frac{\partial u}{\partial t} + \rho u\frac{\partial u}{\partial x}$$

Again retaining only the first power of small quantities and dividing by ρ, we have

$$\frac{\partial u}{\partial t} = -\frac{1}{\rho}\frac{\partial p}{\partial x} \tag{9.7}$$

It is known from the equation of state that p is a function of ρ. Equation (9.7) can then be written

$$\frac{\partial u}{\partial t} = -\frac{1}{\rho}\frac{dp}{d\rho}\frac{\partial\rho}{\partial x} = -\frac{dp}{d\rho}\frac{\partial\ln\rho}{\partial x} \tag{9.8}$$

Differentiating Eq. (9.6) with respect to x and Eq. (9.8) with respect to t, and combining, we have

$$\frac{\partial^2 u}{\partial t^2} = \frac{dp}{d\rho}\frac{\partial^2 u}{\partial x^2} \tag{9.9}$$

The solution of this equation is

$$u = f_1\!\left(x - \sqrt{\frac{dp}{d\rho}}\,t\right) + f_2\!\left(x + \sqrt{\frac{dp}{d\rho}}\,t\right)$$

These functions represent disturbances traveling in opposite directions with the velocity $\sqrt{dp/d\rho}$. This velocity is termed the velocity of sound a.

As stated previously, the changes in pressure occur so rapidly that no heat is considered to be conducted to or away from the elements of gas.

The relation between the pressure and density is therefore considered to be adiabatic [Eq. (9.2)]. Then

$$a = \sqrt{\frac{dp}{d\rho}} = \sqrt{\frac{\gamma p}{\rho}} \tag{9.10}$$

For a perfect gas,

$$a = K\sqrt{T} \tag{9.11}$$

where T is the absolute temperature, and the value of K is given in the following table for various English units, and for dry air, $\gamma = 1.4$.

Units of velocity	Units of temperature	
	°F. absolute	°C. absolute
Feet per second.......	49.02	65.76
Miles per hour........	33.42	44.84
Knots................	29.02	38.94

In applying the concept of the velocity of sound to various problems, it should be kept in mind that the concept as derived is valid only for small disturbances. Very strong shock waves and blasts should not be expected to propagate at speeds corresponding to the local velocity of sound.

c. Bernoulli's Equation for Compressible Flow. In deriving Bernoulli's equation for incompressible flow [Eq. (2.5)], the following expression was obtained:

$$- dp = \frac{\rho}{2} d(V)^2$$

Writing this equation in integral form, we have

$$\frac{1}{2} V^2 + \int \frac{dp}{\rho} = E \tag{9.12}$$

Relating the pressure and density by the adiabatic law of Eq. (9.2) and performing the indicated integration,

$$\frac{1}{2} V^2 + \frac{\gamma}{\gamma - 1} \frac{p}{\rho} = E = \frac{1}{2} V^2 + c_p T \tag{9.13}$$

This equation can also be derived from thermodynamic reasoning. The thermodynamic reasoning indicates that the quantity E in Eq. (9.13) is a constant along any stream tube for adiabatic changes whether reversible or not. For example, E is constant throughout a stream tube containing a normal shock. For reversible changes, Eqs. (9.2) and (9.13) together with a knowledge of the initial conditions permit the calculation of the variation of pressure and density with velocity.

d. Cross-sectional Areas and Pressures in a Stream Tube. Along a stream tube, the equation of continuity for steady flow is

$$\rho S V = \text{constant}$$

where S = cross-sectional area of tube

It is interesting to contrast the manner in which the cross-sectional area varies with velocity for compressible and incompressible flow. For the latter case, the area is seen to vary inversely with the velocity. The determination of the variation of area with velocity for the compressible case requires more extensive analysis. Differentiating the equation of continuity, and dividing by $\rho S V$, we obtain

$$\frac{1}{S}\frac{dS}{dV} + \frac{1}{\rho}\frac{d\rho}{dV} + \frac{1}{V} = 0 \tag{9.14}$$

Differentiating Bernoulli's equation with respect to the velocity,

$$V + \frac{\gamma}{\gamma - 1}\left(\frac{1}{\rho}\frac{dp}{d\rho} - \frac{p}{\rho^2}\right)\frac{d\rho}{dV} = 0$$

or

$$V + \frac{1}{\gamma - 1}\left(\frac{\gamma}{\rho}a^2 - \frac{a^2}{\rho}\right)\frac{d\rho}{dV} = 0$$

or

$$V + \frac{a^2}{\rho}\frac{d\rho}{dV} = 0 \tag{9.15}$$

From Eq. (9.14),

$$\frac{dS}{dV} = -\frac{S}{V}\left(1 + \frac{V}{\rho}\frac{d\rho}{dV}\right)$$

Substituting the value of $d\rho/dV$ from Eq. (9.15), we obtain

$$\frac{dS}{dV} = -\frac{S}{V}\left(1 - \frac{V^2}{a^2}\right) = -\frac{S}{V}(1 - \text{M}^2) \tag{9.16}$$

This relation shows that the stream tube contracts as the velocity increases for Mach numbers less than unity. The area of the stream tube is a minimum for a Mach number of unity and increases with Mach number for supersonic flow.

From consideration of Eqs. (9.2), (9.3), and (9.11), it may be seen that the pressures, densities, temperatures, and velocities of sound are connected by the equations

$$\left(\frac{p}{p_m}\right)^{(\gamma-1)/\gamma} = \left(\frac{\rho}{\rho_m}\right)^{\gamma-1} = \left(\frac{T}{T_m}\right) = \left(\frac{a}{a_m}\right)^2 \tag{9.17}$$

where the subscript m may denote the condition at any point but, in the following development, the subscript will denote the condition at the minimum cross-sectional area of the stream tube where the local Mach number equals unity.

In order to obtain an expression for the cross-sectional area of the stream tube in terms of the local Mach number M, it is convenient to express Bernoulli's equation (9.13) as

$$\frac{1}{2}M^2 + \frac{1}{\gamma - 1} = \frac{\gamma + 1}{2(\gamma - 1)} \frac{a_m{}^2}{a^2} \tag{9.18}$$

where a and a_m represent, respectively, the velocities of sound at any point along the stream tube and at the point of minimum cross-sectional area. From continuity, we can write

$$\frac{S_m}{S} = \frac{\rho V}{\rho_m a_m}$$

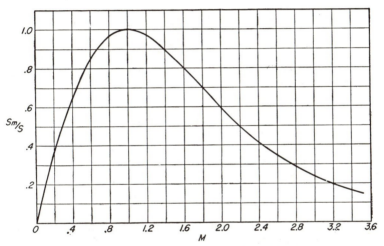

FIG. 144.   Variation of stream-tube area with local Mach number.

where S and S_m are, respectively, the cross-sectional areas of the stream tube at any point and at the point of minimum area.   Substituting the value of ρ/ρ_m from Eq. (9.17),

$$\frac{S_m}{S} = \left(\frac{a}{a_m}\right)^{2/(\gamma-1)} \left(\frac{V}{a_m}\right) = M\left(\frac{a}{a_m}\right)^{(\gamma+1)/(\gamma-1)}$$

and, from Eq. (9.18),

$$\frac{S_m}{S} = M\left(\frac{\gamma + 1}{(\gamma - 1)M^2 + 2}\right)^{(\gamma+1)/2(\gamma-1)} \tag{9.19}$$

A plot of S_m/S against M is given in Fig. 144.

The pressure relations along the stream tube may be obtained by substituting the value of a_m/a from Eq. (9.18) into Eq. (9.17)

$$\frac{p}{p_m} = \left(\frac{\gamma + 1}{(\gamma - 1)M^2 + 2}\right)^{\gamma/(\gamma-1)} \tag{9.20}$$

where p = pressure at any point along stream tube.

It is often more convenient to express the local pressure in terms of the pressure at the stagnation condition,

$$\frac{p}{p_0} = \left(\frac{2}{(\gamma - 1)\mathrm{M}^2 + 2}\right)^{\gamma/(\gamma-1)} \tag{9.21}$$

where p_0 is the pressure of the fluid at rest.

The temperature relations may be obtained from Eqs. (9.17) and (9.18) as

$$\frac{T}{T_m} = \frac{\gamma + 1}{(\gamma - 1)\mathrm{M}^2 + 2}$$

or

$$\frac{T_0}{T} = 1 + \frac{\gamma - 1}{2}\mathrm{M}^2 \tag{9.22}$$

It is interesting to note that there is a limiting velocity which is reached when the gas is expanded indefinitely into a vacuum. Bernoulli's equation (9.13) may be written

$$\frac{1}{2}\frac{v^2}{a_m^2} + \frac{1}{\gamma - 1}\left(\frac{a}{a_m}\right)^2 = \frac{\gamma + 1}{2(\gamma - 1)}$$

With infinite expansion, the temperature approaches zero and the local velocity of sound a approaches zero. Accordingly

$$\frac{V_{\max}}{a_m} = \sqrt{\frac{\gamma + 1}{\gamma - 1}} \tag{9.23}$$

The local Mach number M, of course, continues to increase indefinitely as the expansion is increased.

e. Relations for a Normal Shock. In deriving the relations for the cross-sectional areas, pressures, and temperatures existing in a stream tube, the entire process was considered to be adiabatic and reversible. Experience has shown that this condition is satisfied whenever the velocity is increasing in the direction of flow. However, when an attempt is made to decelerate supersonic flows, discontinuities in velocity, pressure, and temperature generally occur. These discontinuities occur in a very short distance along the direction of motion and are termed "shock waves." The velocity and total pressure of the air decrease and the temperature and static pressure increase in going through the shock. The process, although adiabatic, is irreversible and characterized by an increase of entropy. The shock is stationary for a tube of fixed configuration with constant pressures at the inlet and exit. Since the velocity on the upstream face is supersonic, the shock can hardly be considered to propagate at the velocity of sound. Similar discontinuities occur forward of the nose of a blunt body traveling at supersonic speeds. In the case of a simple tube or of the shock immediately forward of the nose of a blunt body, the plane of the shock is normal to the direction of flow. This type of shock is a fundamental one and is termed a "normal shock," as distinguished from the oblique shocks shown in Fig. 143.

Relations for determining the conditions across a normal shock wave from the conditions on one side are derived[29] from considerations of mass, momentum, and energy. If the subscripts 1 and 2 denote, respectively, the conditions on opposite sides of the shock and if m is the mass flow per unit area and p is the static pressure, the condition for continuity for a shock stationary with respect to the observer is

$$\rho_1 u_1 = \rho_2 u_2 = m \tag{9.24}$$

The equation of conservation of momentum is

$$p_2 - p_1 = m(u_1 - u_2) \tag{9.25}$$

The equation for conservation of energy is

$$p_1 u_1 - p_2 u_2 - \frac{1}{2} m \left(u_2{}^2 - u_1{}^2\right) = \frac{m}{\gamma - 1}\left(\frac{p_2}{\rho_2} - \frac{p_1}{\rho_1}\right) \tag{9.26}$$

The solution[29] for these conditions is

$$\frac{\rho_2}{\rho_1} = \frac{(p_2/p_1)(\gamma + 1)/(\gamma - 1) + 1}{(p_2/p_1) + (\gamma + 1)/(\gamma - 1)} \tag{9.27}$$

This equation relates the density and pressure ratios across the shock when p_1 and ρ_1 are the upstream conditions. This equation corresponds to a change in entropy except in the case where p_1 equals p_2. It can be shown[29] that p_2 must be greater than p_1 for the change in entropy to be positive. It would be convenient to relate the pressure and density ratios across the normal shock to the upstream Mach number. Such an expression may be obtained by expressing Eqs. (9.24) and (9.25) in terms of the local Mach number and combining to give

$$\frac{p_2}{p_1} - 1 = M_1{}^2 \gamma \left(1 - \frac{\rho_1}{\rho_2}\right)$$

Substituting the value of ρ_1/ρ_2 from Eq. (9.27) gives

$$M_1{}^2 = \frac{(p_2/p_1) - 1}{\gamma \left[1 - \dfrac{(p_2/p_1) + (\gamma + 1(/)\gamma - 1)}{(p_2/p_1)(\gamma + 1)/(\gamma - 1) + 1}\right]} \tag{9.28}$$

From Eq. (9.24), we can write

$$\frac{M_2}{M_1} = \sqrt{\frac{p_1}{p_2}\frac{\rho_1}{\rho_2}} \tag{9.29}$$

Values of the ratio p_{0_2}/p_{0_1} may be calculated from Eqs. (9.23), (9.27), (9.28), and (9.29) where p_0 is the total pressure. This ratio is an index of the useful energy remaining in the air downstream of the shock. Values of M_2/M_1, p_2/p_1, ρ_2/ρ_1, and p_{0_2}/p_{0_1} are plotted against M in Fig. 145.

9.4. First-order Compressibility Effects. *a. Glauert-Prandtl rule.* The Glauert-Prandtl rule[36] relates the lift coefficient or slope of the lift curve of a wing section in compressible flow with that for incompressible flow. This relation was derived for the case of small disturbance velocities and

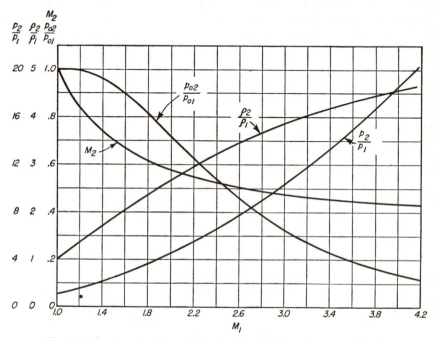

Fig. 145. Pressure, density, and Mach-number relations for a normal shock.

low Mach numbers. These conditions are approximated for thin wing sections with small amounts of camber at low lift coefficients at speeds well below the speed of sound. The Glauert-Prandtl relation is

$$\frac{c_{l_c}}{c_{l_i}} = \frac{a_c}{a_i} = \frac{1}{\sqrt{1 - M^2}} \tag{9.30}$$

where the subscripts c and i denote, respectively, the compressible and incompressible cases. In deriving this formula, Glauert considered all the velocities over the wing section to be increased by the same factor. Equation (9.30) may therefore be applied equally well to the moment coefficient.

The Glauert-Prandtl rule agrees remarkably well with experimental data, considering the assumptions made in its derivation. A comparison of this rule as applied to the slope of the lift curve for three wing sections of 6, 9, and 12 per cent thickness is given[44] in Fig. 146.

Kaplan[54] obtained a first-step improvement of the Glauert-Prandtl rule for the lift of an elliptical cylinder and extended it to arbitrary symmetrical profiles. The Kaplan rule is

$$\frac{c_{l_c}}{c_{l_i}} = \mu + \frac{t}{1+t}\left[\mu\left(\mu - 1\right) + \frac{1}{4}\left(\gamma + 1\right)\left(\mu^2 - 1\right)^2\right] \qquad (9.31)$$

where $\mu = \dfrac{1}{\sqrt{1 - M^2}}$

$t = $ thickness ratio

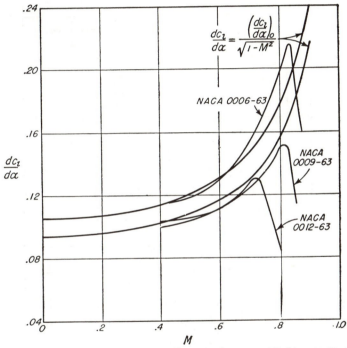

FIG. 146. Lift-curve slope variation with Mach number for the NACA 0012-63, 0009-63, and 0006-63 airfoils.

Equation (9.31) approaches the Glauert-Prandtl rule as the thickness ratio approaches zero. It will also be noted that the Kaplan rule shows a weak dependence on the ratio of the specific heats γ. The Kaplan rule for wing sections 5, 10, 15, and 20 per cent thick in air together with the Glauert-Prandtl rule are plotted in Fig. 147.

b. Effect of Mach number on the Pressure Coefficient. The Glauert-Prandtl rule supplies a first approximation to the variation of pressure coefficient with Mach number.

$$\frac{C_{p_M}}{C_{p_i}} = \frac{1}{\sqrt{1 - M^2}}$$

Numerous attempts have been made to obtain more accurate expressions for the variation of pressure with Mach number, notably by Chaplygin,[25] Temple and Yarwood,[119] and von Kármán and Tsien.[129,140] Garrick and Kaplan[34] succeeded in unifying these results as approximate solutions of

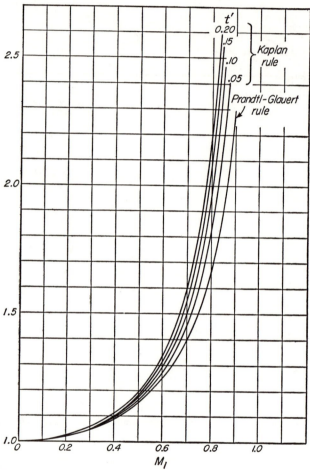

Fig. 147. Ratio of lifts for compressible and incompressible fluids as function of stream Mach number.

the general problem and presented two other approximations. No simple general solution of the problem is known.

The Kármán-Tsien relation is widely used in the United States. Experimental evidence appears to indicate that this relation is as applicable as any of the solutions. This relation is

$$\frac{C_{p_M}}{C_{p_i}} = \frac{1}{\sqrt{1 - M^2} + [M^2/(1 + \sqrt{1 - M^2})](c_{p_i}/2)} \tag{9.32}$$

A plot of the pressure coefficient as a function of the free stream Mach number M is given in Fig. 148 for a number of values of the incompressible

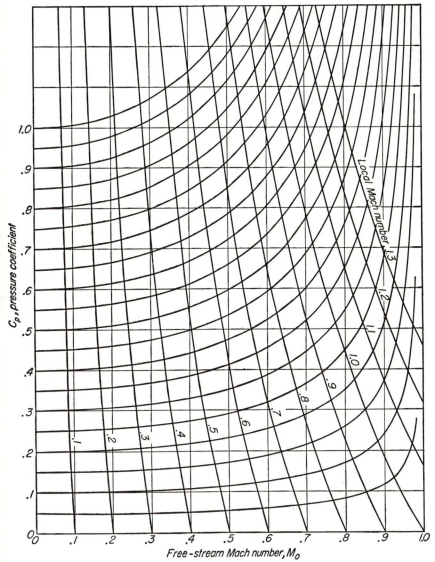

FIG. 148. Variation of local pressure coefficient and local Mach number with free stream Mach number according to Kármán-Tsien.

pressure coefficient. Also shown in Fig. 148 are curves of the local Mach number M_L.

A special use for relations such as Eq. (9.32) is the prediction of the stream Mach number at which the velocity of sound is reached locally over the wing section. Jacobs[44] applied the Glauert-Prandtl rule to this problem. It is now customary in the United States to use the Kármán-

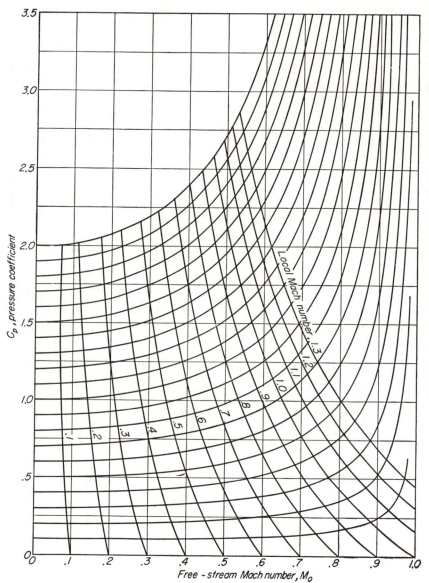

C_p, pressure coefficient

Free - stream Mach number, M_0

Fig. 148. (*Concluded*)

Tsien relation [Eq. (9.32)]. This problem may be solved from Fig. 148 by following the curve for the highest incompressible pressure coefficient on the wing section to the line $M_L = 1$ and reading the corresponding stream

Mach number. Figure 148 may be used conveniently when the maximum local pressure coefficient is known experimentally at some subcritical Mach number. If the incompressible pressure coefficient is known theoretically, it is more convenient to use Fig. 149, which gives the corresponding critical Mach number directly.

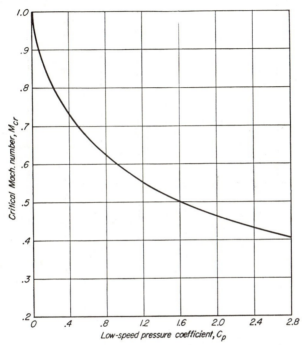

Fig. 149. Critical Mach number chart from Kármán-Tsien relation.

9.5. Flow about Wing Sections at High Speed. Consideration of the effects of compressibility on the flow over wing sections may be conveniently divided into two parts. The first part deals with the characteristics in the range of subcritical Mach numbers. The second part deals with the characteristics at supercritical Mach numbers. The critical Mach number dividing these two ranges is defined as the stream Mach number at which the local velocity of sound is just attained at any point in the field of flow.

a. Flow at Subcritical Mach Numbers. The first-order effects of compressibility on the potential flow at subcritical Mach numbers are well described by the theoretical relations of Sec. 9.4. An indication of the accuracy of the Kármán-Tsien relation is given[115] in Fig. 150b. This figure shows the experimental pressure distribution for the NACA 4412 section at a Mach number of 0.512 and a pressure distribution predicted from the experimental data obtained at a Mach number of 0.191 (Fig.

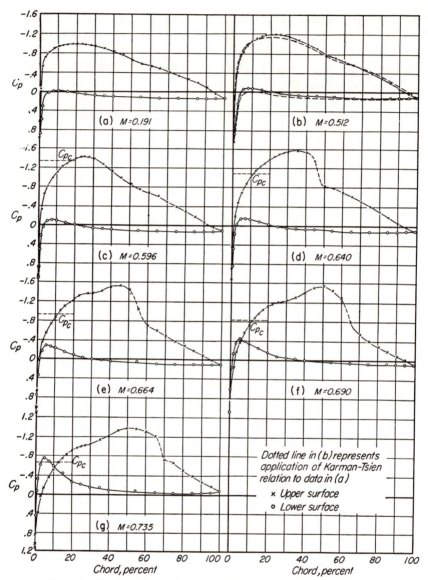

FIG. 150. Pressure distribution for the NACA 4412 airfoil. $\alpha = 1°52.5'$.

150*a*). The accuracy appears to be satisfactory for such applications. The larger area of the pressure diagram at the higher Mach number corresponds to an increased lift coefficient at the same angle of attack. The resulting change of lift-curve slope has been shown (Fig. 146) to agree fairly well with the Glauert-Prandtl rule.

The applicability of these theoretical relations is limited to cases where the effects of separation are not marked. If separation is present, the increased pressure gradients associated with higher Mach numbers tend to increase the separation and thus change the whole field of flow. Moreover, the absence of appreciable separation at low speeds is not necessarily a sufficient criterion, because the increased pressure gradients at the higher Mach numbers may cause separation in marginal cases. The theoretical relations may therefore be applied with confidence only to wing sections of normal shape at low lift coefficients.

b. Flow at Supercritical Mach Numbers. The flow at supercritical Mach numbers is partly subsonic and partly supersonic and is characterized by the presence of shocks. No theory dealing with these mixed subsonic and supersonic flows has been developed to the extent of being useful in the prediction of wing characteristics. The almost complete lack of theoretical treatment requires reliance on experimental data at supercritical Mach numbers.

For the typical case of a wing section operating at a small positive lift coefficient, the velocity of sound is reached first on the upper surface, as is shown in Fig. 150 by the pressure coefficient increasing negatively beyond the critical value C_{p_c}. Figure 150*c* shows no drastic change in the pressure distribution when the local velocity of sound is exceeded by a small amount. Many experimental data indicate that drastic changes in the forces on wing sections do not occur until the critical Mach number is exceeded by a small but appreciable margin. There is some doubt as to whether a shock necessarily occurs when the velocity of sound is locally exceeded by a small margin.[124] In any case, the losses associated with a shock from very low supersonic velocities are very small, as shown by Fig. 145. Such small losses would not be expected to produce drastic changes in the field of flow.

A shock occurs when the critical Mach number is exceeded appreciably, as shown by Fig. 150*d* to *g*. The shocks act, at least qualitatively, like a normal shock to reduce the velocity to subsonic values. The resulting rather sudden changes in the pressure distribution are shown dotted in Fig. 150. Schlieren pictures of the shocks corresponding to the diagrams of Figs. 150*f* and *g* are shown[115] in Fig. 151. The positions of the shocks as shown by the photographs correspond closely with the dotted regions of the pressure diagrams. Figure 152 presents another series of schlieren photographs showing the shocks on a NACA 23015 wing section at an

angle of attack of 3 degrees.   These pictures,[114] which were taken in the
NACA rectangular high-speed wind tunnel, show the progressive intensi-

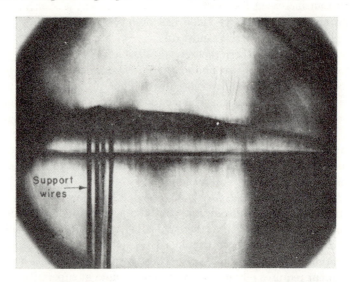

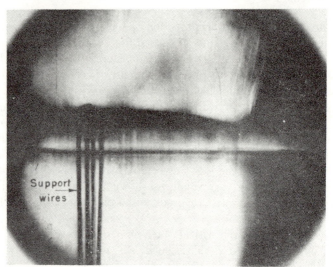

Fig. 151.   Schlieren photograph of flow for the NACA 4412 airfoil. $\alpha = 1°52.5'$.

fication and rearward movement of the shock with increasing Mach number.

The character of the flow with shock is greatly affected by the interaction of the shock and the boundary layer.   It is apparent that the shock
cannot extend to the surface through the region of low velocity in the

boundary layer. If the plane of the shock is perpendicular to the flow, the increase of static pressure through the shock will correspond to that shown in Fig. 145. The relatively high pressure downstream of the shock will tend to propagate upstream through the boundary layer. The pres-

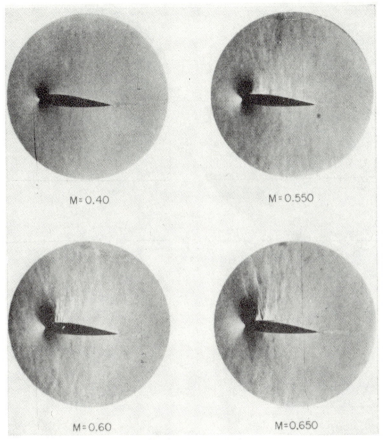

Fig. 152. Schlieren photographs of flow. NACA 23015 airfoil. NACA rectangular high-speed wind tunnel. Airfoil chord, 2 inches. Angle of attack, 3 degrees.

sure distribution measured at the surface cannot, therefore, be expected to show a discontinuity. Figure 150 shows that the sharp pressure rise is, in fact, shown as a steep gradient at the surface. Such propagation of the pressure along the boundary layer necessarily leads to pressure gradients across the boundary layer in the vicinity of the shock. This situation is quite different from the usual boundary-layer conditions at low Mach numbers.

The large adverse pressure gradients associated with the shock always produce a large increase of boundary-layer thickness and frequently cause separation, as shown[114] in Figs. 152 and 153. This separation is an important or even predominant factor in causing the large force changes occurring at supercritical speeds. The relative importance of the shock losses and the effects of separation on the drag of a typical wing section

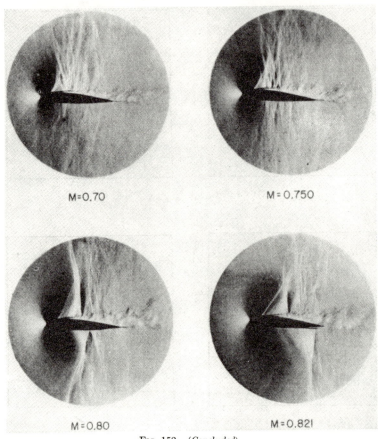

Fig. 152. *(Concluded)*

are shown[114] by Fig. 154. This figure shows the total pressure loss across the wake of the section. These losses are intimately associated with the drag, and, to the first order, the drag coefficient is proportional to the area under the curve. Intense shock has not occurred at the lower Mach number, and the diagram has the typical shape obtained at low speeds. Intense shock has occurred at the higher Mach number, and the drag coefficient has obviously increased. This increase may be considered to be of two parts. The very large losses of pressure in the center of the diagram are similar in shape to the lower speed diagram and are attributed to the usual type of loss associated with skin friction and separation. The smaller pressure losses extending far out into the stream are attributed to the losses across the shock. It is apparent that the drag rise associated with the usual type of loss is as important as that directly associated with the losses across the shocks.

The pressure rise measured[114] on the surface of the wing section in the vicinity of the shock is smaller than that corresponding to a normal shock with the losses of total pressure indicated by the wake surveys. The pressure rise corresponding to a normal shock may also be obtained by means of Fig. 145 from the local Mach number ahead of the shock as indicated by pressure measurements on the surface. In this case, also, the computed pressure rise is greater than that shown by the surface pressures. The complete explanation of this discrepancy is not known, but two factors could tend to reduce the pressure rise on the surface below that corresponding to a normal shock.

The first factor is the pressure gradient through the boundary layer previously discussed. The importance of this factor may be inferred from Figs. 152 and 153. The lines in the schlieren photographs correspond to density gradients. Dark lines are shown in the vicinity of

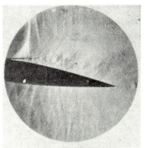

M = 0.691

Fig. 153. Schlieren photograph of separated flow for rear portion of NACA 23015 airfoil. NACA rectangular high-speed wind tunnel. Airfoil chord, 5 inches. Angle of attack, 6 degrees.

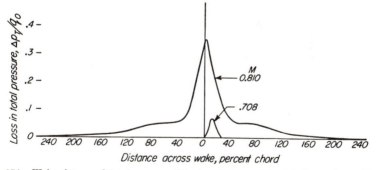

Fig. 154. Wake shape and total pressure defect as influenced by Mach number. NACA 0012 airfoil, $\alpha = 0$ degrees; NACA rectangular high-speed wind tunnel.

the shock making a small angle with the surface. Although these density gradients may be partly associated with temperature gradients, the sharpness and intensity of these lines are thought to indicate the presence of appreciable pressure gradients.

The second factor is the possible reaction of the boundary layer on the shock. The thickening or separation of the boundary layer tends to produce oblique shocks. Such shocks would produce some pressure rise ahead of the normal shock and would accordingly reduce the intensity of the normal shock, especially close to the surface. The extent to which the normal shock is locally softened in this manner is uncertain, but Figs. 152 and 153 suggest that this mechanism may be of considerable importance.

Investigations made by the NACA, by Liepman,[66] and by Ackeret[4] and his associates emphasize the complexity of the interaction of the shock and the boundary layer. These investigations indicate important differences between the shock phenomena for laminar and for turbulent boundary layers. Although exceptions have been noted, laminar boundary layers

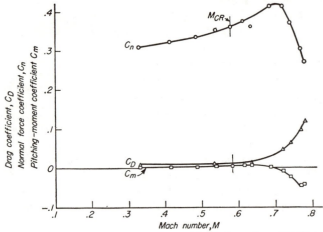

Fig. 155. Force and moment coefficient variation with Mach number. NACA 23015 airfoil, $\alpha = 0$ degrees; NACA rectangular high-speed wind tunnel.

tend to produce multiple shocks of lower intensity, or "lambda-type" shocks, while those associated with turbulent boundary layers tend to resemble intense normal shocks more closely. The understanding of these phenomena awaits further theoretical and experimental investigations.

The typical changes of forces experienced at supercritical speeds by a wing section of moderate thickness are shown[114] in Fig. 155. The Mach number for force break is seen to be appreciably higher than the critical Mach number. At Mach numbers higher than that for force break, the lift coefficient decreases, the drag coefficient increases, and the moment coefficient usually increases negatively.

9.6. Experimental Wing Characteristics at High Speeds. The existing three-dimensional wing theory, which is the basis for the concept of section characteristics, is not valid for supercritical speeds where part of the flow is supersonic. Consequently section data cannot be applied quantitatively to the prediction of the characteristics of wings at such speeds. Moreover, present trends for wings designed for efficient operation at supercritical speeds are toward low aspect ratios and large amounts of sweep. As pointed out previously, section data are of doubtful quantitative significance for such plan forms at any speed. Under these circumstances, the detailed consideration of section characteristics as design data is not warranted.

Nevertheless, the characteristics of wings at supercritical speeds depend to some extent on the wing sections employed. The variation of the wing characteristics with wing section appears to be qualitatively similar to the variation of the wing-section characteristics if the wing plan form is such as to permit any semblance to two-dimensional flow. Consequently, the purpose of this section is to present typical data showing qualitatively the major effects of variation of the wing profile at high subsonic Mach numbers.

a. Lift Characteristics. One of the first and most severe effects of compressibility encountered in flight was the tendency of many airplanes to "tuck under" in high-speed dives. This "tucking under" tendency consisted of a large negative shift in the angle of trim together with a large increase of stability that resisted the efforts of the pilot to trim at the desired positive lift coefficients required for recovery from the dive. The resulting large elevator motions required to recover from the dive, or even to prevent the dive from becoming steeper, corresponded to excessive control forces, thus leading to the impression that the "stick became frozen." These changes in longitudinal stability limited the maximum safe flying speed for many airplanes.

These effects are intimately associated with the changes in lift characteristics of wing sections at Mach numbers above that for force break. The relatively thick cambered wing sections commonly used during the Second World War experienced a positive shift of the angle of zero lift and a reduction of lift-curve slope. Although many effects associated with other characteristics of the airplane are present, the change in angle of zero lift directly affects the angle of trim and the reduction of lift-curve slope directly increases the longitudinal stability.

The shift in the angle of zero lift is associated with the camber of the wing section. A symmetrical section shows, of course, no change in the angle of zero lift[116] (Fig. 156). The rather large positive shift in the angle of zero lift shown[116] in Fig. 157 for the NACA 2409-34 section is typical for cambered sections without reflex. In this case, the change amounts to about 2 degrees for a change of Mach number from 0.80 to only 0.83. The effect of increasing the camber is shown[116] by Fig. 158. Doubling the amount of camber causes a similar shift to occur at a lower Mach number.

The effects of thickness on the lift characteristics of symmetrical sections are shown[30] in Figs. 159 to 161 for the NACA 0006-34, 0008-34, and 0012-34 sections. The 6 per cent thick section shows a large increase of lift coefficient at a given angle of attack with increasing Mach number at values below about 0.85. At Mach numbers above the force break, the lift coefficient decreases rapidly, but it never goes below the low-speed values, at least at Mach numbers up to 0.95. The 12 per cent thick section, however, fails to show much increase of lift coefficient at subcritical Mach numbers, but it shows a large decrease of lift coefficient at Mach numbers just above the force break. At a Mach number of 0.85, the lift coefficient,

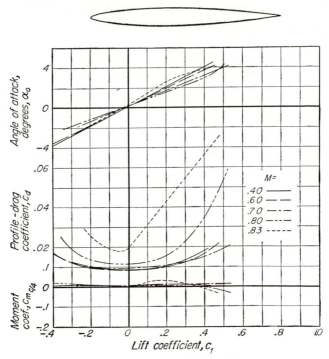

FIG. 156. Aerodynamic characteristics of the NACA 0009-34 airfoil.

and accordingly the lift-curve slope, are only about two-thirds of the low-speed value. At higher Mach numbers, the lift coefficients again increase. The effects of Mach number on the lift characteristics of the 8 per cent thick section are intermediate. These data show that the thin sections experience force breaks at higher Mach numbers than the thicker sections and that the lift force breaks are much less severe for the thin than for the thicker sections.

The effects of thickness for a series of cambered sections are shown[30] in Figs. 162 to 165. These data indicate to an even greater degree the adverse effect of thickness on the character of the lift force break. Comparison of these data with those for the symmetrical sections (Figs. 159 to 161) shows the adverse effects of camber on the Mach number for lift force break and on the shift in the angle of zero lift.

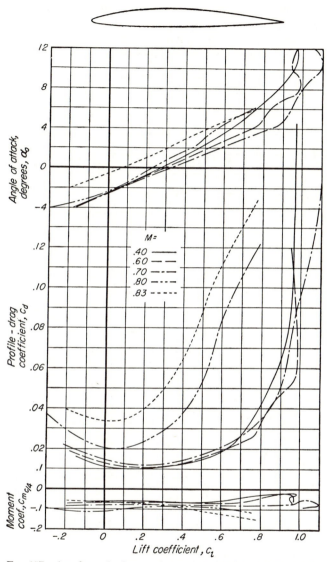

FIG. 157. Aerodynamic characteristics of the NACA 2409-34 airfoil.

Comparatively few data are available to show the effect of thickness distribution on the characteristics above the force break. The trends indicated by Stack and von Doenhoff[116] cannot be considered representative for properly designed families of wing sections because of the arbitrary

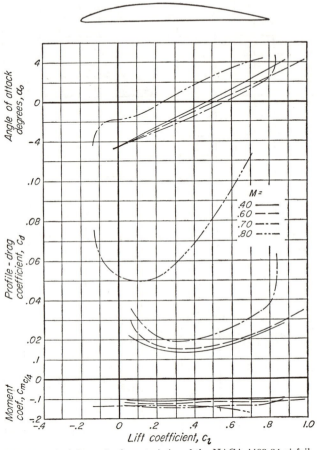

FIG. 158. Aerodynamic characteristics of the NACA 4409-34 airfoil.

geometrical manner in which the sections were varied. If the thickness distributions are varied in such a manner as to avoid peaks in the low-speed pressure distributions, as in the case of the NACA 6-series sections, the indications are that the effect of thickness distribution is minor compared with the effects of thickness ratio and camber. There is some indication[116] that the trailing-edge angle should be kept small to avoid adverse effects on the lift characteristics similar to those for thick sections.

Figures 166 to 168 show the[37b] lift and drag characteristics of the NACA 66-210 wing section. The most outstanding feature of these data as com-

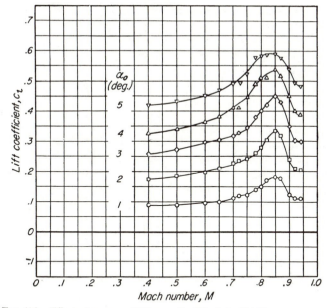

Fig. 159. Effect of compressibility on the lift of the NACA 0006-34 airfoil.

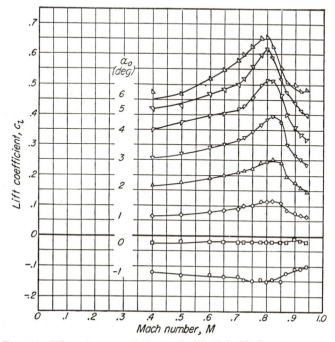

Fig. 160. Effect of compressibility on the lift of the NACA 0008-34 airfoil.

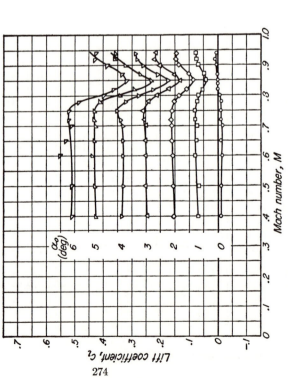

FIG. 162. Effect of compressibility on the lift of the NACA 2306 airfoil.

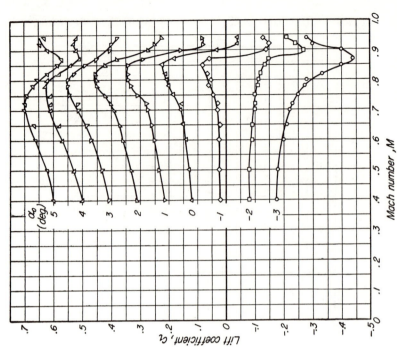

FIG. 161. Effect of compressibility on the lift of the NACA 0012-34 airfoil.

274

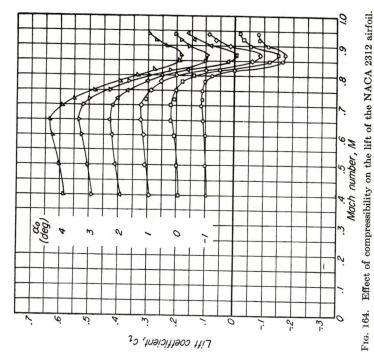

Fig. 164. Effect of compressibility on the lift of the NACA 2312 airfoil.

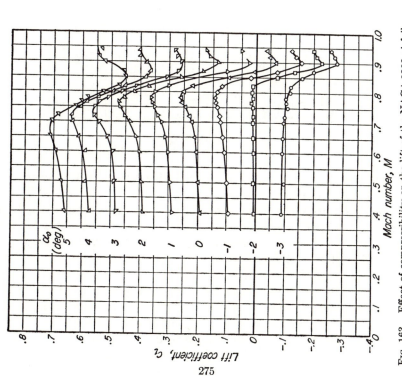

Fig. 163. Effect of compressibility on the lift of the NACA 2309 airfoil.

275

pared with those for the NACA four-digit series sections is the ability of
the 6-series section to carry large lifts at moderately high angles of attack
and high Mach numbers. The lift-curve slope for the NACA 66-210
section at a Mach number of 0.75 is high and substantially constant up to
a lift coefficient of about 0.8 (Fig. 168). At the same Mach number, the
NACA 2309 section (Fig. 163) shows a reduced lift-curve slope at a lift

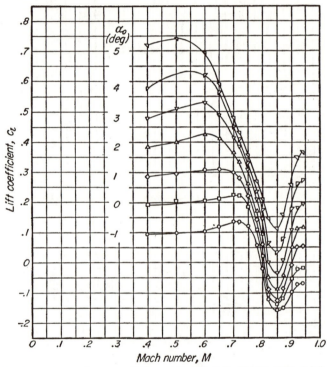

Fig. 165. Effect of compressibility on the lift of the NACA 2315 airfoil.

coefficient of only about 0.55 and almost no increase of lift for a change of
angle of attack from 4 to 5 degrees.

Figure 169 shows[37b] the predicted critical Mach numbers and the Mach
numbers for force divergence as functions of the lift coefficient for the
NACA 66-210 section. The outstanding characteristic of these data is the
wide range of lift coefficients over which high Mach numbers for force
divergence are realized as compared with the range for high critical Mach
numbers. The predicted critical Mach numbers are lower than those for
force divergence but approximate the latter values over the range of lift
coefficients where the speed of sound is first reached near the location of
minimum pressure at the design lift coefficient. It is apparent that critical
Mach numbers predicted on the basis of the attainment of the velocity of

sound near the leading edge are useless as an indication of the force characteristics of NACA 6-series sections.

The general effects of Mach number on the maximum lift coefficients of wing sections may be inferred from the data[79] of Fig. 170. These data were

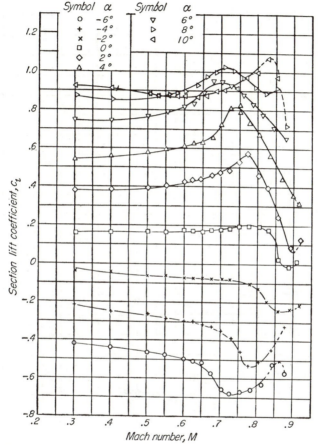

Fig. 166. The variation of section lift coefficient with Mach number at various angles of attack for the NACA 66-210 airfoil.

obtained from tests in the NACA Langley 16-foot high-speed wind tunnel of a wing having an aspect ratio of 6 and having NACA 23016 sections at the root and NACA 23009 sections at the tip. At Mach numbers above 0.30, the maximum lift coefficient decreases rapidly with increasing Mach number up to values of about 0.55 where the maximum lift coefficient is approximately 1.0. At higher Mach numbers, the actual maximum lift coefficient decreases more slowly; but the angle of attack, or lift coefficient, at which the lift curve departs radically from its normal slope continues to decrease to the limits of the tests. The resulting variation of maximum lift coefficient with Mach number is shown[79] in Fig. 171.

The increase of maximum lift coefficient with Mach number at values below 0.30 is associated with the variation of Reynolds number for these

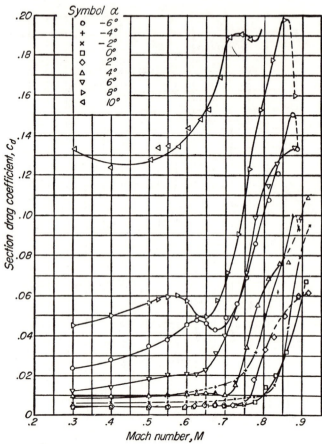

FIG. 167. The variation of section drag coefficient with Mach number at various angles of attack for the NACA 66-210 airfoil.

tests. This effect was studied more extensively in the NACA 19-foot-pressure wind tunnel where the effects of Mach and Reynolds numbers could be separated[33] for low Mach numbers. These data indicated, at least qualitatively, that the peak maximum lift coefficient is determined by the critical Mach number at maximum lift which, of course, is attained at relatively low free-stream Mach numbers. At subcritical Mach numbers, the maximum lift coefficient is primarily a function of the Reynolds number as at very low Mach numbers, although the effect of Mach number is

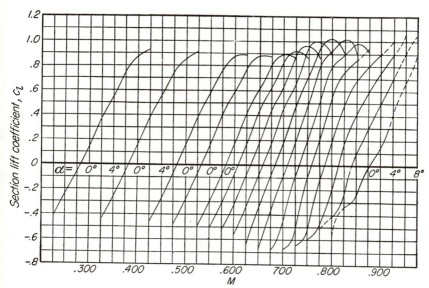

FIG. 168. The variation of section lift coefficient with angle of attack at various Mach numbers for the NACA 66-210 airfoil.

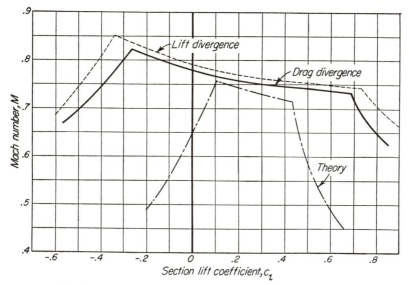

FIG. 169. Critical Mach numbers for the NACA 66-210 airfoil.

not negligible. At supercritical Mach numbers, the value of the maximum lift coefficient decreases rapidly with increasing speed, and the effect of Reynolds number is secondary. For the wing tested, the critical Mach

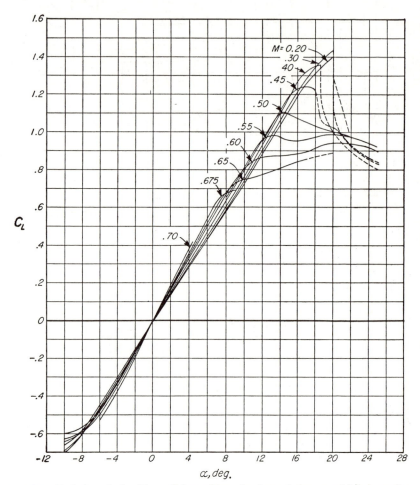

Fig. 170. Variation of wing lift coefficient with angle of attack for several Mach numbers.

number was of the order of 0.25 to 0.30 for the flaps-retracted condition and of 0.17 to 0.20 with 0.20c split flaps deflected 60 degrees.

It should be noted that the maximum usable lift coefficient for an airplane at high Mach numbers may not be the same as the actual maximum lift coefficient. Referring to Fig. 170, the airplane may be unable to fly on the relatively flat portions of the lift curves above the force breaks, because of the serious buffeting associated with the partly separated flow and, perhaps also, because of stability and control difficulties.

At supercritical speeds, changing the camber near the trailing edge, as by small deflections of a flap or aileron, cannot affect the flow over the front part of the section because pressures cannot propagate upstream through the supersonic region. Consequently, the flap effectiveness must be greatly reduced at supercritical Mach numbers. This effect is clearly shown[114] by Fig. 172. This loss of flap effectiveness is related to the positive

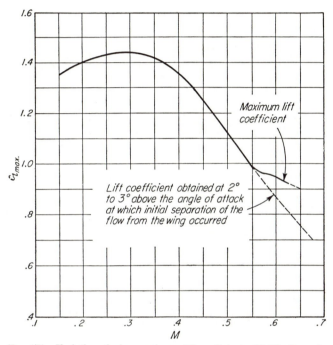

FIG. 171. Variation of wing maximum lift coefficient with Mach number.

shift of the angle of zero lift for cambered sections. The rearward portion of a cambered section may be considered to be a special case of a deflected flap. The loss of flap effectiveness at the force break leaves the forward portion of the section at an effectively lower angle of attack, and the resulting change in lift coefficient corresponds to a positive shift of the angle of zero lift. Conversely, a strongly reflexed section would show the opposite effect. It follows that the change in lift characteristics of cambered sections may be reduced by incorporation of reflex or of an upwardly deflected flap.

b. Drag Characteristics. The effect of thickness ratio on the drag characteristics of symmetrical wing sections is illustrated[30] by Figs. 173 to 175. These data show the typical rapid increase of drag coefficient at high Mach numbers. This rapid rise (force break) occurs at smaller Mach numbers as the angle of attack and thickness are increased. Similar data

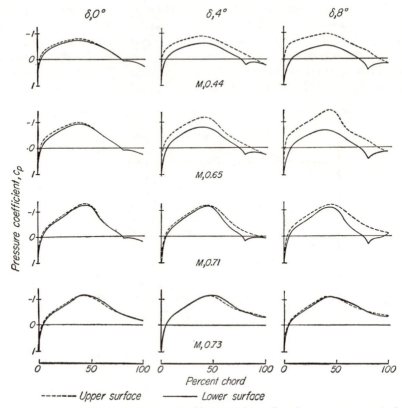

FIG. 172. Variation of aileron action with Mach number. Two-dimensional pressure distribution results. Symmetrical airfoil, 19 per cent thick, $\alpha = 0$ degrees; NACA rectangular high-speed wind tunnel.

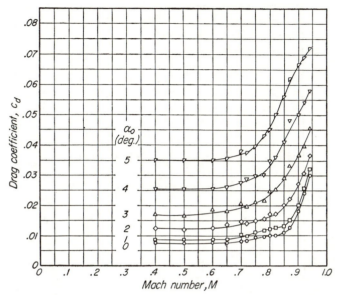

FIG. 173. Effect of compressibility on the drag of the NACA 0006-34 airfoil.

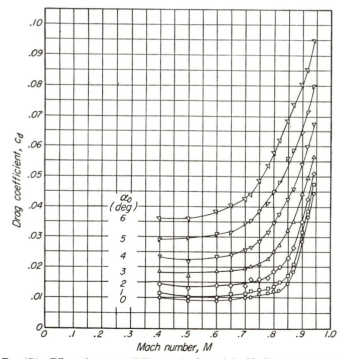

FIG. 174. Effect of compressibility on the drag of the NACA 0008-34 airfoil.

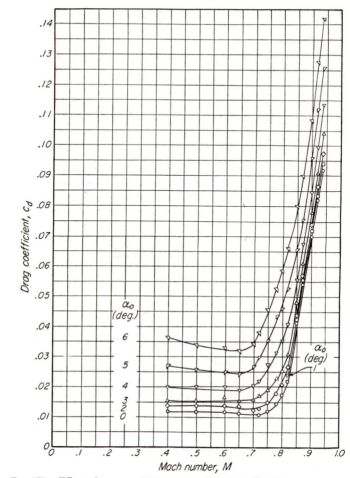

FIG. 175.   Effect of compressibility on the drag of the NACA 0012-34 airfoil.

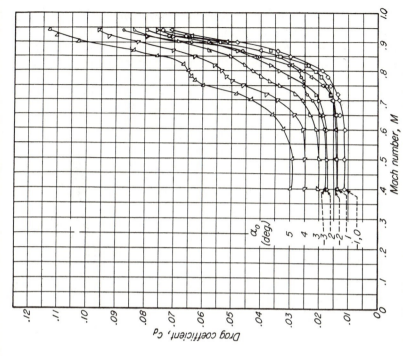

Fig. 177. Effect of compressibility on the drag of the NACA 2309 airfoil.

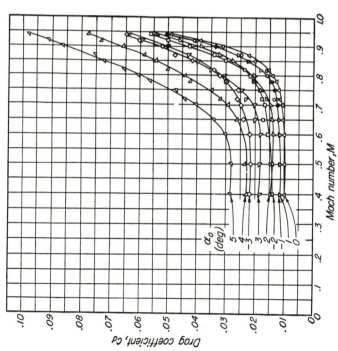

Fig. 176. Effect of compressibility on the drag of the NACA 2306 airfoil.

for the cambered sections, NACA 2306, 2309, 2312, and 2315, are shown[30] in Figs. 176 to 179. The effect of camber is to reduce the Mach number at which the drag force break occurs as compared with the data of Figs. 173 to 175. Although the thickness distribution is not the same for these

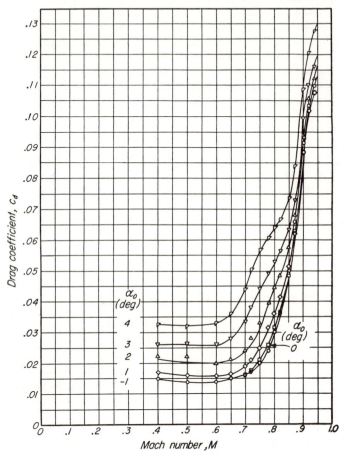

Fɪɢ. 178.   Effect of compressibility on the drag of the NACA 2312 airfoil.

symmetrical and cambered sections, the differences attributable to this fact are very small, at least for the lower thickness ratios.

Comparisons of the predicted critical Mach numbers and the Mach numbers for drag force break are presented[37b] in Fig. 169 for the NACA 66-210 wing section. For the range of lift coefficients corresponding to high predicted critical speeds, the Mach number for drag force break is about 0.01 to 0.03 greater than the critical. The Mach number for drag force break is intermediate between that for lift force break and the critical value. The slope of the boundary for drag force break is similar to that for the critical Mach number, but high Mach numbers without force break

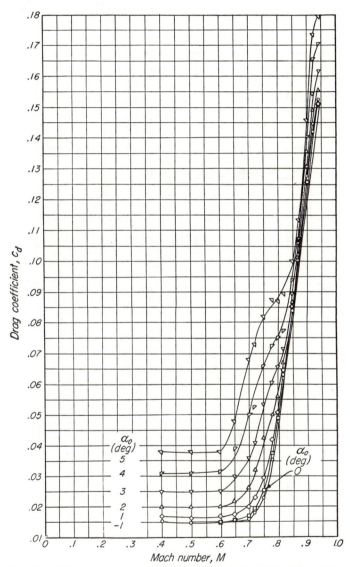

FIG. 179. Effect of compressibility on the drag of the NACA 2315 airfoil.

may be realized over a much wider range of lift coefficient than is indicated by the theoretically predicted critical speed.

c. Moment Characteristics. Figures 157 and 158 illustrate[116] the negative shift of the moment coefficient with increasing Mach number that is typical for cambered wing sections. In addition to this shift, there is a tendency at the highest Mach numbers shown for the moment coefficient to increase negatively with increasing lift coefficient as is shown[37b] more

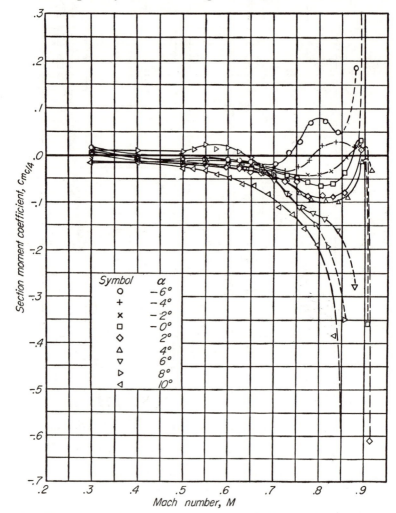

Fig. 180. The variation of section moment coefficient with Mach number at various angles of attack for the NACA 66-210 airfoil.

clearly for the NACA 66-210 section in Fig. 180. This rearward shift of the aerodynamic center at high Mach numbers is stabilizing, and the more negative values of the moment coefficient tend to make the airplane trim at lower angles of attack. These moment changes thus add to the previously discussed effects of the lift coefficient in producing the "tucking under" tendency of airplanes with thick cambered wing sections.

The moment characteristics of the NACA 0006-34, 0008-34, 0012-34, 2306, 2309, 2312, and 2315 wing sections are shown[30] in Figs. 181 to 187. These data indicate that the changes in moment characteristics with Mach

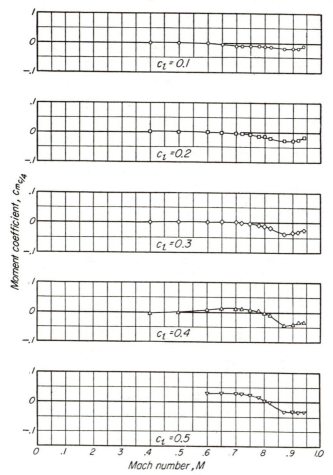

FIG. 181.   Effect of compressibility on the pitching moment of the NACA 0006-34 airfoil.

number are minimized for symmetrical sections, although a rearward shift of the aerodynamic center at high Mach numbers persists.

9.7. Wings for High-speed Applications.   As discussed in the previous section, all wing sections experience undesirable variations in their characteristics at high subsonic Mach numbers.   Although these undesirable variations can be minimized by the use of thin symmetrical sections, it does not appear to be possible to eliminate them in two-dimensional flow.

In three-dimensional flow, however, it is obvious that the shocks cannot extend to the wing tips without modification.   The presence of the shock at

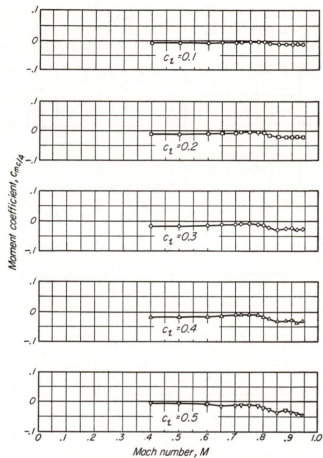

Fig. 182. Effect of compressibility on the pitching moment of the NACA 0008-34 airfoil.

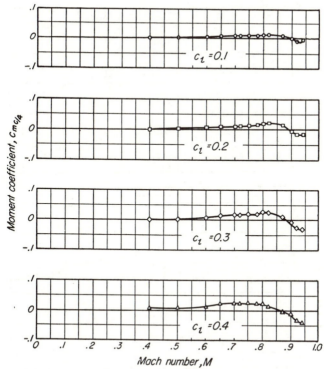

Fɪɢ. 183.   Effect of compressibility on the pitching moment of the NACA 0012-34 airfoil.

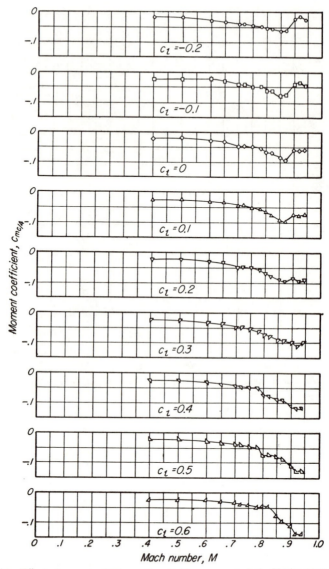

FIG. 184. Effect of compressibility on the pitching moment of the NACA 2306 airfoil.

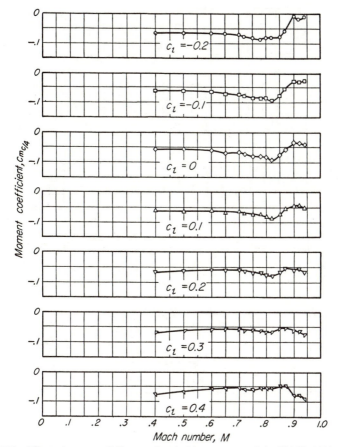

FIG. 185. Effect of compressibility on the pitching moment of the NACA 2309 airfoil.

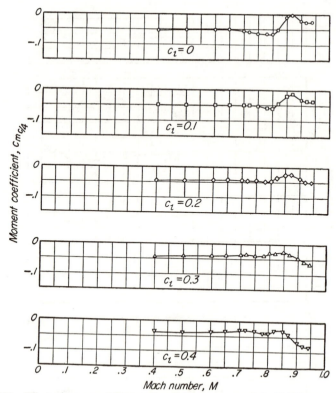

FIG. 186. Effect of compressibility on the pitching moment of the NACA 2312 airfoil.

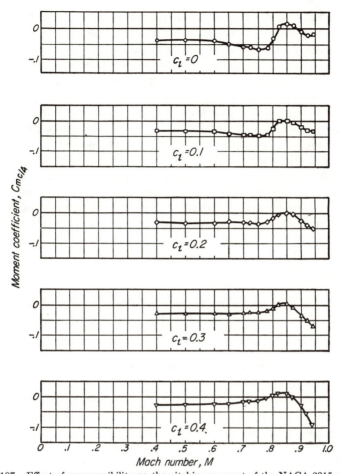

FIG. 187. Effect of compressibility on the pitching moment of the NACA 2315 airfoil.

the wing tip with the accompanying sudden pressure rise would induce a three-dimensional flow that would tend to reduce the severity of the shock. The existence of such "tip relief" became apparent in early tests of propellers at high tip speeds. The data showed that losses of efficiency did not occur until tip Mach numbers were reached that were well in excess of those at which the sections used at the tips suffered large increases of drag in two-dimensional flow. Corresponding advantages for wings may be expected if the aspect ratio is sufficiently low. Experimental data have confirmed the advantages of low aspect ratio at high Mach numbers.

The theoretical approach for wings of very low aspect ratio is entirely different from that for normal wings of high aspect ratio, and the concept of wing-section characteristics is not applicable. Jones[51] showed that, for wings with aspect ratios approaching zero, the lift depends on the angle of attack and on the positive rate of increase of span in the direction of the air flow, or

$$c_l = \pi\alpha \frac{db}{dx}$$

where c_l = local lift coefficient of section perpendicular to direction of air flow

α = angle of attack

b = span

x = distance in the direction of air flow

db/dx is positive

This simplified theory indicates that $c_l = 0$ for those portions of the wing where db/dx is zero or negative. This theory shows that the center of lift for a triangular wing with the apex forward is at the center of area. The theory indicates that, for the particular case of a slender triangular plan form, the lift coefficient and center of pressure are independent of Mach number at both subsonic and supersonic speeds so long as the wing lies well within the Mach cone. The applicability of the theory at a Mach number of 1.75 is shown by Fig. 188. Low-aspect-ratio wings of triangular plan form thus appear to provide one solution to the stability difficulties in the transonic speed range.

Another solution to the problems of the transonic speed range is the use of large amounts of sweep. The manner in which sweep is effective is clearly seen from consideration of a wing of infinite span with sideslip (Fig. 189). It is obvious that the spanwise component of velocity V_s will produce no forces on the wing, neglecting viscosity, and that the force on the wing will be determined by the normal component of velocity V_n. The resulting velocity V may be transonic or even supersonic, while the normal component V_n is less than the critical speed if the angle of sideslip, or sweep, β is sufficiently large. This effect of sweep in avoiding compres-

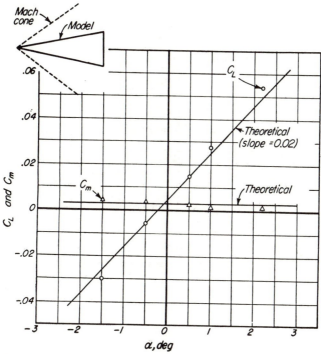

Fig. 188. Test of triangular airfoil in Langley model supersonic tunnel. Mach number, 1.75; Reynolds number, 1,600,000.

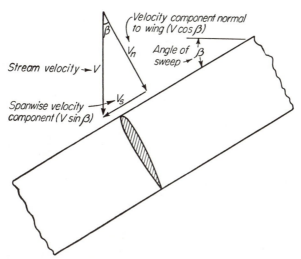

Fig. 189. Velocity components on a swept wing.

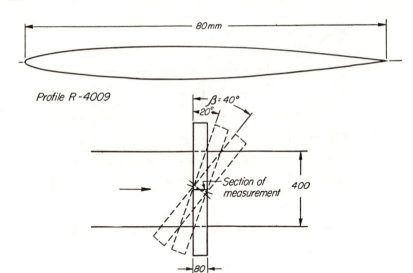

Profile R-4009

FIG. 190. Wing placed obliquely across a two-dimensional wind tunnel.

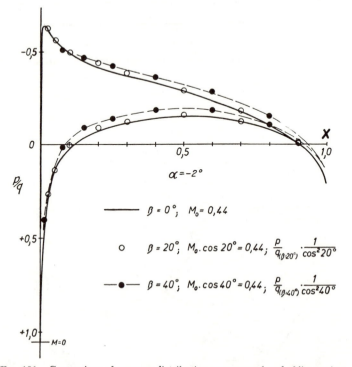

FIG. 191. Comparison of pressure distribution over normal and oblique wing.

sibility troubles was independently discovered by German investigators and by Jones[52] during the Second World War.

Application of this principle results in wings with large amounts of sweepback or sweepforward. Such plan forms obviously violate the simple two-dimensional considerations at the root and tip even if the wing has no taper. These violations, however, do not appear to be sufficient to destroy the value of sweep if the wing is of reasonably high aspect ratio, as indicated by tests[67] of a wing placed obliquely across a two-dimensional wing tunnel (Fig. 190). Figure 191 shows that the pressure distribution over the center section of the oblique wing is reasonably close to that expected from the simple theory, despite the fact that the tunnel walls violate the assumed end conditions. Experimental studies have shown that swept wings of all useful aspect ratios are effective in postponing and alleviating compressibility troubles. Consequently, the theory of swept wings of arbitrary plan forms is being intensively investigated.[18, 53, 90, 117]

It thus appears that the wings of the high-speed airplanes of the future may be quite different from those of comparatively slow subsonic airplanes. These highly swept low-aspect-ratio wings, however, have serious disadvantages for the low-speed flight necessary for landing and take-off. Some of these disadvantages are low lift-curve slopes, high angles of stall, and poor stability characteristics. These disadvantages seriously compromise the design of airplanes using such plan forms and may force the use of more conventional plan forms with very thin supersonic sections for some applications. In any case, these disadvantages at slow speeds are expected to dictate the use of conventional plan forms for airplanes designed for such speeds that it is possible to avoid serious compressibility difficulties by the use of thin wing sections.

REFERENCES

1. ABBOTT, FRANK T., JR., and TURNER, HAROLD R., JR.: The Effects of Roughness at High Reynolds Numbers on the Lift and Drag Characteristics of Three Thick Airfoils. NACA ACR No. L4H21, 1944 (Wartime Rept. No. L-46).
2. ABBOTT, IRA H., and GREENBERG, HARRY: Tests in the Variable-Density Wind Tunnel of the NACA 23012 Airfoil with Plain and Split Flaps. NACA Rept. No. 661, 1939.
3. ABBOTT, IRA H., VON DOENHOFF, ALBERT E., and STIVERS, LOUIS S.: Summary of Airfoil Data. NACA Rept. No. 824, 1945.
4. ACKERET, J., FELDMAN, F., and ROTT, N.: Investigations of Compression Shocks and Boundary Layers in Gases Moving at High Speed. NACA TM No. 1113, 1947.
5. AIKEN, WILLIAM, JR.: Standard Nomenclature for Airspeeds with Tables and Charts for Use in Calculation of Airspeed. NACA TN No. 1120, 1946.
6. ALLEN, H. JULIAN: Calculation of the Chordwise Load Distribution over Airfoil Sections with Plain, Split or Serially-hinged Trailing Edge Flaps. NACA Rept. No. 634, 1938.
7. ALLEN, H. JULIAN: A Simplified Method for the Calculation of Airfoil Pressure Distribution. NACA TN No. 708, 1939.
8. ALLEN, H. JULIAN: General Theory of Airfoil Sections Having Arbitrary Shape or Pressure Distribution. NACA ACR No. 3G29, 1943.
9. ALLEN, H. JULIAN: Notes on the Effect of Surface Distortion on the Drag and Critical Mach Number of Airfoils. NACA ACR No. 3I29, 1943.
10. ANDERSON, RAYMOND F.: Determination of the Characteristics of Tapered Wings. NACA Rept. No. 572, 1936.
12. BAIRSTOW, L.: Skin Friction. *J. Roy. Aeronaut. Soc.*, January, 1925, pp. 3–23.
13. BAMBER. MILLARD J.: Wind-tunnel Tests on Airfoil Boundary Layer Control Using a Backward-opening Slot. NACA Rept. No. 385, 1931.
14. BLASIUS, H.: Grenzschichten in Flüssigkeiten mit kleiner Reibung. *Z. Math. Physik*, vol. 56, pp. 1–37, 1908.
15. BLASIUS, H.: *Ver. deut. Ing. Forschungsheft* 131, 1911.
16. BOSHAR, JOHN: The Determination of Span Load Distribution at High Speeds by the Use of High-speed Wind Tunnel Section Data. NACA ACR No. 4B22, 1944 (Wartime Rept. No. L-436).
17. BRASLOW, ALBERT L.: Investigation of Effects of Various Camouflage Paints and Painting Procedures on the Drag Characteristics of an NACA $65_{(421)}$-420 $a = 1.0$ Airfoil Section. NACA CB No. L4G17, 1944 (Wartime Rept. No. L-141).
18. BROWN, CLINTON E.: Theoretical Lift and Drag of Thin Triangular Wings at Supersonic Speeds. NACA TN No. 1183, 1946.
19. BURGERS, J. M.: The Motion of a Fluid in the Boundary Layer along a Plane Smooth Surface. *Proc. First Intern. Congr. Appl. Mech.*, Delft, 1924, pp. 113–128.
20. CAHILL, JONES F.: Two-dimensional Wind-tunnel Investigation of Four Types of High-lift Flap on an NACA 65-210 Airfoil Section. NACA TN No. 1191, 1947.
21. CAHILL, JONES F., and RACISZ, STANLEY: Wind-tunnel Development of Optimum Double-slotted-flap Configurations for Seven Thin NACA Airfoil Sections. NACA RM No. L7B17, 1947, also TN No. 1545.

22. CAHILL, JONES F.: Aerodynamic Data for a Wing Section of the Republic XF–12 Airplane Equipped with a Double Slotted Flap. NACA MR No. L6A08a, 1946 (Wartime Rept. No. L–544).
23. CAHILL, JONES F.: Summary of Section Data on Trailing-edge High-lift Devices. NACA RM No. L8D09, 1948.
25. CHAPLYGIN, SERGEI: Gas Jets. NACA TM No. 1063, 1944 (from Scientific Memoirs, Moscow University, 1902, pp. 1–121).
26. CHARTERS, ALEX C.: Transition between Laminar and Turbulent Flow by Transverse Contamination. NACA TN No. 891, 1943.
27. DRYDEN, H. L., and KUETHE, A. M.: Effect of Turbulence in Wind Tunnel Measurements. NACA Rept. No. 342, 1929.
28. DURAND, W. F.: Aerodynamic Theory, vol. 3, Div. G. "The Mechanics of Viscous Fluids" by L. Prandtl. Verlag Julius Springer, Berlin, or Durand Reprinting Comm., Cal. Inst. of Tech.
29. DURAND, W.F.: Aerodynamic Theory, vol. 3, Div. H. "The Mechanics of Compressible Fluids" by G. I. Taylor and J. W. Maccoll. Verlag Julius Springer, Berlin, or Durand Reprinting Comm., Cal. Inst. of Tech.
30. FERRI, ANTONIO: Completed Tabulation in the United States of Tests of 24 Airfoils at High Mach Numbers. (Derived from interrupted work at Guidonia Italy, in the 1.31- by 1.74-foot High-speed Tunnel) NACA ACR No. L5E21, 1945 (Wartime Rept. No. L–143).
31. FISCHEL, JACK, and RIEBE, JOHN M.: Wind-tunnel Investigation of a NACA 23021 Airfoil with a 0.32-airfoil-chord Double Slotted Flap. NACA ARR No. L4J05, 1944 (Wartime Rept. No. L–7).
32. FULLMER, FELICIEN F., JR.: Two-dimensional Wind-tunnel Investigation of the NACA 64₁-012 Airfoil Equipped with Two Types of Leading-edge Flap. NACA TN No. 1277, 1947.
33. FURLONG, G. CHESTER, and FITZPATRICK, JAMES E.: Effects of Mach Number and Reynolds Number on the Maximum Lift Coefficient of a Wing of NACA 230-series Airfoil Sections. NACA MR No. L6F04, 1946, also TN No. 1299.
34. GARRICK, I. E., and KAPLAN, CARL: On the Flow of a Compressible Fluid by the Hodograph Method. I. Unification and Extension of Present-day Results. NACA ACR No. L4C24, 1944, also Rept. No. 789.
35. GLAUERT, H.: Theoretical Relationships for an Aerofoil with Hinged Flap. R. & M. No. 1095, British ARC, 1927.
36. GLAUERT, H.: The Effect of Compressibility on the Lift of an Aerofoil. R. & M. No. 1135, British ARC, 1927.
37. GLAUERT, H.: "The Elements of Aerofoil and Airscrew Theory," Cambridge University Press, London, 1926.
37a. GOLDSTEIN, SIDNEY: Low-drag and Suction Airfoils, Eleventh Wright Brothers Lecture, J. Inst. Aeronaut. Sci., vol. 15, No. 4, pp. 189–214, April, 1948.
37b. GRAHAM, DONALD J.: High-speed Tests of an Airfoil Section Cambered to Have Critical Mach Numbers Higher than Those Attainable with a Uniform-load Mean Line. NACA TN No. 1396, 1947.
38. GRUSCHWITZ, E.: Die turbulente Reibungsschicht in ebener Strömung bei Druckabfall und Druckanstieg, Ing. Archiv, Bd. 11, Heft 3, pp. 321–346, September, 1931.
39. HARRIS, THOMAS A.: Wind-tunnel Investigation of an NACA Airfoil with Two Arrangements of a Wide-chord Slotted Flap. NACA TN No. 715, 1939.
40. HARRIS, THOMAS A., and RECANT, ISIDORE G.: Wind-tunnel Investigation of NACA 23012, 23021, and 23030 Airfoils Equipped with 40-percent-chord Double Slotted Flaps. NACA Rept. No. 723, 1941.

41. HARRIS, THOMAS A., and LOWRY, JOHN G.: Pressure Distribution over an NACA 23012 Airfoil with a Fixed Slot and a Slotted Flap. NACA Rept. No. 732, 1942.

42. HOOD, MANLEY J.: The Effects of Some Common Surface Irregularities on Wing Drag. NACA TN No. 695, 1939.

43. HOOD, MANLEY J., and GAYDOS, M. EDWARD: Effects of Propellers and of Vibration on the Extent of Laminar Flow on the NACA 27-212 Airfoil. NACA ACR, October 1939, (Wartime Rept. No. L–784).

44. JACOBS, EASTMAN N.: Methods Employed in America for the Experimental Investigation of Aerodynamic Phenomena at High Speeds. NACA Misc. Paper No. 42, 1936. Paper presented at Volta meeting in Italy, Sept. 30 to Oct. 6, 1935.

45. JACOBS, EASTMAN N.: Preliminary Report on Laminar Flow Airfoils and New Methods Adopted for Airfoil and Boundary-layer Investigations. NACA ACR, June, 1939, (Wartime Rept. No. L–345).

45a. JACOBS, EASTMAN N., and ABBOTT, IRA H.: Airfoil Section Data Obtained in the NACA Variable-density Tunnel as Affected by Support Interference and Other Corrections. NACA Rept. No. 669, 1939.

46. JACOBS, EASTMAN N., and PINKERTON, ROBERT M.: Tests in the Variable-density Wind Tunnel of Related Airfoils Having the Maximum Camber Unusually Far Forward. NACA Rept. No. 537, 1935.

47. JACOBS, EASTMAN N., and PINKERTON, ROBERT M.: Pressure Distribution over a Symmetrical Airfoil Section with Trailing Edge Flap. NACA Rept. No. 360, 1930.

48. JACOBS, EASTMAN N., PINKERTON, ROBERT M., and GREENBERG, HARRY: Tests of Related Forward-camber Airfoils in the Variable-density Wind Tunnel. NACA Rept. No. 610, 1937.

49. JACOBS, EASTMAN N., WARD, KENNETH E., and PINKERTON, ROBERT M.: The Characteristics of 78 Related Airfoil Sections from Tests in the Variable-density Wind Tunnel. NACA Rept. No. 460, 1932.

50. JONES, ROBERT T.: Correction of the Lifting-line Theory for the Effect of the Chord. NACA TN No. 617, 1941.

51. JONES, ROBERT T.: Properties of Low-aspect Ratio Pointed Wings at Speeds below and above the Speed of Sound. NACA TN No. 1032, 1946.

52. JONES, ROBERT T.: Wing Plan Forms for High-speed Flight. NACA TN No. 1033, 1946.

53. JONES, ROBERT T.: Thin Oblique Airfoils at Supersonic Speed. NACA TN No. 1107, 1946.

54. KAPLAN, CARL: Effect of Compressibility at High Subsonic Velocities on the Lifting Force Acting on an Elliptic Cylinder. NACA TN No. 1118, 1946.

55. KEENAN, JOSEPH H., and NEUMANN, ERNEST P.: Friction in Pipes at Supersonic and Subsonic Velocities. NACA TN No. 963, 1945.

56. KENNARD, EARLE H.: "Kinetic Theory of Gases," McGraw-Hill Book Company, Inc., New York, 1938.

57. KNIGHT, MONTGOMERY, and BAMBER, MILLARD J.: Wind Tunnel Tests on Airfoil Boundary Layer Control Using a Backward Opening Slot. NACA TN No. 323, 1929.

58. KOSTER, H.: Messungen an Profile 0_00 12-0, 55 45 mit Spreitz- und Nasanspreizklappe, Deutsche Versuchsanstalt für Luftfahrt, UM 1317, July 29, 1944.

59. KRÜGER, W.: Systematische Windkanalmessungen an einem Laminarflügel mit Nasanklappe, Aerodynamische Versuchsanstalt Göttingen, Fb Nr. 1948, June 13, 1944.

60. KRÜGER, W.: Wind-tunnel Investigation on a Changed Mustang Profile with

Nose Flap. Force and Pressure Distribution Measurements. NACA TM No. 1177, 1947.

61. LAMB, HORACE: "Hydrodynamics," Cambridge University Press, London, 1932.
62. LEES, LESTER: The Stability of a Laminar Boundary Layer in a Compressible Fluid. NACA TN No. 1115, 1946.
63. LEMME, H. G.: Kraftmessungen und Drukverteilungmessungen an einem Flügel mit Knicknase, Vorflügel, Wölbungs- und Spreizklappe, Aerodynamische Versuchsanstalt Göttingen, Fb. Nr. 1676, Oct. 15, 1942.
64. LEMME, H. G.: Kraftmessungen und Druckverteilungmessungen an einem Rechteckflügel mit Spaltknicknase, Wölbungs- und Spreizklappe oder Rollklappe, Aerodynamische Versuchsanstalt Göttingen, Fb. Nr. 1676/2, May 14, 1943 (also NACA TM No. 1108, 1947).
65. LEMME, H. G.: Kraftmessungen und Druckverteilungmessungen an einem Rechteckflügel mit Doppelknicknase, Aerodynamische Versuchsanstalt Göttingen, Fb. Nr. 1676/3, June 2, 1944.
66. LIEPMAN, HANS WOLFGANG: The Interaction between Boundary Layer and Shock Waves in Transonic Flow, *J. Aeronaut. Sci.*, vol. 13, No. 12, pp. 623–637, December, 1946.
66a. LIN, C. C.: On the Stability of Two-dimensional Parallel Flows, Part I, *Quart. Appl. Math.*, vol. 3, No. 2, pp. 117–142, July, 1945; Part II, vol. 3, No. 3, pp. 218–234, October 1945; and Part III, vol. 3, No. 4, pp. 277–301, January, 1946.
67. LIPPISCH, A., and BEUSCHAUSEN, W.: Pressure Distribution Measurements at High Speed and Oblique Incidence of Flow. NACA TM No. 1115, 1947.
68. LOFTIN, LAURENCE K., JR.: Effects of Specific Types of Surface Roughness on Boundary-layer Transition. NACA ACR L5J29a, 1945 (Wartime Rept. No. L–48).
68a. LOFTIN, LAURENCE K., JR., and COHEN, KENNETH S.: Aerodynamic Characteristics of a Number of Modified NACA Four-digit-series Airfoil Sections. NACA RM No. L7I22, 1947.
69. LOWRY, JOHN G.: Wind-tunnel Investigation of an NACA 23012 Airfoil with Several Arrangements of Slotted Flap with Extended Lips. NACA TN No. 808, 1941.
70. MUNK, MAX M.: The Determination of the Angles of Attack of Zero Lift and Zero Moment, Based on Munk's Integrals. NACA TN No. 122, 1923.
71. MUNK, MAX M.: Elements of the Wing Section Theory and of the Wing Theory. NACA Rept. No. 191, 1924.
72. MUNK, MAX M.: Calculation of Span Lift Distribution (Part 2), *Aero Digest,* vol. 48, No. 3, p. 84, Feb. 1, 1945.
73. NAIMAN, IRVEN: Numerical Evaluation of the ϵ-Integral Occurring in the Theodorsen Arbitrary Airfoil Potential Theory. NACA ARR No. L4D27a, April, 1944 (Wartime Rept. No. L–136).
73a. NAIMAN, IRVEN: Numerical Evaluation by Harmonic Analysis of the ϵ-Function of the Theodorsen Arbitrary Potential Theory. NACA ARR No. L5H18, September, 1945 (Wartime Rept. No. L–153).
74. NEELY, ROBERT H., BOLLECH, THOMAS V., WESTRICK, GERTRUDE C., and GRAHAM, ROBERT R.: Experimental and Calculated Characteristics of Several NACA 44-series Wings with Aspect Ratios of 8, 10, and 12, and Taper Ratios of 2.5 and 3.5. NACA TN No. 1270, 1947.
75. NIKURADSE, J.: Gesetzmässigkeiten der turbulenten Strömung in glatten Rohren, *Forsch. Gebiete Ingenieurw.*, 1932, *Forschungsheft* 356.
75a. NIKURADSE, J.: Vörträge aus dem Gebiete der Aerodynamik, p. 63, Aachen, 1929.
76. NITZBERG, GERALD E.: A Concise Theoretical Method for Profile-drag Calculation. NACA ACR No. 4B05, 1944.

77. Noyes, Richard W.: Wind-tunnel Tests of a Wing with a Trailing-edge Auxiliary Airfoil Used as a Flap. NACA TN No. 524, 1935.

78. Pankhurst, R. C.: A Method for the Rapid Evaluation of Glauert's Expressions for the Angle of Zero Lift and the Moment at Zero Lift. R. & M. No. 1914, British ARC, 1944.

79. Pearson, E. O., Jr., Evans, A. J., and West, F. E.: Effects of Compressibility on the Maximum Lift Characteristics, and Spanwise Load Distribution of a 12-foot-span Fighter-type Wing of NACA 230-series Airfoil Sections. NACA ACR No. L5G10, 1945 (Wartime Rept. No. L–51).

80. Peirce, B. O.: "A Short Table of Integrals," Ginn & Company, Boston, 1929.

80a. Pierce, H.: On the Dynamic Response of Airplane Wings Due to Gusts. NACA TN No. 1320, 1947.

81. Pinkerton, Robert M.: Analytical Determination of the Load on a Trailing Edge Flap. NACA TN No. 353, 1930.

82. Pinkerton, Robert M.: Calculated and Measured Pressure Distributions over the Midspan Section of the NACA 4412 Airfoil. NACA Rept. No. 563, 1936.

83. Platt, Robert C.: Aerodynamic Characteristics of Wings with Cambered External-airfoil Flaps, Including Lateral Control with a Full-span Flap. NACA Rept. No. 541, 1935.

84. Platt, Robert C., and Abbott, Ira H.: Aerodynamic Characteristics of NACA 23012 and 23021 Airfoils with 20-percent-chord External-airfoil Flaps of NACA 23012 Section. NACA Rept. No. 573, 1936.

85. Platt, Robert C., and Shortal, Joseph A.: Wind-tunnel Investigation of Wings with Ordinary Ailerons and Full-span External Airfoil Flaps. NACA Rept. No. 603, 1937.

86. Pohlhausen, K.: Zur näherungsweisen Integration der Differentialgleichung der laminaren Grenzschicht, Abhandl. aero. Inst. Aachen, 1. Lieferung, 1921. See also Z. angew. Math. Mech., Bd. 1, Heft 4, pp. 252–258, 1921.

87. Prandtl, L.: Motion of Fluids with Very Little Viscosity. NACA TN No. 452, 1928 (originally presented to the Third International Mathematical Congress, Heidelberg, 1904).

88. Prandtl, L.: Applications of Modern Hydrodynamics to Aeronautics. NACA Rept. No. 116, 1921.

89. Prandtl, L., and Tietjens, O. G.: "Applied Hydro- and Aeromechanics," McGraw-Hill Book Company, Inc., New York, 1934.

90. Puckett, Allen E.: Supersonic Wave Drag of Thin Airfoils, J. Aeronaut. Sci., vol. 13, No. 9, pp. 475–484, September, 1946.

91. Purser, Paul E., Fischel, Jack, and Riebe, John M.: Wind-tunnel Investigation of an NACA 23012 Airfoil with a 0.30-airfoil-chord Double Slotted Flap. NACA ARR No. 3L10, 1943 (Wartime Rept. No. L–469).

92. Purser, Paul E., and Johnson, Harold S.: Effects of Trailing-edge Modifications on Pitching-moment Characteristics of Airfoils. NACA CB No. L4I30, 1944 (Wartime Rept. No. L–664).

93. Quinn, John H., Jr.: Tests of the NACA 65_3-018 Airfoil Section with Boundary-layer Control by Suction. NACA CB No. L4H10, 1944 (Wartime Rept. No. L–209).

94. Quinn, John H., Jr.: Wind-tunnel Investigation of Boundary-layer Control by Suction on the NACA 65_3-418, $a = 1.0$ Airfoil Section with a 0.29 Airfoil-chord Double Slotted Flap. NACA TN No. 1071, 1946.

95. Quinn, John H., Jr.: Tests of the NACA 64_1A212 Airfoil Section with a Slat, a Double Slotted Flap, and Boundary-layer Control by Suction. NACA TN No. 1293, 1947.

96. RECANT, I. G.: Wind-tunnel Investigation of an NACA 23030 Airfoil with Various Arrangements of Slotted Flap. NACA TN No. 755, 1940.

97. REID, E. G., and BAMBER, M. J.: Preliminary Investigation on Boundary Layer Control by Means of Suction and Pressure with the U.S.A. 27 Airfoil. NACA TN No. 286, 1928.

98. SCHILLER, L.: Die Entwicklung der laminaren Geschwindigkeitsverteilung und ihr Bedeutung für Zähigkeitsmessungen, Z. angew. Math. Mech., Bd. 2, Heft 2, pp. 96–106, 1922. See also Forsch. Gebiete Ingenieurw., Heft 248, Aachen III, 3, 1922.

99. SCHLICHTING, H.: Amplitudenverteilung und Energiebilanz der kleinen Störungen bei der Plattenströmung, Nachr. Ges. Wiss. Göttingen, Math. physik. Klasse, Bd. 1, 1935.

100 SCHLICHTING, H.: Zur Entstehung der Turbulenz bei der Plattenströmung, Nachr. Ges. Wiss. Göttingen, Math. physik. Klasse, 1933, pp. 181–209.

101. SCHLICHTING, H., and ULRICH, A.: Zur Berechnung des Umschlages Laminar Turbulent, Jahrbuch 1942 der deutschen Luftfahrtforschung, pp. 18–135, R. Oldenbourg, Munich.

102. SCHRENK, OSKAR: Experiments with a Wing Model from Which the Boundary Layer Is Removed by Suction. NACA TM No. 534, 1929.

103. SCHRENK, OSKAR: Experiments with a Wing from Which the Boundary Layer Is Removed by Suction. NACA TM No. 634, 1931.

104. SCHRENK, OSKAR: Experiments with Suction-type Wings. NACA TM No. 773, 1935.

105. SCHUBAUR, G. B.: Air Flow in the Boundary Layer of an Elliptical Cylinder. NACA Rept. No. 652, 1939.

106. SCHUBAUR, G. B., and SKRAMSTAD, H. K.: Laminar-boundary-layer Oscillations and Transition on a Flat Plate. NACA ACR, April, 1943 (Wartime Rept. No. 8).

107. SCHULDENFREI, MARVIN J.: Wind-tunnel Investigation of an NACA 23012 Airfoil with a Handley Page Slot and Two Flap Arrangements. NACA ARR, February, 1942 (Wartime Rept. No. L–261).

108. SHERMAN, ALBERT, and HARRIS, T. A.: The Effects of Equal Pressure Fixed Slots on the Characteristics of a Clark-Y Airfoil. NACA TN No. 507, 1934.

109. SHERMAN, ALBERT: A Simple Method of Obtaining Span Load Distribution. NACA TN No. 732, 1939.

109a. SILVERSTEIN, ABE, KATZOFF, SAMUEL, and HOOTMAN, JAMES: Comparative Flight and Full-scale Wind-tunnel Measurements of the Maximum Lift of an Airplane. NACA Rept. No. 618, 1938.

110. SIVELLS, JAMES C.: Experimental Characteristics of Three Wings of NACA 64-210 and 65-210 Airfoil Sections with and without 2° Washout. NACA TN No. 1422, 1947.

111. SIVELLS, JAMES C., and NEELY, ROBERT H.: Method of Calculating Wing Characteristics by Lifting-line Theory Using Nonlinear Section Lift Data. NACA TN No. 1269, 1947.

112. SQUIRE, H. B., and YOUNG, A. D.: The Calculation of the Profile Drag of Aerofoils. R. & M. No. 1838, British ARC, 1938.

113. STACK, JOHN: Tests of Airfoils Designed to Delay the Compressibility Burble. NACA TN No. 976, 1944.

114. STACK, JOHN: Compressible Flows in Aeronautics, Eighth Wright Brothers Lecture, J. Aeronaut. Sci., vol. 12, No. 2, April, 1945.

115. STACK, JOHN, LINDSAY, W. F., and LITTELL, ROBERT E.: The Compressibility Burble and the Effects of Compressibility on Pressures and Forces Acting on an Airfoil. NACA Rept. No. 646, 1938.

116. STACK, JOHN, and VON DOENHOFF, ALBERT E.: Tests of 16 Related Airfoils at High Speeds. NACA Rept. No. 492, 1934.

117. STEWART, H. J.: Lift of a Delta Wing at Supersonic Speeds, *Quart. Appl. Math.*, vol. IV, No. 3, pp. 246–254, October, 1946.

118. TANI, ITIRO: A Simple Method of Calculating the Induced Velocity of a Monoplane Wing, *Aeronaut. Research Inst. Tokyo Imp. Univ. Rept.* 111, vol. IX, p. 3, August, 1934.

119. TEMPLE, G., and YARWOOD, J.: The Approximate Solution of the Hodograph Equations for Compressible Flow. Rept. No. S.M.E. 3201, RAE, June, 1942.

120. TETERVIN, NEAL: A Method for the Rapid Estimation of Turbulent Boundary-layer Thickness for Calculating Profile Drag. NACA ACR No. L4G14, July, 1944 (Wartime Rept. No. L–16).

121. THEODORSEN, THEODORE: On the Theory of Wing Sections with Particular Reference to the Lift Distribution. NACA Rept. No. 383, 1931.

122. THEODORSEN, THEODORE: Theory of Wing Sections of Arbitrary Shape. NACA Rept. No. 411, 1931.

123. THEODORSEN, THEODORE: Airfoil Contour Modification Based on ε-Curve Method of Calculating Pressure Distribution. NACA ARR No. L4G05, 1944 (Wartime Rept. No. L–135).

124. THEODORSEN, THEODORE: A Condition on the Initial Shock. NACA TN No. 1029, 1946.

125. THEODORSEN, THEODORE, and GARRICK, I. E.: General Potential Theory of Arbitrary Wing Sections. NACA Rept. No. 452, 1933.

126. THEODORSEN, THEODORE, and REGIER, ARTHUR: Experiments on Drag of Revolving Disks, Cylinders and Streamline Rods at High Speeds. NACA ACR No. L4F16, 1944 (Rept. No. 793).

127. TOLLMIEN, W.: The Production of Turbulence. NACA TM No. 609, 1931.

128. TOLLMIEN, W.: General Instability Criterion of Laminar Velocity Distributions. NACA TM No. 792, 1936.

129. TSIEN, Hsue-Shen: Two-dimensional Subsonic Flow of Compressible Fluids, *J. Aeronaut. Sci.*, vol. 6, No. 10, pp. 399–407, August, 1939.

132. VON DOENHOFF, ALBERT E.: A Method of Rapidly Estimating the Position of the Laminar Separation Point. NACA TN No. 671, 1938.

133. VON DOENHOFF, ALBERT E., and ABBOTT, FRANK T., JR.: The Langley Two-dimensional Low-turbulence Pressure Tunnel. NACA TN No. 1283, 1947.

134. VON DOENHOFF, ALBERT E., and TETERVIN, NEAL: Investigation of the Variation of Lift Coefficient with Reynolds Number at a Moderate Angle of Attack on a Low-drag Airfoil. NACA CB, 1942 (Wartime Rept. No. L–661).

135. VON DOENHOFF, A. E., and TETERVIN, NEAL: Determination of General Relations for the Behavior of Turbulent Boundary Layers. NACA ACR No. 3G13, 1943 (Rept. No. 772).

136. VON KÁRMÁN, T.: Über laminare und turbulente Reibung, *Z. angew. Math. Mech.*, Bd. 1, Heft 4, pp. 233–252, 1921.

137. VON KÁRMÁN, T.: Gastheoretische Deutung der Reynoldsschen Kennzahl, *Abhandl. Aero. Inst. Aachen*, Heft 4, 1925, Verlag Julius Springer, Berlin.

138. VON KÁRMÁN, T.: Mechanical Similitude and Turbulence. NACA TM No. 611, 1931.

139. VON KÁRMÁN, T.: Turbulence and Skin Friction, *J. Aeronaut. Sci.*, vol. 1, No. 1, pp. 1–20, January, 1934.

140. VON KÁRMÁN, T.: Compressibility Effects in Aerodynamics, *J. Aeronaut. Sci.*, vol. 8, No. 9, pp. 337–356, July, 1941.

141. von Kármán, T., and Millikan, C. B.: On the Theory of Laminar Boundary Layers Involving Separation. NACA Rept. No. 504, 1934.

142. von Kármán, T., and Tsien, H. S.: Boundary Layer in Compressible Fluids, J. Aeronaut. Sci., vol. 5, No. 6, April, 1938.

143. Warfield, Calvin N. for the NACA Special Subcommittee on the Upper Atmosphere: Tentative Tables for the Upper Atmosphere. NACA TN No. 1200, 1947.

144. Weick, Fred E., and Harris, Thomas A.: The Aerodynamic Characteristics of a Model Wing Having a Split Flap Deflected Downward and Moved to the Rear. NACA TN No. 422, 1932.

145. Weick, Fred E., and Platt, Robert C.: Wind-tunnel Tests on a Model Wing with Fowler Flap and Specially Developed Leading-edge Slot. NACA TN No. 459, 1933.

146. Weick, Fred E., and Sanders, Robert: Wind-tunnel Tests of a Wing with Fixed Auxiliary Airfoils Having Various Chords and Profiles. NACA Rept. No. 472, 1933.

147. Weick, Fred E., and Shortal, Joseph A.: The Effect of Multiple Fixed Slots and a Trailing-edge Flap on the Lift and Drag of a Clark Y Airfoil. NACA Rept. No. 427, 1932.

148. Wenzinger, Carl J.: Wind-tunnel Investigation of Ordinary and Split Flaps on Airfoils of Different Profile. NACA Rept. No. 554, 1936.

149. Wenzinger, Carl J.: Pressure Distribution over an Airfoil Section with a Flap and Tab. NACA Rept. No. 574, 1936.

150. Wenzinger, Carl J.: Wind-tunnel Tests of a Clark Y Wing Having Split Flaps with Gaps. NACA TN No. 650, 1938.

151. Wenzinger, Carl J.: Pressure Distribution over an NACA 23012 Airfoil with an NACA 23012 External-airfoil Flap. NACA Rept. No. 614, 1938.

152. Wenzinger, Carl J., and Delano, James B.: Pressure Distribution over an NACA 23012 Airfoil with a Slotted and a Plain Flap. NACA Rept. No. 633, 1938.

153. Wenzinger, Carl J., and Harris, Thomas A.: Wind-tunnel Investigation of an NACA 23012 Airfoil with Various Arrangements of Slotted Flaps. NACA Rept. No. 664, 1939.

154. Wenzinger, Carl J., and Harris, Thomas A.: Wind-tunnel Investigation of NACA 23012, 23021, and 23030 Airfoils with Various Sizes of Split Flap. NACA Rept. No. 668, 1939.

155. Wenzinger, Carl J., and Harris, Thomas A.: Wind-tunnel Investigation of an NACA 23021 Airfoil with Various Arrangements of Slotted Flaps. NACA Rept. No. 677, 1939.

156. Wenzinger, Carl J., and Rogallo, Francis M.: Résumé of Airload Data on Slats and Flaps. NACA TN No. 690, 1939.

156a. Wragg, C. A.: Wragg Compound Aerofoil. U. S. Air Service, March, 1921, pp. 32-34.

157. Young, A. D.: Note on the Effect of Slipstream on Boundary Layer Flow. Rept. No. B.A. 1404, British RAE, May, 1937.

158. Young, A. D.: Further Note on the Effect of Slipstream on Boundary Layer Flow. Rept. No. B.A. 1404a, British RAE, October, 1938.

159. Young, A. D., and Morris, D. E.: Further Note on the Effect of Slipstream on Boundary Layer Flow. Rept. No. B.A. 1404b, British RAE, September, 1939.

160. Zalovick, John A.: Profile Drag Coefficients of Conventional and Low-drag Airfoils as Obtained in Flight. NACA ACR No. L4E31, 1944 (Wartime Rept. No. L-139).

APPENDIXES

APPENDIX I

BASIC THICKNESS FORMS

CONTENTS

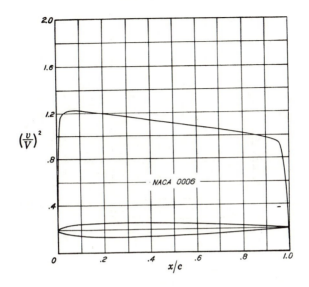

x (per cent c)	y (per cent c)	$(v/V)^2$	v/V	$\Delta v_a/V$
0	0	0	0	3.992
0.5	. . . . .	0.880	0.938	2.015
1.25	0.947	1.117	1.057	1.364
2.5	1.307	1.186	1.089	0.984
5.0	1.777	1.217	1.103	0.696
7.5	2.100	1.225	1.107	0.562
10	2.341	1.212	1.101	0.478
15	2.673	1.206	1.098	0.378
20	2.869	1.190	1.091	0.316
25	2.971	1.179	1.086	0.272
30	3.001	1.162	1.078	0.239
40	2.902	1.136	1.066	0.189
50	2.647	1.109	1.053	0.152
60	2.282	1.086	1.042	0.123
70	1.832	1.057	1.028	0.097
80	1.312	1.026	1.013	0.073
90	0.724	0.980	0.990	0.047
95	0.403	0.949	0.974	0.032
100	0.063	0	0	0
L.E. radius: 0.40 per cent c				

NACA 0006 Basic Thickness Form

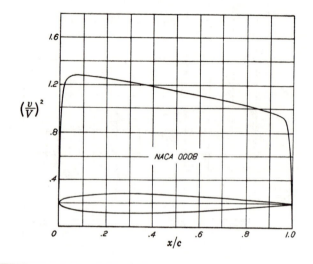

x (per cent c)	y (per cent c)	$(v/V)^2$	v/V	$\Delta v_a/V$
0	0	0	0	2.900
0.5	.....	0.792	0.890	1.795
1.25	1.263	1.103	1.050	1.310
2.5	1.743	1.221	1.105	0.971
5.0	2.369	1.272	1.128	0.694
7.5	2.800	1.284	1.133	0.561
10	3.121	1.277	1.130	0.479
15	3.564	1.272	1.128	0.379
20	3.825	1.259	1.122	0.318
25	3.961	1.241	1.114	0.273
30	4.001	1.223	1.106	0.239
40	3.869	1.186	1.089	0.188
50	3.529	1.149	1.072	0.152
60	3.043	1.111	1.054	0.121
70	2.443	1.080	1.039	0.096
80	1.749	1.034	1.017	0.071
90	0.965	0.968	0.984	0.047
95	0.537	0.939	0.969	0.031
100	0.084	.....	.....	0
L.E. radius: 0.70 per cent c				

NACA 0008 Basic Thickness Form

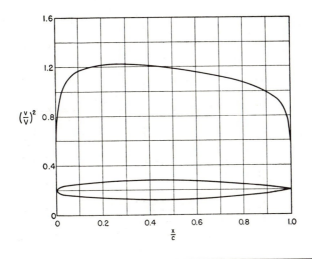

x (per cent c)	y (per cent c)	$(v/V)^2$	v/V	$\Delta v_a/V$
0	0	0	0	4.839
1.25	0.756	0.917	0.958	1.338
2.5	1.120	1.023	1.011	0.966
5.0	1.662	1.092	1.045	0.691
7.5	2.089	1.137	1.066	0.564
10	2.436	1.162	1.078	0.485
15	2.996	1.188	1.090	0.387
20	3.396	1.206	1.098	0.326
30	3.867	1.217	1.103	0.248
40	4.000	1.202	1.096	0.197
50	3.884	1.185	1.089	0.157
60	3.547	1.163	1.079	0.128
70	2.987	1.127	1.062	0.100
80	2.213	1.067	1.033	0.074
90	1.244	0.993	0.996	0.047
95	0.684	0.932	0.965	0.031
100	0.080	0	0	0

L.E. radius: 0.174 per cent c

NACA 0008-34 Basic Thickness Form

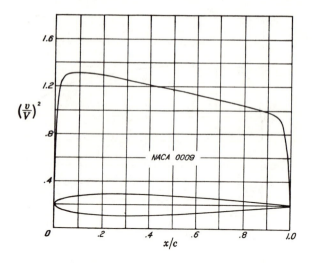

x (per cent c)	y (per cent c)	$(v/V)^2$	v/V	$\Delta v_a/V$
0	0	0	0	0.595
0.5	. . . . .	0.750	0.866	1.700
1.25	1.420	1.083	1.041	1.283
2.5	1.961	1.229	1.109	0.963
5.0	2.666	1.299	1.140	0.692
7.5	3.150	1.310	1.145	0.560
10	3.512	1.309	1.144	0.479
15	4.009	1.304	1.142	0.380
20	4.303	1.293	1.137	0.318
25	4.456	1.275	1.129	0.273
30	4.501	1.252	1.119	0.239
40	4.352	1.209	1.100	0.188
50	3.971	1.170	1.082	0.151
60	3.423	1.126	1.061	0.120
70	2.748	1.087	1.043	0.095
80	1.967	1.037	1.018	0.070
90	1.086	0.984	0.982	0.046
95	0.605	0.933	0.966	0.030
100	0.095	0	0	0

L.E. radius: 0.89 per cent c

NACA 0009 Basic Thickness Form

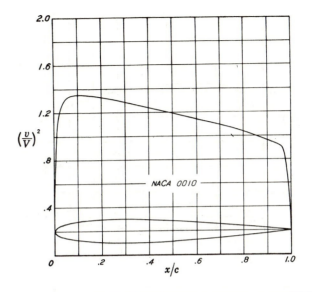

x (per cent c)	y (per cent c)	$(v/V)^2$	v/V	$\Delta v_a/V$
0	0	0	0	2.372
0.5	.....	0.712	0.844	1.618
1.25	1.578	1.061	1.030	1.255
2.5	2.178	1.237	1.112	0.955
5.0	2.962	1.325	1.151	0.690
7.5	3.500	1.341	1.158	0.559
10	3.902	1.341	1.158	0.479
15	4.455	1.341	1.158	0.380
20	4.782	1.329	1.153	0.318
25	4.952	1.309	1.144	0.273
30	5.002	1.284	1.133	0.239
40	4.837	1.237	1.112	0.188
50	4.412	1.190	1.091	0.150
60	3.803	1.138	1.067	0.119
70	3.053	1.094	1.046	0.094
80	2.187	1.040	1.020	0.069
90	1.207	0.960	0.980	0.045
95	0.672	0.925	0.962	0.030
100	0.105	.....	.....	0
L.E. radius: 1.10 per cent c				

NACA 0010 Basic Thickness Form

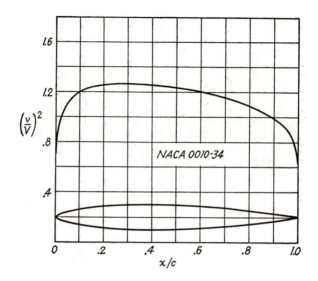

x (per cent c)	y (per cent c)	$(v/V)^2$	v/V	$\Delta v_a/V$
0	0	0	0	3.857
1.25	0.944	0.892	0.944	1.282
2.5	1.400	1.011	1.005	0.950
5.0	2.078	1.113	1.055	0.688
7.5	2.611	1.167	1.080	0.564
10	3.044	1.200	1.095	0.486
15	3.744	1.238	1.113	0.389
20	4.244	1.256	1.121	0.327
30	4.833	1.265	1.124	0.249
40	5.000	1.253	1.119	0.197
50	4.856	1.235	1.111	0.159
60	4.433	1.205	1.098	0.127
70	3.733	1.157	1.076	0.100
80	2.767	1.089	1.044	0.073
90	1.556	0.990	0.995	0.045
95	0.856	0.910	0.954	0.030
100	0.100	0	0	0
L.E. radius: 0.272 per cent c				

NACA 0010-34 Basic Thickness Form

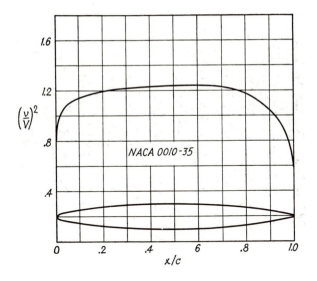

x (per cent c)	y (per cent c)	$(v/V)^2$	v/V	$\Delta v_a/V$
0	0	0	0	4.068
1.25	0.878	0.954	0.977	1.309
2.5	1.267	1.032	1.016	0.952
5.0	1.844	1.087	1.043	0.679
7.5	2.289	1.122	1.059	0.555
10	2.667	1.141	1.068	0.476
15	3.289	1.172	1.083	0.382
20	3.789	1.194	1.093	0.323
30	4.478	1.214	1.102	0.247
40	4.878	1.229	1.109	0.198
50	5.000	1.235	1.111	0.162
60	4.867	1.240	1.114	0.131
70	4.389	1.227	1.108	0.104
80	3.500	1.176	1.084	0.076
90	2.100	1.046	1.023	0.048
95	1.178	0.920	0.959	0.030
100	0.100	0	0	0

L.E. radius: 0.272 per cent c

NACA 0010-35 Basic Thickness Form

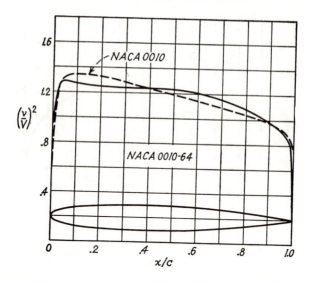

x (per cent c)	y (per cent c)	$(v/V)^2$	v/V	$\Delta v_a/V$
0	0	0	0	2.324
1.25	1.511	1.108	1.053	1.286
2.5	2.044	1.245	1.116	0.966
5.0	2.722	1.286	1.134	0.690
7.5	3.178	1.277	1.130	0.556
10	3.533	1.269	1.127	0.475
15	4.056	1.261	1.123	0.377
20	4.411	1.248	1.117	0.316
30	4.856	1.244	1.116	0.241
40	5.000	1.242	1.115	0.193
50	4.856	1.231	1.110	0.155
60	4.433	1.211	1.101	0.126
70	3.733	1.155	1.074	0.098
80	2.767	1.089	1.043	0.072
90	1.556	0.980	0.990	0.045
95	0.856	0.912	0.955	0.030
100	0.100	0	0	0

L.E. radius: 1.10 per cent c

NACA 0010-64 Basic Thickness Form

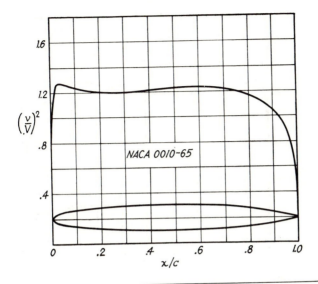

x (per cent c)	y (per cent c)	$(v/V)^2$	v/V	$\Delta v_a/V$
0	0	0	0	2.584
1.25	1.467	1.140	1.068	1.295
2.5	1.967	1.273	1.128	0.970
5.0	2.589	1.271	1.127	0.684
7.5	2.989	1.252	1.119	0.551
10	3.300	1.236	1.112	0.470
15	3.756	1.213	1.101	0.372
20	4.089	1.200	1.095	0.312
30	4.578	1.196	1.094	0.239
40	4.889	1.212	1.101	0.193
50	5.000	1.229	1.109	0.158
60	4.867	1.234	1.111	0.128
70	4.389	1.226	1.107	0.103
80	3.500	1.173	1.083	0.076
90	2.100	1.049	1.024	0.046
95	1.178	0.915	0.957	0.029
100	0.100	0	0	0

L.E. radius: 1.10 per cent c

NACA 0010-65 Basic Thickness Form

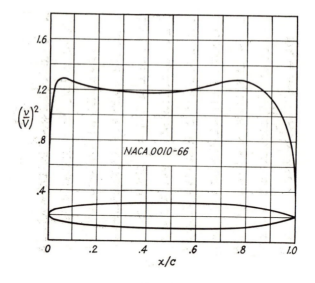

x (per cent c)	y (per cent c)	$(v/V)^2$	v/V	$\Delta v_a/V$
0	0	0	0	2.434
1.25	1.489	1.130	1.063	1.289
2.5	2.011	1.246	1.116	0.959
5.0	2.656	1.286	1.134	0.687
7.5	3.089	1.282	1.132	0.554
10	3.400	1.258	1.122	0.471
15	3.856	1.225	1.107	0.372
20	4.178	1.209	1.100	0.310
30	4.578	1.189	1.090	0.236
40	4.822	1.178	1.085	0.190
50	4.956	1.184	1.088	0.153
60	5.000	1.214	1.102	0.129
70	4.889	1.265	1.125	0.104
80	4.300	1.278	1.130	0.080
90	2.833	1.135	1.065	0.049
95	1.656	0.960	0.980	0.030
100	0.100	0	0	0
L.E. radius: 1.10 per cent c				

NACA 0010-66 Basic Thickness Form

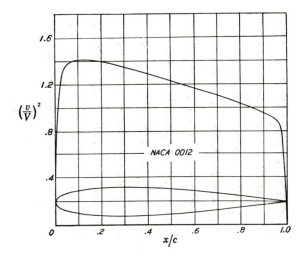

x (per cent c)	y (per cent c)	$(v/V)^2$	v/V	$\Delta v_a/V$
0	0	0	0	1.988
0.5	.....	0.640	0.800	1.475
1.25	1.894	1.010	1.005	1.199
2.5	2.615	1.241	1.114	0.934
5.0	3.555	1.378	1.174	0.685
7.5	4.200	1.402	1.184	0.558
10	4.683	1.411	1.188	0.479
15	5.345	1.411	1.188	0.381
20	5.737	1.399	1.183	0.319
25	5.941	1.378	1.174	0.273
30	6.002	1.350	1.162	0.239
40	5.803	1.288	1.135	0.187
50	5.294	1.228	1.108	0.149
60	4.563	1.166	1.080	0.118
70	3.664	1.109	1.053	0.092
80	2.623	1.044	1.022	0.068
90	1.448	0.956	0.978	0.044
95	0.807	0.906	0.952	0.029
100	0.126	0	0	0
L.E. radius: 1.58 per cent c				

NACA 0012 Basic Thickness Form

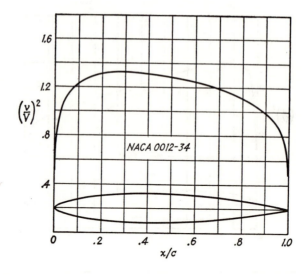

x (per cent c)	y (per cent c)	$(v/V)^2$	v/V	$\Delta v_a/V$
0	0	0	0	3.154
1.25	1.133	0.865	0.930	1.251
2.5	1.680	0.997	0.999	0.933
5.0	2.493	1.122	1.059	0.683
7.5	3.133	1.186	1.089	0.560
10	3.653	1.229	1.109	0.484
15	4.493	1.282	1.132	0.389
20	5.093	1.310	1.145	0.329
30	5.800	1.329	1.153	0.250
40	6.000	1.311	1.145	0.198
50	5.827	1.284	1.133	0.158
60	5.320	1.249	1.118	0.128
70	4.480	1.192	1.092	0.098
80	3.320	1.112	1.055	0.071
90	1.867	0.985	0.992	0.045
95	1.027	0.894	0.946	0.029
100	0.120	0	0	0
L.E. radius: 0.391 per cent c				

NACA 0012-34 Basic Thickness Form

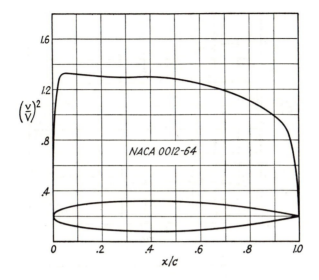

x (per cent c)	y (per cent c)	$(v/V)^2$	v/V	$\Delta v_a/V$
0	0	0	0	2.019
1.25	1.813	1.072	1.035	1.236
2.5	2.453	1.270	1.127	0.952
5.0	3.267	1.330	1.153	0.685
7.5	3.813	1.325	1.151	0.554
10	4.240	1.322	1.150	0.474
15	4.867	1.313	1.146	0.372
20	5.293	1.303	1.141	0.315
30	5.827	1.297	1.139	0.241
40	6.000	1.300	1.140	0.199
50	5.827	1.280	1.131	0.154
60	5.320	1.244	1.115	0.126
70	4.480	1.189	1.090	0.096
80	3.320	1.102	1.050	0.070
90	1.867	0.993	0.996	0.044
95	1.027	0.889	0.943	0.028
100	0.120	0	0	0

L.E. radius: 1.582 per cent c

NACA 0012-64 Basic Thickness Form

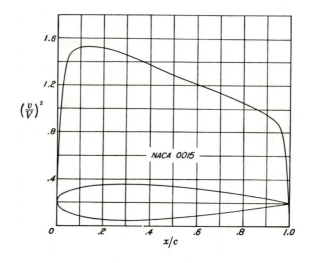

x (per cent c)	y (per cent c)	$(v/V)^2$	v/V	$\Delta v_a/V$
0	0	0	0	1.600
0.5	.....	0.546	0.739	1.312
1.25	2.367	0.933	0.966	1.112
2.5	3.268	1.237	1.112	0.900
5.0	4.443	1.450	1.204	0.675
7.5	5.250	1.498	1.224	0.557
10	5.853	1.520	1.233	0.479
15	6.682	1.520	1.233	0.381
20	7.172	1.510	1.229	0.320
25	7.427	1.484	1.218	0.274
30	7.502	1.450	1.204	0.239
40	7.254	1.369	1.170	0.185
50	6.617	1.279	1.131	0.146
60	5.704	1.206	1.098	0.115
70	4.580	1.132	1.064	0.090
80	3.279	1.049	1.024	0.065
90	1.810	0.945	0.972	0.041
95	1.008	0.872	0.934	0.027
100	0.158	0	0	0

L.E. radius: 2.48 per cent c

NACA 0015 Basic Thickness Form

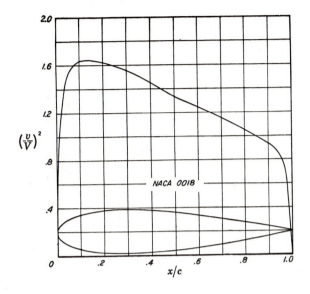

x (per cent c)	y (per cent c)	$(v/V)^2$	v/V	$\Delta v_a/V$
0	0	0	0	1.342
0.5	. . . . .	0.465	0.682	1.178
1.25	2.841	0.857	0.926	1.028
2.5	3.922	1.217	1.103	0.861
5.0	5.332	1.507	1.228	0.662
7.5	6.300	1.598	1.264	0.555
10	7.024	1.628	1.276	0.479
15	8.018	1.633	1.278	0.381
20	8.606	1.625	1.275	0.320
25	8.912	1.592	1.262	0.274
30	9.003	1.556	1.247	0.238
40	8.705	1.453	1.205	0.184
50	7.941	1.331	1.154	0.144
60	6.845	1.246	1.116	0.113
70	5.496	1.153	1.074	0.087
80	3.935	1.051	1.025	0.063
90	2.172	0.933	0.966	0.039
95	1.210	0.836	0.914	0.025
100	0.189	0	0	0

L.E. radius: 3.56 per cent c

NACA 0018 Basic Thickness Form

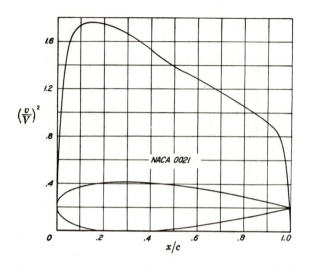

x (per cent c)	y (per cent c)	$(v/V)^2$	v/V	$\Delta v_a/V$
0	0	0	0	1.167
0.5	. . . . . .	0.397	0.630	1.065
1.25	3.315	0.787	0.887	0.946
2.5	4.576	1.182	1.087	0.818
5.0	6.221	1.543	1.242	0.648
7.5	7.350	1.682	1.297	0.550
10	8.195	1.734	1.317	0.478
15	9.354	1.756	1.325	0.381
20	10.040	1.742	1.320	0.320
25	10.397	1.706	1.306	0.274
30	10.504	1.664	1.290	0.238
40	10.156	1.538	1.240	0.183
50	9.265	1.388	1.178	0.142
60	7.986	1.284	1.133	0.111
70	6.412	1.177	1.085	0.084
80	4.591	1.055	1.027	0.061
90	2.534	0.916	0.957	0.037
95	1.412	0.801	0.895	0.023
100	0.221	0	0	0
L.E. radius: 4.85 per cent c				

NACA 0021 Basic Thickness Form

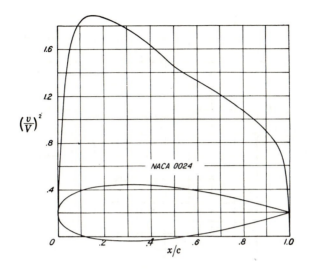

x (per cent c)	y (per cent c)	$(v/V)^2$	v/V	$\Delta v_a/V$
0	0	0	0	1.050
0.5	. . . . . .	0.335	0.579	0.964
1.25	3.788	0.719	0.848	0.870
2.5	5.229	1.130	1.063	0.771
5.0	7.109	1.548	1.244	0.632
7.5	8.400	1.748	1.322	0.542
10	9.365	1.833	1.354	0.476
15	10.691	1.888	1.374	0.383
20	11.475	1.871	1.368	0.321
25	11.883	1.822	1.350	0.274
30	12.004	1.777	1.333	0.238
40	11.607	1.631	1.277	0.181
50	10.588	1.450	1.204	0.140
60	9.127	1.325	1.151	0.109
70	7.328	1.203	1.097	0.082
80	5.247	1.065	1.032	0.059
90	2.896	0.891	0.944	0.035
95	1.613	0.773	0.879	0.022
100	0.252	0	0	0

L.E. radius: 6.33 per cent c

NACA 0024 Basic Thickness Form

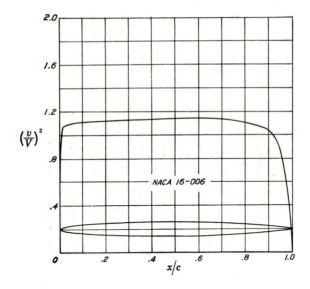

x (per cent c)	y (per cent c)	$(v/V)^2$	v/V	$\Delta v_a/V$
0	0	0	0	5.471
1.25	0.646	1.059	1.029	1.376
2.5	0.903	1.085	1.042	0.980
5.0	1.255	1.097	1.047	0.689
7.5	1.516	1.105	1.051	0.557
10	1.729	1.108	1.053	0.476
15	2.067	1.112	1.055	0.379
20	2.332	1.116	1.057	0.319
30	2.709	1.123	1.060	0.244
40	2.927	1.132	1.064	0.196
50	3.000	1.137	1.066	0.160
60	2.917	1.141	1.068	0.130
70	2.635	1.132	1.064	0.104
80	2.099	1.104	1.051	0.077
90	1.259	1.035	1.017	0.049
95	0.707	0.962	0.981	0.032
100	0.060	0	0	0

L.E. radius: 0.176 per cent c

NACA 16-006 Basic Thickness Form

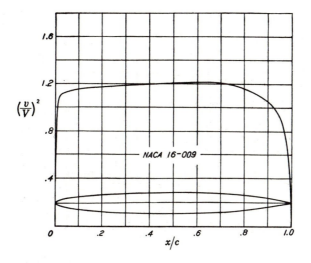

x (per cent c)	y (per cent c)	$(v/V)^2$	v/V	$\Delta v_a/V$
0	0	0	0	3.644
1.25	0.969	1.042	1.021	1.330
2.5	1.354	1.109	1.053	0.964
5.0	1.882	1.139	1.067	0.684
7.5	2.274	1.152	1.073	0.554
10	2.593	1.158	1.076	0.475
15	3.101	1.168	1.081	0.378
20	3.498	1.177	1.085	0.319
30	4.063	1.190	1.091	0.245
40	4.391	1.202	1.096	0.197
50	4.500	1.211	1.100	0.160
60	4.376	1.214	1.106	0.131
70	3.952	1.206	1.099	0.103
80	3.149	1.156	1.075	0.076
90	1.888	1.043	1.022	0.047
95	1.061	0.939	0.969	0.030
100	0.090	0	0	0

L.E. radius: 0.396 per cent c

NACA 16-009 Basic Thickness Form

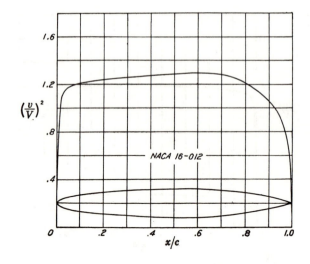

x (per cent c)	y (per cent c)	$(v/V)^2$	v/V	$\Delta v_a/V$
0	0	0	0	2.624
1.25	1.292	1.002	1.001	1.268
2.5	1.805	1.109	1.053	0.942
5.0	2.509	1.173	1.083	0.677
7.5	3.032	1.197	1.094	0.551
10	3.457	1.208	1.099	0.473
15	4.135	1.223	1.106	0.378
20	4.664	1.237	1.112	0.319
30	5.417	1.257	1.121	0.245
40	5.855	1.271	1.128	0.197
50	6.000	1.286	1.134	0.161
60	5.835	1.293	1.137	0.131
70	5.269	1.275	1.129	0.102
80	4.199	1.203	1.097	0.075
90	2.517	1.051	1.025	0.045
95	1.415	0.908	0.953	0.027
100	0.120	0	0	0
L.E. radius: 0.703 per cent c				

NACA 16-012 Basic Thickness Form

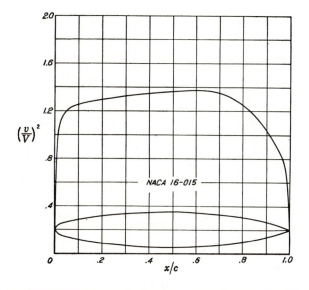

x (per cent c)	y (per cent c)	$(v/V)^2$	v/V	$\Delta v_a/V$
0	0	0	0	2.041
1.25	1.615	0.956	0.978	1.209
2.5	2.257	1.105	1.051	0.916
5.0	3.137	1.200	1.095	0.668
7.5	3.790	1.239	1.113	0.547
10	4.322	1.256	1.121	0.471
15	5.168	1.278	1.130	0.377
20	5.830	1.297	1.139	0.318
30	6.772	1.327	1.152	0.245
40	7.318	1.349	1.161	0.197
50	7.500	1.364	1.168	0.161
60	7.293	1.374	1.172	0.131
70	6.587	1.348	1.161	0.102
80	5.248	1.254	1.120	0.074
90	3.147	1.053	1.026	0.043
95	1.768	0.875	0.935	0.025
100	0.150	0	0	0
L.E. radius: 1.100 per cent c				

NACA 16-015 Basic Thickness Form

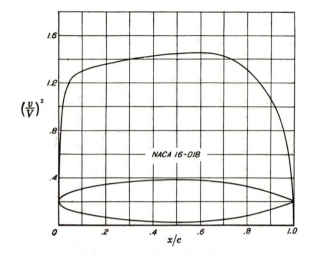

x (per cent c)	y (per cent c)	$(v/V)^2$	v/V	$\Delta v_a/V$
0	0	0	0	1.744
1.25	1.938	0.903	0.950	1.140
2.5	2.708	1.092	1.045	0.883
5.0	3.764	1.217	1.103	0.657
7.5	4.548	1.271	1.128	0.541
10	5.186	1.302	1.141	0.468
15	6.202	1.332	1.154	0.376
20	6.996	1.357	1.165	0.318
30	8.126	1.399	1.183	0.245
40	8.782	1.426	1.194	0.198
50	9.000	1.447	1.203	0.162
60	8.752	1.452	1.205	0.131
70	7.904	1.421	1.192	0.102
80	6.298	1.306	1.143	0.073
90	3.776	1.051	1.025	0.042
95	2.122	0.837	0.915	0.024
100	0.180	0	0	0

L.E. radius: 1.584 per cent c

NACA 16-018 Basic Thickness Form

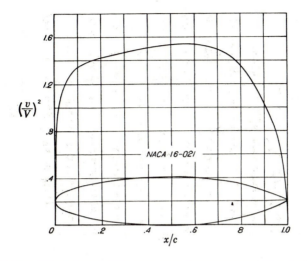

x (per cent c)	y (per cent c)	$(v/V)^2$	v/V	$\Delta v_a/V$
0	0	0	0	1.574
1.25	2.261	0.826	0.909	1.069
2.5	3.159	1.062	1.031	0.828
5.0	4.391	1.221	1.105	0.640
7.5	5.306	1.295	1.138	0.534
10	6.050	1.342	1.159	0.463
15	7.236	1.391	1.179	0.374
20	8.162	1.419	1.191	0.317
30	9.480	1.474	1.214	0.245
40	10.246	1.506	1.227	0.198
50	10.500	1.535	1.239	0.162
60	10.211	1.536	1.239	0.131
70	9.221	1.495	1.223	0.102
80	7.348	1.361	1.166	0.072
90	4.405	1.039	1.019	0.041
95	2.476	0.801	0.895	0.023
100	0.210	0	0	0
L.E. radius: 2.156 per cent c				

NACA 16-021 Basic Thickness Form

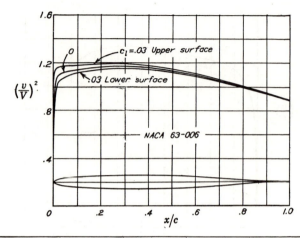

x (per cent c)	y (per cent c)	$(v/V)^2$	v/V	$\Delta v_a/V$
0	0	0	0	4.483
0.5	0.503	0.973	0.986	2.110
0.75	0.609	1.050	1.025	1.778
1.25	0.771	1.080	1.039	1.399
2.5	1.057	1.110	1.054	0.981
5.0	1.462	1.130	1.063	0.692
7.5	1.766	1.142	1.069	0.562
10	2.010	1.149	1.072	0.484
15	2.386	1.159	1.077	0.384
20	2.656	1.165	1.079	0.321
25	2.841	1.170	1.082	0.279
30	2.954	1.174	1.084	0.245
35	3.000	1.170	1.082	0.218
40	2.971	1.164	1.079	0.196
45	2.877	1.151	1.073	0.176
50	2.723	1.137	1.066	0.158
55	2.517	1.118	1.057	0.141
60	2.267	1.096	1.047	0.125
65	1.982	1.074	1.036	0.111
70	1.670	1.046	1.023	0.098
75	1.342	1.020	1.010	0.085
80	1.008	0.994	0.997	0.073
85	0.683	0.965	0.982	0.060
90	0.383	0.936	0.967	0.047
95	0.138	0.910	0.954	0.032
100	0	0.886	0.941	0

L.E. radius: 0.297 per cent c

NACA 63-006 Basic Thickness Form

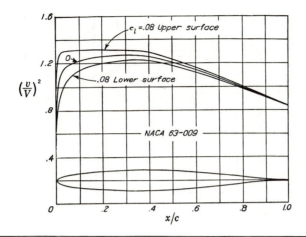

x (per cent c)	y (per cent c)	$(v/V)^2$	v/V	$\Delta v_a/V$
0	0	0	0	3.058
0.5	0.749	0.885	0.941	1.889
0.75	0.906	1.002	1.001	1.647
1.25	1.151	1.051	1.025	1.339
2.5	1.582	1.130	1.063	0.961
5.0	2.196	1.180	1.086	0.689
7.5	2.655	1.205	1.098	0.560
10	3.024	1.221	1.105	0.484
15	3.591	1.241	1.114	0.386
20	3.997	1.255	1.120	0.324
25	4.275	1.264	1.124	0.281
30	4.442	1.269	1.126	0.248
35	4.500	1.265	1.125	0.220
40	4.447	1.255	1.120	0.196
45	4.296	1.235	1.111	0.175
50	4.056	1.208	1.099	0.156
55	3.739	1.175	1.084	0.140
60	3.358	1.141	1.068	0.124
65	2.928	1.104	1.051	0.109
70	2.458	1.065	1.032	0.095
75	1.966	1.025	1.012	0.082
80	1.471	0.984	0.992	0.069
85	0.990	0.942	0.971	0.057
90	0.550	0.903	0.950	0.044
95	0.196	0.868	0.932	0.030
100	0	0.838	0.915	0

L.E. radius: 0.631 per cent c

NACA 63-009 Basic Thickness Form

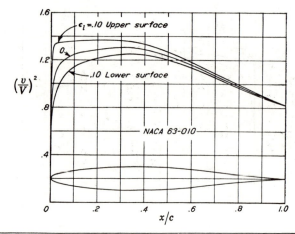

x (per cent c)	y (per cent c)	$(v/V)^2$	v/V	$\Delta v_a/V$
0	0	0	0	2.775
0.5	0.829	0.841	0.917	1.825
0.75	1.004	0.978	0.989	1.603
1.25	1.275	1.037	1.018	1.316
2.5	1.756	1.131	1.063	0.952
5.0	2.440	1.193	1.092	0.687
7.5	2.950	1.223	1.106	0.560
10	3.362	1.245	1.116	0.484
15	3.994	1.270	1.127	0.386
20	4.445	1.285	1.134	0.325
25	4.753	1.296	1.138	0.282
30	4.938	1.302	1.141	0.248
35	5.000	1.299	1.140	0.220
40	4.938	1.286	1.134	0.196
45	4.766	1.262	1.123	0.175
50	4.496	1.231	1.110	0.156
55	4.140	1.193	1.092	0.139
60	3.715	1.154	1.074	0.123
65	3.234	1.113	1.055	0.108
70	2.712	1.069	1.034	0.094
75	2.166	1.025	1.012	0.081
80	1.618	0.979	0.989	0.069
85	1.088	0.935	0.967	0.056
90	0.604	0.893	0.945	0.043
95	0.214	0.853	0.924	0.030
100	0	0.822	0.907	0

L.E. radius: 0.770 per cent c

NACA 63-010 Basic Thickness Form

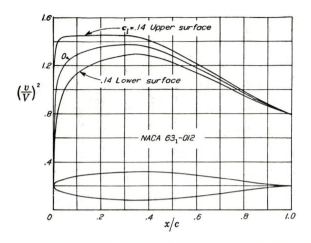

x (per cent c)	y (per cent c)	$(v/V)^2$	v/V	$\Delta v_a/V$
0	0	0	0	2.336
0.5	0.985	0.750	0.866	1.695
0.75	1.194	0.925	0.962	1.513
1.25	1.519	1.005	1.003	1.266
2.5	2.102	1.129	1.063	0.933
5.0	2.925	1.217	1.103	0.682
7.5	3.542	1.261	1.123	0.559
10	4.039	1.294	1.138	0.484
15	4.799	1.330	1.153	0.387
20	5.342	1.349	1.161	0.326
25	5.712	1.362	1.167	0.283
30	5.930	1.370	1.170	0.249
35	6.000	1.366	1.169	0.221
40	5.920	1.348	1.161	0.196
45	5.704	1.317	1.148	0.174
50	5.370	1.276	1.130	0.155
55	4.935	1.229	1.109	0.137
60	4.420	1.181	1.087	0.121
65	3.840	1.131	1.063	0.106
70	3.210	1.076	1.037	0.091
75	2.556	1.023	1.011	0.079
80	1.902	0.969	0.984	0.067
85	1.274	0.920	0.959	0.055
90	0.707	0.871	0.933	0.042
95	0.250	0.826	0.909	0.029
100	0	0.791	0.889	0

L.E. radius: 1.087 per cent c

NACA 63_1-012 Basic Thickness Form

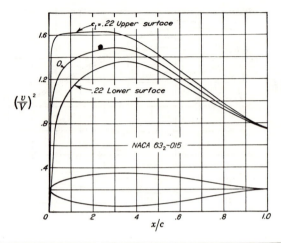

x (per cent c)	y (per cent c)	$(v/V)^2$	v/V	$\Delta v_a/V$
0	0	0	0	1.918
0.5	1.204	0.600	0.775	1.513
0.75	1.462	0.822	0.907	1.379
1.25	1.878	0.938	0.969	1.182
2.5	2.610	1.105	1.051	0.903
5.0	3.648	1.244	1.115	0.674
7.5	4.427	1.315	1.147	0.557
10	5.055	1.360	1.166	0.484
15	6.011	1.415	1.190	0.388
20	6.693	1.446	1.202	0.330
25	7.155	1.467	1.211	0.286
30	7.421	1.481	1.217	0.251
35	7.500	1.475	1.214	0.222
40	7.386	1.446	1.202	0.196
45	7.099	1.401	1.184	0.174
50	6.665	1.345	1.160	0.153
55	6.108	1.281	1.132	0.135
60	5.453	1.220	1.105	0.118
65	4.721	1.155	1.075	0.102
70	3.934	1.085	1.042	0.088
75	3.119	1.019	1.009	0.076
80	2.310	0.953	0.976	0.063
85	1.541	0.894	0.946	0.051
90	0.852	0.839	0.916	0.039
95	0.300	0.789	0.888	0.026
100	0	0.750	0.866	0

L.E. radius: 1.594 per cent c

NACA 63_2-015 Basic Thickness Form

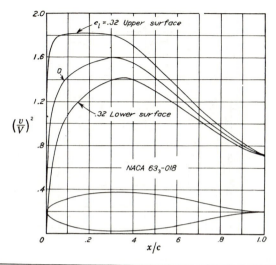

x (per cent c)	y (per cent c)	$(v/V)^2$	v/V	$\Delta v_a/V$
0	0	0	0	1.639
0.5	1.404	0.441	0.664	1.361
0.75	1.713	0.700	0.837	1.258
1.25	2.217	0.848	0.921	1.105
2.5	3.104	1.065	1.032	0.871
5.0	4.362	1.260	1.122	0.663
7.5	5.308	1.360	1.166	0.553
10	6.068	1.424	1.193	0.484
15	7.225	1.500	1.225	0.390
20	8.048	1.547	1.244	0.333
25	8.600	1.579	1.257	0.289
30	8.913	1.598	1.264	0.253
35	9.000	1.585	1.259	0.223
40	8.845	1.550	1.245	0.197
45	8.482	1.490	1.221	0.173
50	7.942	1.411	1.188	0.152
55	7.256	1.330	1.153	0.133
60	6.455	1.252	1.119	0.115
65	5.567	1.170	1.082	0.099
70	4.622	1.087	1.043	0.084
75	3.650	1.009	1.004	0.072
80	2.691	0.933	0.966	0.059
85	1.787	0.868	0.932	0.048
90	0.985	0.807	0.898	0.036
95	0.348	0.753	0.868	0.024
100	0	0.712	0.844	0

L.E. radius: 2.120 per cent c

NACA 63_3-018 Basic Thickness Form

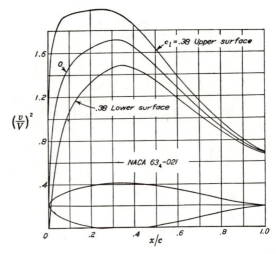

x (per cent c)	y (per cent c)	$(v/V)^2$	v/V	$\Delta v_a/V$
0	0	0	0	1.439
0.5	1.583	0.275	0.524	1.236
0.75	1.937	0.564	0.751	1.156
1.25	2.527	0.725	0.851	1.034
2.5	3.577	1.010	1.005	0.842
5.0	5.065	1.260	1.122	0.653
7.5	6.182	1.394	1.181	0.550
10	7.080	1.487	1.219	0.484
15	8.441	1.592	1.262	0.392
20	9.410	1.655	1.286	0.335
25	10.053	1.698	1.303	0.291
30	10.412	1.721	1.312	0.255
35	10.500	1.709	1.307	0.225
40	10.298	1.654	1.286	0.198
45	9.854	1.578	1.256	0.173
50	9.206	1.479	1.216	0.150
55	8.390	1.380	1.175	0.130
60	7.441	1.281	1.132	0.112
65	6.396	1.180	1.086	0.096
70	5.290	1.084	1.041	0.081
75	4.160	0.994	0.997	0.068
80	3.054	0.911	0.954	0.057
85	2.021	0.839	0.916	0.046
90	1.113	0.774	0.880	0.035
95	0.392	0.721	0.849	0.023
100	0	0.676	0.822	0
L.E. radius: 2.650 per cent c				

NACA 63_4-021 Basic Thickness Form

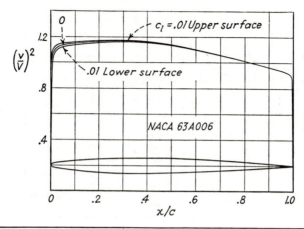

x (per cent c)	y (per cent c)	$(v/V)^2$	v/V	$\Delta v_a/V$
0	0	0	0	4.560
0.5	0.495	0.900	0.949	2.079
0.75	0.595	1.063	1.031	1.794
1.25	0.754	1.086	1.042	1.370
2.5	1.045	1.112	1.055	0.976
5.0	1.447	1.134	1.065	0.693
7.5	1.747	1.142	1.069	0.563
10	1.989	1.150	1.072	0.485
15	2.362	1.159	1.077	0.383
20	2.631	1.165	1.079	0.321
25	2.820	1.168	1.081	0.278
30	2.942	1.170	1.082	0.244
35	2.996	1.169	1.081	0.217
40	2.985	1.162	1.078	0.195
45	2.914	1.151	1.073	0.175
50	2.788	1.138	1.067	0.158
55	2.613	1.120	1.058	0.140
60	2.396	1.100	1.049	0.126
65	2.143	1.079	1.039	0.112
70	1.859	1.057	1.028	0.098
75	1.556	1.035	1.017	0.085
80	1.248	1.010	1.005	0.072
85	0.939	0.986	0.993	0.060
90	0.630	0.964	0.982	0.047
95	0.322	0.939	0.969	0.033
100	0.013	0	0	0

L.E. radius: 0.265 per cent c
T.E. radius: 0.014 per cent c

NACA 63A006 Basic Thickness Form

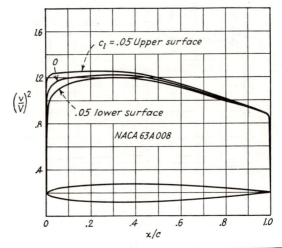

x (per cent c)	y (per cent c)	$(v/V)^2$	v/V	$\Delta v_a/V$
0	0	0	0	3.465
0.5	0.658	0.850	0.922	1.961
0.75	0.791	1.034	1.017	1.674
1.25	1.003	1.080	1.039	1.344
2.5	1.391	1.132	1.064	0.967
5.0	1.930	1.168	1.081	0.689
7.5	2.332	1.185	1.089	0.562
10	2.656	1.198	1.095	0.484
15	3.155	1.212	1.101	0.383
20	3.515	1.221	1.105	0.322
25	3.766	1.227	1.108	0.279
30	3.926	1.230	1.109	0.246
35	3.995	1.228	1.108	0.218
40	3.978	1.219	1.104	0.195
45	3.878	1.204	1.097	0.174
50	3.705	1.183	1.088	0.156
55	3.468	1.159	1.077	0.138
60	3.176	1.132	1.064	0.123
65	2.837	1.104	1.051	0.109
70	2.457	1.073	1.036	0.096
75	2.055	1.042	1.021	0.083
80	1.647	1.010	1.005	0.070
85	1.240	0.980	0.990	0.058
90	0.833	0.951	0.975	0.045
95	0.425	0.919	0.959	0.030
100	0.018	0	0	0

L.E. radius: 0.473 per cent c
T.E. radius: 0.020 per cent c

NACA 63A008 Basic Thickness Form

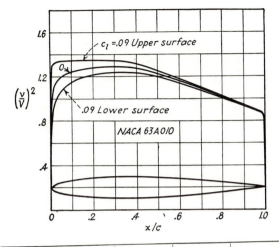

x (per cent c)	y (per cent c)	$(v/V)^2$	v/V	$\Delta v_a/V$
0	0	0	0	2.805
0.5	0.816	0.774	0.880	1.833
0.75	0.983	0.985	0.992	1.594
1.25	1.250	1.043	1.021	1.307
2.5	1.737	1.140	1.068	0.957
5.0	2.412	1.200	1.095	0.684
7.5	2.917	1.225	1.107	0.560
10	3.324	1.245	1.116	0.483
15	3.950	1.268	1.126	0.383
20	4.400	1.282	1.132	0.324
25	4.714	1.290	1.136	0.280
30	4.913	1.294	1.138	0.247
35	4.995	1.291	1.136	0.218
40	4.968	1.279	1.131	0.195
45	4.837	1.258	1.122	0.174
50	4.613	1.230	1.109	0.155
55	4.311	1.196	1.094	0.137
60	3.943	1.162	1.078	0.122
65	3.517	1.125	1.061	0.108
70	3.044	1.086	1.042	0.094
75	2.545	1.048	1.024	0.081
80	2.040	1.009	1.004	0.068
85	1.535	0.972	0.986	0.057
90	1.030	0.938	0.969	0.044
95	0.525	0.900	0.949	0.030
100	0.021	0	0	0

L.E. radius: 0.742 per cent c
T.E. radius: 0.023 per cent c

NACA 63A010 Basic Thickness Form

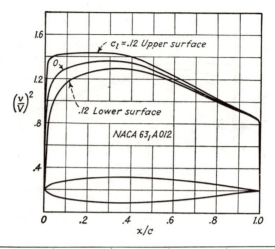

x (per cent c)	y (per cent c)	$(v/V)^2$	v/V	$\Delta v_a/V$
0	0	0	0	2.361
0.5	0.973	0.686	0.828	1.701
0.75	1.173	0.924	0.961	1.515
1.25	1.492	0.985	0.992	1.258
2.5	2.078	1.136	1.066	0.935
5.0	2.895	1.229	1.109	0.679
7.5	3.504	1.265	1.125	0.559
10	3.994	1.291	1.136	0.482
15	4.747	1.324	1.151	0.384
20	5.287	1.344	1.159	0.325
25	5.664	1.355	1.164	0.281
30	5.901	1.360	1.166	0.248
35	5.995	1.357	1.165	0.219
40	5.957	1.340	1.158	0.196
45	5.792	1.312	1.145	0.174
50	5.517	1.275	1.129	0.154
55	5.148	1.234	1.111	0.136
60	4.700	1.191	1.091	0.120
65	4.186	1.145	1.070	0.106
70	3.621	1.098	1.048	0.092
75	3.026	1.051	1.025	0.079
80	2.426	1.007	1.003	0.066
85	1.826	0.964	0.982	0.055
90	1.225	0.925	0.962	0.042
95	0.625	0.880	0.938	0.029
100	0.025	0	0	0

L.E. radius: 1.071 per cent c
T.E. radius: 0.028 per cent c

NACA 63_1A012 Basic Thickness Form

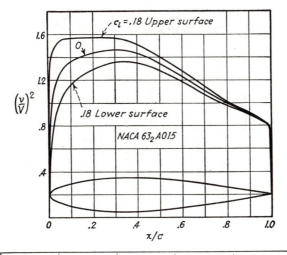

x (per cent c)	y (per cent c)	$(v/V)^2$	v/V	$\Delta v_a/V$
0	0	0	0	1.930
0.5	1.203	0.550	0.742	1.504
0.75	1.448	0.825	0.908	1.370
1.25	1.844	0.882	0.939	1.176
2.5	2.579	1.120	1.058	0.905
5.0	3.618	1.257	1.121	0.669
7.5	4.382	1.323	1.150	0.555
10	4.997	1.361	1.167	0.482
15	5.942	1.408	1.187	0.384
20	6.619	1.437	1.199	0.326
25	7.091	1.455	1.206	0.282
30	7.384	1.464	1.210	0.250
35	7.496	1.458	1.207	0.220
40	7.435	1.435	1.198	0.196
45	7.215	1.396	1.182	0.174
50	6.858	1.349	1.161	0.152
55	6.387	1.296	1.138	0.134
60	5.820	1.237	1.112	0.118
65	5.173	1.175	1.084	0.104
70	4.468	1.115	1.056	0.090
75	3.731	1.055	1.027	0.077
80	2.991	1.000	1.000	0.064
85	2.252	0.950	0.975	0.052
90	1.512	0.900	0.949	0.040
95	0.772	0.850	0.922	0.028
100	0.032	0	0	0

L.E. radius: 1.630 per cent c
T.E. radius: 0.037 per cent c

NACA 63_2A015 Basic Thickness Form

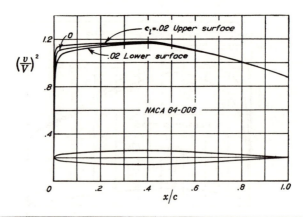

x (per cent c)	y (per cent c)	$(v/V)^2$	v/V	$\Delta v_a/V$
0	0	0	0	4.623
0.5	0.494	0.995	0.997	2.175
0.75	0.596	1.058	1.029	1.780
1.25	0.754	1.085	1.042	1.418
2.5	1.024	1.108	1.053	0.982
5.0	1.405	1.119	1.058	0.692
7.5	1.692	1.128	1.062	0.560
10	1.928	1.134	1.065	0.483
15	2.298	1.146	1.071	0.385
20	2.572	1.154	1.074	0.321
25	2.772	1.160	1.077	0.279
30	2.907	1.164	1.079	0.246
35	2.981	1.168	1.081	0.220
40	2.995	1.171	1.082	0.198
45	2.919	1.160	1.077	0.178
50	2.775	1.143	1.069	0.158
55	2.575	1.124	1.060	0.142
60	2.331	1.102	1.050	0.126
65	2.050	1.079	1.039	0.112
70	1.740	1.054	1.027	0.098
75	1.412	1.028	1.014	0.085
80	1.072	1.000	1.000	0.072
85	0.737	0.970	0.985	0.060
90	0.423	0.939	0.969	0.047
95	0.157	0.908	0.953	0.031
100	0	0.876	0.936	0

L.E. radius: 0.256 per cent c

NACA 64-006 Basic Thickness Form

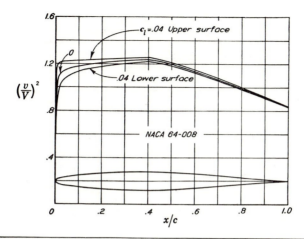

x (per cent c)	y (per cent c)	$(v/V)^2$	v/V	$\Delta v_a/V$
0	0	0	0	3.544
0.5	0.658	0.912	0.955	1.994
0.75	0.794	1.016	1.008	1.686
1.25	1.005	1.084	1.041	1.367
2.5	1.365	1.127	1.062	0.969
5.0	1.875	1.152	1.073	0.688
7.5	2.259	1.167	1.080	0.560
10	2.574	1.179	1.086	0.480
15	3.069	1.195	1.093	0.385
20	3.437	1.208	1.099	0.323
25	3.704	1.217	1.103	0.279
30	3.884	1.225	1.107	0.246
35	3.979	1.230	1.109	0.220
40	3.992	1.235	1.111	0.198
45	3.883	1.220	1.105	0.176
50	3.684	1.191	1.091	0.158
55	3.411	1.163	1.078	0.141
60	3.081	1.133	1.064	0.125
65	2.704	1.102	1.050	0.110
70	2.291	1.069	1.034	0.096
75	1.854	1.033	1.016	0.083
80	1.404	0.995	0.997	0.071
85	0.961	0.957	0.978	0.059
90	0.550	0.918	0.958	0.046
95	0.206	0.878	0.937	0.031
100	0	0.839	0.916	0

L.E. radius: 0.455 per cent c

NACA 64-008 Basic Thickness Form

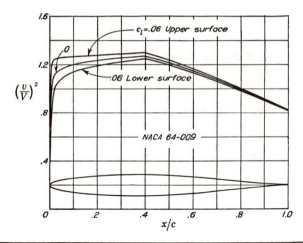

x (per cent c)	y (per cent c)	$(v/V)^2$	v/V	$\Delta v_a/V$
0	0	0	0	3.130
0.5	0.739	0.872	0.934	1.905
0.75	0.892	0.990	0.995	1.637
1.25	1.128	1.075	1.037	1.340
2.5	1.533	1.131	1.063	0.963
5.0	2.109	1.166	1.080	0.686
7.5	2.543	1.186	1.089	0.560
10	2.898	1.200	1.095	0.479
15	3.455	1.221	1.105	0.383
20	3.868	1.236	1.112	0.323
25	4.170	1.246	1.116	0.281
30	4.373	1.255	1.120	0.248
35	4.479	1.262	1.123	0.221
40	4.490	1.267	1.126	0.198
45	4.364	1.246	1.116	0.176
50	4.136	1.217	1.103	0.158
55	3.826	1.183	1.088	0.140
60	3.452	1.149	1.072	0.125
65	3.026	1.112	1.055	0.109
70	2.561	1.073	1.036	0.095
75	2.069	1.033	1.016	0.082
80	1.564	0.992	0.996	0.070
85	1.069	0.950	0.975	0.057
90	0.611	0.907	0.952	0.044
95	0.227	0.865	0,930	0.030
100	0	0.822	0.907	0̇

L.E. radius: 0.579 per cent c

NACA 64-009 Basic Thickness Form

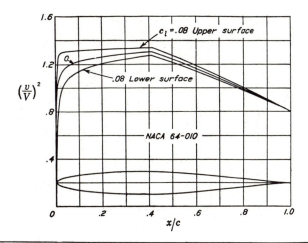

x (per cent c)	y (per cent c)	$(v/V)^2$	v/V	$\Delta v_a/V$
0	0	0	0	2.815
0.5	0.820	0.834	0.913	1.817
0.75	0.989	0.962	0.981	1.586
1.25	1.250	1.061	1.030	1.313
2.5	1.701	1.130	1.063	0.957
5.0	2.343	1.181	1.087	0.684
7.5	2.826	1.206	1.098	0.559
10	3.221	1.221	1.105	0.480
15	3.842	1.245	1.116	0.386
20	4.302	1.262	1.123	0.325
25	4.639	1.275	1.129	0.280
30	4.864	1.286	1.134	0.246
35	4.980	1.295	1.138	0.220
40	4.988	1.300	1.140	0.199
45	4.843	1.279	1.131	0.176
50	4.586	1.241	1.114	0.158
55	4.238	1.201	1.096	0.139
60	3.820	1.161	1.077	0.124
65	3.345	1.120	1.058	0.109
70	2.827	1.080	1.039	0.095
75	2.281	1.036	1.018	0.081
80	1.722	0.990	0.995	0.069
85	1.176	0.944	0.972	0.057
90	0.671	0.900	0.949	0.044
95	0.248	0.850	0.922	0.030
100	0	0.805	0.897	0
L.E. radius: 0.720 per cent c				

NACA 64-010 Basic Thickness Form

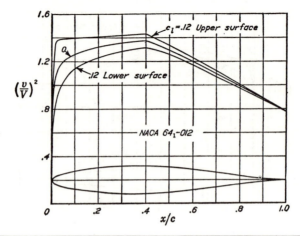

x (per cent c)	y (per cent c)	$(v/V)^2$	v/V	$\Delta v_a/V$
0	0	0	0	2.379
0.5	0.978	0.750	0.866	1.663
0.75	1.179	0.885	0.941	1.508
1.25	1.490	1.020	1.010	1.271
2.5	2.035	1.129	1.063	0.943
5.0	2.810	1.204	1.097	0.685
7.5	3.394	1.240	1.114	0.559
10	3.871	1.264	1.124	0.482
15	4.620	1.296	1.139	0.388
20	5.173	1.320	1.149	0.328
25	5.576	1.338	1.156	0.281
30	5.844	1.351	1.162	0.247
35	5.978	1.362	1.167	0.221
40	5.981	1.372	1.171	0.199
45	5.798	1.335	1.156	0.177
50	5.480	1.289	1.136	0.158
55	5.056	1.243	1.115	0.138
60	4.548	1.195	1.093	0.122
65	3.974	1.144	1.070	0.103
70	3.350	1.091	1.044	0.088
75	2.695	1.037	1.018	0.074
80	2.029	0.981	0.990	0.063
85	1.382	0.928	0.963	0.052
90	0.786	0.874	0.935	0.045
95	0.288	0.825	0.908	0.028
100	0	0.775	0.880	0

L.E. radius: 1.040 per cent c

NACA 64_1-012 Basic Thickness Form

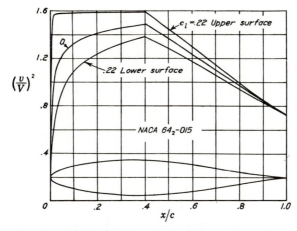

x (per cent c)	y (per cent c)	$(v/V)^2$	v/V	$\Delta v_a/V$
0	0	0	0	1.939
0.5	1.208	0.670	0.819	1.476
0.75	1.456	0.762	0.873	1.354
1.25	1.842	0.896	0.947	1.188
2.5	2.528	1.113	1.055	0.916
5.0	3.504	1.231	1.109	0.670
7.5	4.240	1.284	1.133	0.559
10	4.842	1.323	1.150	0.482
15	5.785	1.375	1.172	0.389
20	6.480	1.410	1.187	0.326
25	6.985	1.434	1.198	0.285
30	7.319	1.454	1.206	0.250
35	7.482	1.470	1.213	0.225
40	7.473	1.485	1.218	0.202
45	7.224	1.426	1.195	0.179
50	6.810	1.365	1.168	0.158
55	6.266	1.300	1.140	0.135
60	5.620	1.233	1.110	0.121
65	4.895	1.167	1.080	0.105
70	4.113	1.101	1.049	0.090
75	3.296	1.033	1.016	0.078
80	2.472	0.967	0.983	0.065
85	1.677	0.902	0.950	0.054
90	0.950	0.841	0.917	0.041
95	0.346	0.785	0.886	0.031
100	0	0.730	0.855	0

L.E. radius: 1.590 per cent c

NACA 64_2-015 Basic Thickness Form

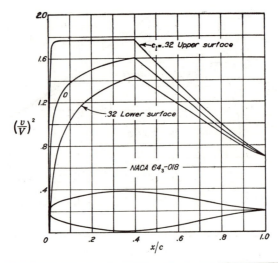

x (per cent c)	y (per cent c)	$(v/V)^2$	v/V	$\Delta v_a/V$
0	0	0	0	1.646
0.5	1.428	0.546	0.739	1.360
0.75	1.720	0.705	0.840	1.269
1.25	2.177	0.862	0.920	1.128
2.5	3.005	1.079	1.039	0.904
5.0	4.186	1.244	1.115	0.669
7.5	5.076	1.327	1.152	0.558
10	5.803	1.380	1.175	0.486
15	6.942	1.450	1.204	0.391
20	7.782	1.497	1.224	0.331
25	8.391	1.535	1.239	0.288
30	8.789	1.562	1.250	0.255
35	8.979	1.585	1.259	0.228
40	8.952	1.600	1.265	0.200
45	8.630	1.518	1.232	0.177
50	8.114	1.436	1.198	0,154
55	7.445	1.354	1.164	0.134
60	6.658	1.272	1.128	0,117
65	5.782	1.190	1.091	0.102
70	4.842	1.109	1.053	0.088
75	3.866	1.028	1.014	0.074
80	2.888	0.952	0.976	0.063
85	1.951	0.879	0.937	0.051
90	1.101	0.812	0.901	0.039
95	0.400	0.747	0.864	0.027
100	0	0.695	0.834	0

L.E. radius: 2.208 per cent c

NACA 64_3-018 Basic Thickness Form

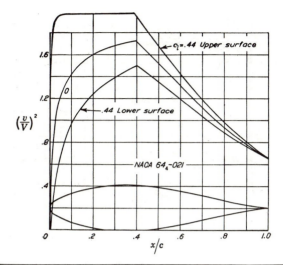

x (per cent c)	y (per cent c)	$(v/V)^2$	v/V	$\Delta v_a/V$
0	0	0	0	1.458
0.5	1.646	0.462	0.680	1.274
0.75	1.985	C.603	0.776	1.203
1.25	2.517	0.759	0.871	1.084
2.5	3.485	1.010	1.005	0.878
5.0	4.871	1.248	1.117	0.665
7.5	5.915	1.358	1.165	0.557
10	6.769	1.431	1.196	0.486
15	8.108	1.527	1.236	0.395
20	9.095	1.593	1.262	0.335
25	9.807	1.654	1.281	0.293
30	10.269	1.681	1.297	0.259
35	10.481	1.712	1.308	0.232
40	10.431	1.709	1.307	0.202
45	10.030	1.607	1.268	0.178
50	9.404	1.507	1.228	0.155
55	8.607	1.406	1.186	0.134
60	7.678	1.307	1.143	0.116
65	6.649	1.209	1.099	0.099
70	5.549	1.112	1.055	0.084
75	4.416	1.020	1.010	0.071
80	3.287	0.932	0.965	0.059
85	2.213	0.851	0.923	0.047
90	1.245	0.778	0.882	0.036
95	0.449	0.711	0.844	0.022
100	0	0.653	0.808	0

L.E. radius: 2.884 per cent c

NACA 64$_4$-021 Basic Thickness Form

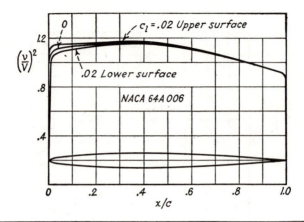

x (per cent c)	y (per cent c)	$(v/V)^2$	v/V	$\Delta v_a/V$
0	0	0	0	4.688
0.5	0.485	1.019	1.009	2.101
0.75	0.585	1.046	1.023	1.798
1.25	0.739	1.076	1.037	1.422
2.5	1.016	1.106	1.052	0.980
5.0	1.399	1.118	1.057	0.694
7.5	1.684	1.126	1.061	0.564
10	1.919	1.132	1.064	0.482
15	2.283	1.141	1.068	0.382
20	2.557	1.149	1.072	0.321
25	2.757	1.154	1.074	0.278
30	2.896	1.158	1.076	0.246
35	2.977	1.162	1.078	0.219
40	2.999	1.165	1.079	0.197
45	2.945	1.156	1.075	0.177
50	2.825	1.142	1.069	0.159
55	2.653	1.125	1.061	0.143
60	2.438	1.107	1.052	0.126
65	2.188	1.087	1.043	0.112
70	1.907	1.066	1.032	0.099
75	1.602	1.043	1.021	0.087
80	1.285	1.018	1.009	0.074
85	0.967	0.992	0.996	0.061
90	0.649	0.964	0.982	0.047
95	0.331	0.935	0.967	0.033
100	0.013	0	0	0

L.E. radius: 0.246 per cent c
T.E. radius: 0.014 per cent c

NACA 64A006 Basic Thickness Form

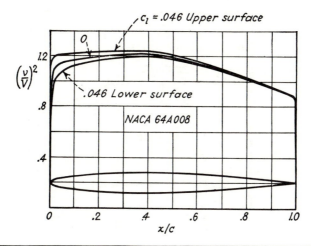

x (per cent c)	y (per cent c)	$(v/V)^2$	v/V	$\Delta v_a/V$
0	0	0	0	3.546
0.5	0.646	0.947	0.973	1.972
0.75	0.778	1.005	1.002	1.697
1.25	0.983	1.068	1.033	1.352
2.5	1.353	1.122	1.059	0.971
5.0	1.863	1.151	1.073	0.692
7.5	2.245	1.165	1.079	0.564
10	2.559	1.176	1.084	0.481
15	3.047	1.191	1.091	0.382
20	3.414	1.201	1.096	0.323
25	3.681	1.209	1.100	0.279
30	3.866	1.217	1.103	0.247
35	3.972	1.221	1.105	0.221
40	3.998	1.225	1.107	0.198
45	3.921	1.211	1.100	0.177
50	3.757	1.191	1.091	0.158
55	3.524	1.167	1.080	0.141
60	3.234	1.141	1.068	0.125
65	2.897	1.113	1.055	0.111
70	2.521	1.084	1.041	0.098
75	2.117	1.053	1.026	0.084
80	1.698	1.020	1.010	0.072
85	1.278	0.987	0.993	0.059
90	0.858	0.951	0.975	0.045
95	0.438	0.914	0.956	0.032
100	0.018	0	0	0

L.E. radius: 0.439 per cent c
T.E. radius: 0.020 per cent c

NACA 64A008 Basic Thickness Form

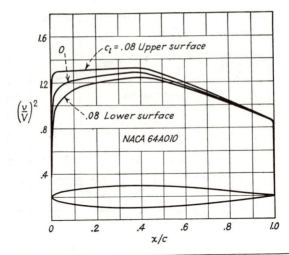

x (per cent c)	y (per cent c)	$(v/V)^2$	v/V	$\Delta v_a/V$
0	0	0	0	2.868
0.5	0.804	0.868	0.932	1.845
0.75	0.969	0.952	0.976	1.603
1.25	1.225	1.042	1.021	1.300
2.5	1.688	1.130	1.063	0.957
5.0	2.327	1.178	1.085	0.688
7.5	2.905	1.201	1.096	0.562
10	3.199	1.217	1.103	0.480
15	3.813	1.238	1.113	0.382
20	4.272	1.254	1.120	0.324
25	4.606	1.266	1.125	0.280
30	4.837	1.275	1.129	0.248
25	4.968	1.282	1.132	0.221
40	4.995	1.288	1.135	0.199
45	4.894	1.268	1.126	0.177
50	4.684	1.240	1.114	0.158
55	4.388	1.208	1.099	0.140
60	4.021	1.174	1.084	0.124
65	3.597	1.139	1.067	0.109
70	3.127	1.102	1.050	0.096
75	2.623	1.063	1.031	0.083
80	2.103	1.023	1.011	0.070
85	1.582	0.981	0.990	0.058
90	1.062	0.938	0.969	0.044
95	0.541	0.893	0.945	0.031
100	0.021	0	0	0

L.E. radius: 0.687 per cent c
T.E. radius: 0.023 per cent c

NACA 64_3-018 Basic Thickness Form

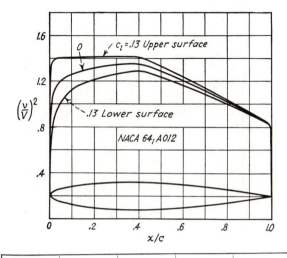

x (per cent c)	y (per cent c)	$(v/V)^2$	v/V	$\Delta v_a/V$
0	0	0	0	2.408
0.5	0.961	0.792	0.890	1.720
0.75	1.158	0.893	0.945	1.515
1.25	1.464	1.006	1.003	1.254
2.5	2.018	1.127	1.062	0.941
5.0	2.788	1.201	1.096	0.681
7.5	3.364	1.235	1.111	0.560
10	3.839	1.257	1.121	0.478
15	4.580	1.288	1.135	0.383
20	5.132	1.308	1.144	0.325
25	5.534	1.324	1.151	0.281
30	5.809	1.336	1.156	0.249
35	5.965	1.346	1.160	0.221
40	5.993	1.354	1.164	0.199
45	5.863	1.326	1.152	0.177
50	5.605	1.289	1.135	0.157
55	5.244	1.250	1.118	0.139
60	4.801	1.207	1.099	0.123
65	4.289	1.164	1.079	0.108
70	3.721	1.118	1.057	0.094
75	3.118	1.071	1.035	0.080
80	2.500	1.023	1.011	0.068
85	1.882	0.974	0.987	0.056
90	1.263	0.925	0.962	0.042
95	0.644	0.873	0.934	0.029
100	0.025	0	0	0

L.E. radius: 0.994 per cent c
T.E. radius: 0.028 per cent c

NACA 64₁A012 Basic Thickness Form

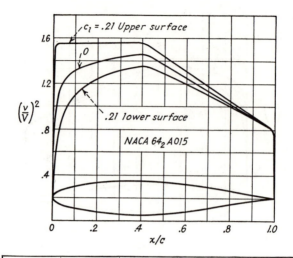

x (per cent c)	y (per cent c)	$(v/V)^2$	v/V	$\Delta v_a/V$
0	0	0	0	1.956
0.5	1.193	0.678	0.823	1.552
0.75	1.436	0.789	0.888	1.404
1.25	1.815	0.936	0.967	1.189
2.5	2.508	1.110	1.054	0.912
5.0	3.477	1.226	1.107	0.671
7.5	4.202	1.280	1.131	0.552
10	4.799	1.314	1.146	0.478
15	5.732	1.360	1.166	0.384
20	6.423	1.390	1.179	0.326
25	6.926	1.413	1.189	0.283
30	7.270	1.430	1.196	0.249
35	7.463	1.445	1.202	0.222
40	7.487	1.458	1.207	0.201
45	7.313	1.414	1.189	0.177
50	6.978	1.364	1.168	0.156
55	6.517	1.311	1.145	0.137
60	5.956	1.255	1.120	0.121
65	5.311	1.198	1.095	0.106
70	4.600	1.139	1.067	0.091
75	3.847	1.079	1.039	0.078
80	3.084	1.020	1.010	0.065
85	2.321	0.961	0.980	0.053
90	1.558	0.901	0.949	0.041
95	0.795	0.843	0.918	0.027
100	0.032	0	0	0

L.E. radius: 1.561 per cent c
T.E. radius: 0.037 per cent c

NACA 64₂A015 Basic Thickness Form

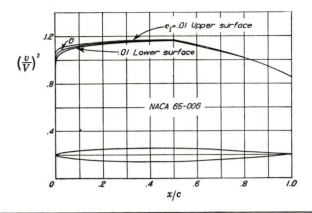

x (per cent c)	y (per cent c)	$(v/V)^2$	v/V	$\Delta v_a/V$
0	0	0	0	4.815
0.5	0.476	1.044	1.022	2.110
0.75	0.574	1.055	1.027	1.780
1.25	0.717	1.063	1.031	1.390
2.5	0.956	1.081	1.040	0.965
5.0	1.310	1.100	1.049	0.695
7.5	1.589	1.112	1.055	0.560
10	1.824	1.120	1.058	0.474
15	2.197	1.134	1.065	0.381
20	2.482	1.143	1.069	0.322
25	2.697	1.149	1.072	0.281
30	2.852	1.155	1.075	0.247
35	2.952	1.159	1.077	0.220
40	2.998	1.163	1.078	0.198
45	2.983	1.166	1.080	0.178
50	2.900	1.165	1.079	0.160
55	2.741	1.145	1.070	0.144
60	2.518	1.124	1.060	0.128
65	2.246	1.100	1.049	0.114
70	1.935	1.073	1.036	0.100
75	1.594	1.044	1.022	0.086
80	1.233	1.013	1.006	0.074
85	0.865	0.981	0.990	0.060
90	0.510	0.944	0.972	0.046
95	0.195	0.902	0.950	0.031
100	0	0.858	0.926	0
L.E. radius: 0.240 per cent c				

NACA 65-006 Basic Thickness Form

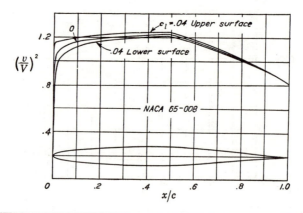

x (per cent c)	y (per cent c)	$(v/V)^2$	v/V	$\Delta v_a/V$
0	0	0	0	3.695
0.5	0.627	0.978	0.989	2.010
0.75	0.756	1.010	1.005	1.696
1.25	0.945	1.043	1.021	1.340
2.5	1.267	1.086	1.042	0.956
5.0	1.745	1.125	1.061	0.689
7.5	2.118	1.145	1.070	0.560
10	2.432	1.158	1.076	0.477
15	2.931	1.178	1.085	0.382
20	3.312	1.192	1.092	0.323
25	3.599	1.203	1.097	0.281
30	3.805	1.210	1.100	0.248
35	3.938	1.217	1.103	0.221
40	3.998	1.222	1.105	0.199
45	3.974	1.226	1.107	0.178
50	3.857	1.222	1.105	0.160
55	3.638	1.193	1.092	0.145
60	3.337	1.163	1.078	0.128
65	2.971	1.130	1.063	0.113
70	2.553	1.094	1.046	0.098
75	2.096	1.055	1.027	0.084
80	1.617	1.014	1.007	0.072
85	1.131	0.971	0.985	0.059
90	0.664	0.923	0.961	0.044
95	0.252	0.873	0.934	0.031
100	0	0.817	0.904	0

L.E. radius: 0.434 per cent c

NACA 65-008 Basic Thickness Form

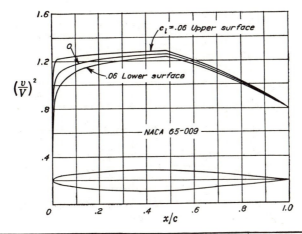

x (per cent c)	y (per cent c)	$(v/V)^2$	v/V	$\Delta v_a/V$
0	0	0	0	3.270
0.5	0.700	0.945	0.972	1.962
0.75	0.845	0.985	0.992	1.655
1.25	1.058	1.037	1.018	1.315
2.5	1.421	1.089	1.044	0.950
5.0	1.961	1.134	1.065	0.687
7.5	2.383	1.159	1.077	0.560
10	2.736	1.177	1.085	0.477
15	3.299	1.200	1.095	0.382
20	3.727	1.216	1.103	0.323
25	4.050	1.229	1.109	0.280
30	4.282	1.238	1.113	0.248
35	4.431	1.246	1.116	0.220
40	4.496	1.252	1.119	0.198
45	4.469	1.258	1.122	0.178
50	4.336	1.250	1.118	0.160
55	4.086	1.220	1.105	0.144
60	3.743	1.185	1.089	0.128
65	3.328	1.145	1.070	0.111
70	2.856	1.103	1.050	0.097
75	2.342	1.059	1.029	0.084
80	1.805	1.013	1.006	0.071
85	1.260	0.963	0.981	0.059
90	0.738	0.912	0.955	0.044
95	0.280	0.856	0.925	0.030
100	0	0.797	0.893	0

L.E. radius: 0.552 per cent c

NACA 65-009 Basic Thickness Form

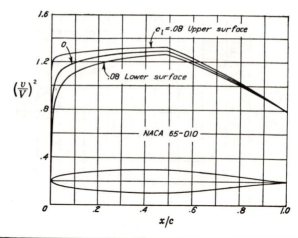

x (per cent c)	y (per cent c)	$(v/V)^2$	v/V	$\Delta v_a/V$
0	0	0	0	2.967
0.5	0.772	0.911	0.954	1.911
0.75	0.932	0.960	0.980	1.614
1.25	1.169	1.025	1.012	1.292
2.5	1.574	1.085	1.042	0.932
5.0	2.177	1.143	1.069	0.679
7.5	2.647	1.177	1.085	0.558
10	3.040	1.197	1.094	0.480
15	3.666	1.224	1.106	0.383
20	4.143	1.242	1.114	0.321
25	4.503	1.257	1.121	0.280
30	4.760	1.268	1.126	0.248
35	4.924	1.277	1.130	0.222
40	4.996	1.284	1.133	0.199
45	4.963	1.290	1.136	0.179
50	4.812	1.284	1.133	0.160
55	4.530	1.244	1.115	0.141
60	4.146	1.202	1.096	0.126
65	3.682	1.158	1.076	0.110
70	3.156	1.112	1.055	0.097
75	2.584	1.062	1.031	0.082
80	1.987	1.011	1.005	0.070
85	1.385	0.958	0.979	0.058
90	0.810	0.903	0.950	0.045
95	0.306	0.844	0.919	0.030
100	0	0.781	0.884	0

L.E. radius: 0.687 per cent c

NACA 65-010 Basic Thickness Form

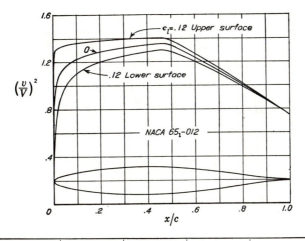

x (per cent c)	y (per cent c)	$(v/V)^2$	v/V	$\Delta v_a/V$
0	0	0	0	2.444
0.5	0.923	0.848	0.921	1.776
0.75	1.109	0.935	0.967	1.465
1.25	1.387	1.000	1.000	1.200
2.5	1.875	1.082	1.040	0.931
5.0	2.606	1.162	1.078	0.702
7.5	3.172	1.201	1.096	0.568
10	3.647	1.232	1.110	0.480
15	4.402	1.268	1.126	0.389
20	4.975	1.295	1.138	0.326
25	5.406	1.316	1.147	0.282
30	5.716	1.332	1.154	0.251
35	5.912	1.343	1.159	0.223
40	5.997	1.350	1.162	0.204
45	5.949	1.357	1.165	0.188
50	5.757	1.343	1.159	0.169
55	5.412	1.295	1.138	0.145
60	4.943	1.243	1.115	0.127
65	4.381	1.188	1.090	0.111
70	3.743	1.134	1.065	0.094
75	3.059	1.073	1.036	0.074
80	2.345	1.010	1.005	0.062
85	1.630	0.949	0.974	0.049
90	0.947	0.884	0.940	0.038
95	0.356	0.819	0.905	0.025
100	0	0.748	0.865	0

L.E. radius: 1.000 per cent c

NACA 65$_1$-012 Basic Thickness Form

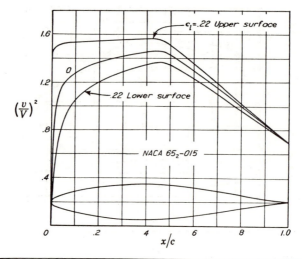

x (per cent c)	y (per cent c)	$(v/V)^2$	v/V	$\Delta v_a/V$
0	0	0	0	2.038
0.5	1.124	0.654	0.809	1.729
0.75	1.356	0.817	0.904	1.390
1.25	1.702	0.939	0.969	1.156
2.5	2.324	1.063	1.031	0.920
5.0	3.245	1.184	1.088	0.682
7.5	3.959	1.241	1.114	0.563
10	4.555	1.281	1.132	0.487
15	5.504	1.336	1.156	0.393
20	6.223	1.374	1.172	0.334
25	6.764	1.397	1.182	0.290
30	7.152	1.418	1.191	0.255
35	7.396	1.438	1.199	0.227
40	7.498	1.452	1.205	0.203
45	7.427	1.464	1.210	0.184
50	7.168	1.433	1.197	0.160
55	6.720	1.369	1.170	0.143
60	6.118	1.297	1.139	0.127
65	5.403	1.228	1.108	0.109
70	4.600	1.151	1.073	0.096
75	3.744	1.077	1.038	0.078
80	2.858	1.002	1.001	0.068
85	1.977	0.924	0.961	0.052
90	1.144	0.846	0.920	0.038
95	0.428	0.773	0.879	0.026
100	0	0.697	0.835	0

L.E. radius: 1.505 per cent c

NACA 65_2-015 Basic Thickness Form

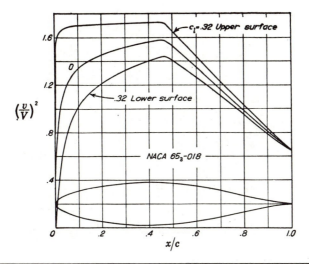

x (per cent c)	y (per cent c)	$(v/V)^2$	v/V	$\Delta v_a/V$
0	0	0	0	1.746
0.5	1.337	0.625	0.791	1.437
0.75	1.608	0.702	0.838	1.302
1.25	2.014	0.817	0.904	1.123
2.5	2.751	1.020	1.010	0.858
5.0	3.866	1.192	1.092	0.650
7.5	4.733	1.275	1.129	0.542
10	5.457	1.329	1.153	0.474
15	6.606	1.402	1.184	0.385
20	7.476	1.452	1.205	0.327
25	8.129	1.488	1.220	0.285
30	8.595	1.515	1.231	0.251
35	8.886	1.539	1.241	0.225
40	8.999	1.561	1.249	0.203
45	8.901	1.578	1.256	0.182
50	8.568	1.526	1.235	0.157
55	8.008	1.440	1.200	0.137
60	7.267	1.353	1.163	0.118
65	6.395	1.262	1.123	0.104
70	5.426	1.170	1.082	0.087
75	4.396	1.076	1.037	0.074
80	3.338	0.985	0.992	0.062
85	2.295	0.896	0.947	0.050
90	1.319	0.813	0.902	0.039
95	0.490	0.730	0.854	0.026
100	0	0.657	0.811	0

L.E. radius: 1.96 per cent c

NACA 65_3-018 Basic Thickness Form

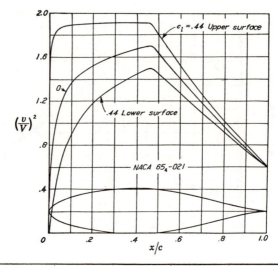

x (per cent c)	y (per cent c)	$(v/V)^2$	v/V	$\Delta v_a/V$
0	0	0	0	1.531
0.5	1.522	0.514	0.717	1.333
0.75	1.838	0.607	0.779	1.215
1.25	2.301	0.740	0.860	1.062
2.5	3.154	0.960	0.980	0.838
5.0	4.472	1.186	1.089	0.649
7.5	5.498	1.293	1.137	0.544
10	6.352	1.371	1.171	0.478
15	7.700	1.469	1.212	0.388
20	8.720	1.533	1.238	0.330
25	9.487	1.580	1.257	0.289
30	10.036	1.621	1.273	0.255
35	10.375	1.654	1.286	0.229
40	10.499	1.680	1.296	0.206
45	10.366	1.700	1.304	0.184
50	9.952	1.633	1.278	0.158
55	9.277	1.508	1.228	0.139
60	8.390	1.397	1.182	0.120
65	7.360	1.286	1.134	0.101
70	6.224	1.177	1.085	0.087
75	5.024	1.073	1.036	0.073
80	3.800	0.970	0.985	0.058
85	2.598	0.872	0.934	0.047
90	1.484	0.778	0.882	0.035
95	0.546	0.694	0.833	0.020
100	0	0.616	0.785	0

L.E. radius: 2.50 per cent c

NACA 65_4-021 Basic Thickness Form

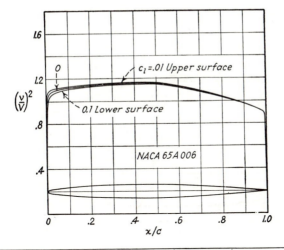

x (per cent c)	y (per cent c)	$(v/V)^2$	v/V	$\Delta v_a/V$
0	0	0	0	4.879
0.5	0.464	1.034	1.017	2.145
0.75	0.563	1.043	1.021	1.763
1.25	0.718	1.058	1.029	1.365
2.5	0.981	1.080	1.039	0.966
5.0	1.313	1.101	1.049	0.688
7.5	1.591	1.112	1.055	0.562
10	1.824	1.120	1.058	0.480
15	2.194	1.131	1.063	0.382
20	2.474	1.139	1.067	0.323
25	2.687	1.145	1.070	0.278
30	2.842	1.149	1.072	0.246
35	2.945	1.153	1.074	0.219
40	2.996	1.157	1.076	0.198
45	2.992	1.159	1.077	0.178
50	2.925	1.157	1.076	0.161)
55	2.793	1.141	1.068	0.143
60	2.602	1.124	1.060	0.127
65	2.364	1.106	1.052	0.112
70	2.087	1.083	1.041	0.099
75	1.775	1.059	1.029	0.087
80	1.437	1.032	1.016	0.076
85	1.083	1.003	1.001	0.061
90	0.727	0.973	0.986	0.047
95	0.370	0.936	0.967	0.033
100	0.013	0	0	0

L.E. radius: 0.229 per cent c
T.E. radius: 0.014 per cent c

NACA 65A006 Basic Thickness Form

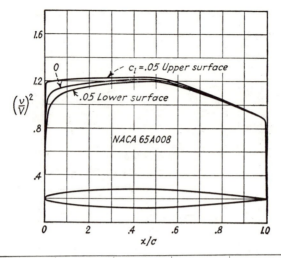

x (per cent c)	y (per cent c)	$(v/V)^2$	v/V	$\Delta v_a/V$
0	0	0	0	3.698
0.5	0.615	0.973	0.986	2.010
0.75	0.746	1.001	1.000	1.693
1.25	0.951	1.038	1.019	1.333
2.5	1.303	1.088	1.043	0.954
5.0	1.749	1.127	1.062	0.685
7.5	2.120	1.145	1.070	0.561
10	2.432	1.157	1.076	0.479
15	2.926	1.175	1.084	0.382
20	3.301	1.186	1.089	0.322
25	3.585	1.195	1.093	0.279
30	3.791	1.202	1.096	0.247
35	3.928	1.207	1.099	0.219
40	3.995	1.213	1.101	0.198
45	3.988	1.217	1.103	0.178
50	3.895	1.214	1.102	0.161
55	3.714	1.191	1.091	0.144
60	3.456	1.167	1.080	0.128
65	3.135	1.139	1.067	0.112
70	2.763	1.108	1.053	0.098
75	2.348	1.076	1.037	0.086
80	1.898	1.041	1.020	0.073
85	1.430	1.002	1.001	0.060
90	0.960	0.961	0.980	0.046
95	0.489	0.916	0.957	0.031
100	0.018	0	0	0

L.E. radius: 0.408 per cent c
T.E. radius: 0.020 per cent c

NACA 65A008 Basic Thickness Form

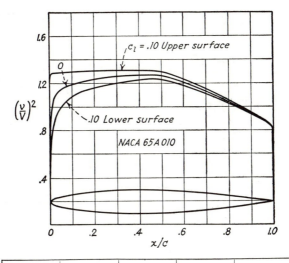

x (per cent c)	y (per cent c)	$(v/V)^2$	v/V	$\Delta v_a/V$
0	0	0	0	2.987
0.5	0.765	0.897	0.947	1.878
0.75	0.928	0.948	0.974	1.619
1.25	1.183	1.010	1.005	1.303
2.5	1.623	1.089	1.044	0.936
5.0	2.182	1.148	1.071	0.679
7.5	2.650	1.176	1.084	0.559
10	3.040	1.194	1.093	0.478
15	3.658	1.218	1.104	0.382
20	4.127	1.234	1.111	0.323
25	4.483	1.247	1.117	0.281
30	4.742	1.257	1.121	0.249
35	4.912	1.265	1.125	0.222
40	4.995	1.272	1.128	0.198
45	4.983	1.277	1.130	0.178
50	4.863	1.271	1.127	0.161
55	4.632	1.241	1.114	0.144
60	4.304	1.208	1.099	0.127
65	3.899	1.172	1.083	0.111
70	3.432	1.133	1.064	0.097
75	2.912	1.091	1.045	0.084
80	2.352	1.047	1.023	0.071
85	1.771	0.999	0.999	0.058
90	1.188	0.949	0.974	0.045
95	0.604	0.893	0.945	0.029
100	0.021	0	0	0

L.E. radius: 0.639 per cent c
T.E. radius: 0.023 per cent c

NACA 65A010 Basic Thickness Form

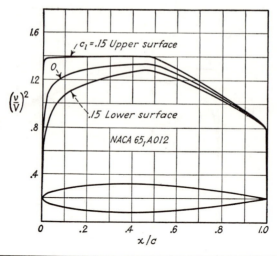

x (per cent c)	y (per cent c)	$(v/V)^2$	v/V	$\Delta v_a/V$
0	0	0	0	2.520
0.5	0.913	0.824	0.908	1.757
0.75	1.106	0.883	0.940	1.543
1.25	1.414	0.969	0.984	1.263
2.5	1.942	1.081	1.040	0.914
5.0	2.614	1.166	1.080	0.672
7.5	3.176	1.204	1.097	0.557
10	3.647	1.228	1.108	0.477
15	4.392	1.263	1.124	0.382
20	4.956	1.285	1.134	0.324
25	5.383	1.301	1.141	0.281
30	5.693	1.313	1.146	0.250
35	5.897	1.324	1.151	0.224
40	5.995	1.332	1.154	0.198
45	5.977	1.338	1.157	0.178
50	5.828	1.329	1.153	0.161
55	5.544	1.292	1.137	0.143
60	5.143	1.251	1.118	0.126
65	4.654	1.204	1.097	0.111
70	4.091	1.156	1.075	0.096
75	3.467	1.104	1.051	0.082
80	2.798	1.051	1.025	0.069
85	2.106	0.994	0.997	0.057
90	1.413	0.936	0.967	0.043
95	0.719	0.871	0.933	0.027
100	0.025	0	0	0

L.E. radius: 0.922 per cent c
T.E. radius: 0.029 per cent c

NACA 65_1A012 Basic Thickness Form

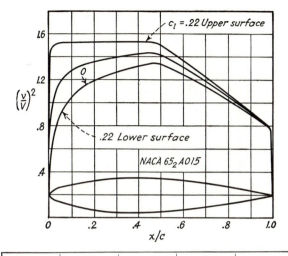

x (per cent c)	y (per cent c)	$(v/V)^2$	v/V	$\Delta v_a/V$
0	0	0	0	2.048
0.5	1.131	0.714	0.845	1.586
0.75	1.371	0.781	0.884	1.417
1.25	1.750	0.891	0.944	1.195
2.5	2.412	1.059	1.029	0.880
5.0	3.255	1.187	1.089	0.660
7.5	3.962	1.243	1.115	0.553
10	4.553	1.280	1.131	0.476
15	5.488	1.328	1.152	0.382
20	6.198	1.359	1.166	0.326
25	6.734	1.383	1.176	0.282
30	7.122	1.401	1.184	0.252
35	7.376	1.416	1.190	0.227
40	7.496	1.427	1.195	0.204
45	7.467	1.437	1.199	0.181
50	7.269	1.419	1.191	0.161
55	6.903	1.368	1.170	0.142
60	6.393	1.311	1.145	0.124
65	5.772	1.249	1.118	0.109
70	5.063	1.186	1.089	0.094
75	4.282	1.123	1.060	0.080
80	3.451	1.056	1.028	0.067
85	2.598	0.986	0.993	0.055
90	1.743	0.913	0.956	0.041
95	0.887	0.841	0.917	0.026
100	0.032	0	0	0

L.E. radius: 1.446 per cent c
T.E. radius: 0.038 per cent c

NACA 65₂A015 Basic Thickness Form

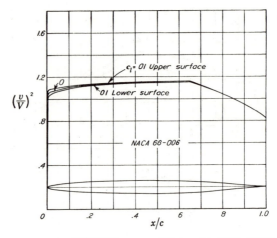

x (per cent c)	y (per cent c)	$(v/V)^2$	v/V	$\Delta v_a/V$
0	0	0	0	4.941
0.5	0.461	1.052	1.026	2.500
0.75	0.554	1.057	1.028	2.020
1.25	0.693	1.062	1.031	1.500
2.5	0.918	1.071	1.035	0.967
5.0	1.257	1.086	1.042	0.695
7.5	1.524	1.098	1.048	0.554
10	1.752	1.107	1.052	0.474
15	2.119	1.119	1.058	0.379
20	2.401	1.128	1.062	0.320
25	2.618	1.133	1.064	0.278
30	2.782	1.138	1.067	0.245
35	2.899	1.142	1.069	0.219
40	2.971	1.145	1.070	0.197
45	3.000	1.148	1.071	0.178
50	2.985	1.151	1.073	0.161
55	2.925	1.153	1.074	0.145
60	2.815	1.155	1.075	0.130
65	2.611	1.154	1.074	0.116
70	2.316	1.118	1.057	0.102
75	1.953	1.081	1.040	0.089
80	1.543	1.040	1.020	0.075
85	1.107	0.996	0.998	0.061
90	0.665	0.948	0.974	0.047
95	0.262	0.890	0.943	0.030
100	0	0.822	0.907	0
L.E. radius: 0.223 per cent c				

NACA 66-006 Basic Thickness Form

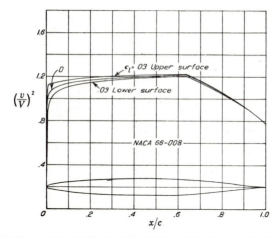

x (per cent c)	y (per cent c)	$(v/V)^2$	v/V	$\Delta v_a/V$
0	0	0	0	3.794
0.5	0.610	0.968	0.984	2.220
0.75	0.735	1.023	1.011	1.825
1.25	0.919	1.046	1.023	1.388
2.5	1.219	1.078	1.038	0.949
5.0	1.673	1.107	1.052	0.689
7.5	2.031	1.128	1.062	0.552
10	2.335	1.141	1.068	0.474
15	2.826	1.158	1.076	0.379
20	3.201	1.171	1.082	0.321
25	3.490	1.178	1.085	0.278
30	3.709	1.186	1.089	0.246
35	3.865	1.191	1.091	0.220
40	3.962	1.196	1.094	0.198
45	4.000	1.201	1.096	0.178
50	3.978	1.205	1.098	0.161
55	3.896	1.208	1.099	0.145
60	3.740	1.213	1.101	0.130
65	3.459	1.202	1.096	0.115
70	3.062	1.156	1.075	0.101
75	2.574	1.103	1.050	0.087
80	2.027	1.048	1.024	0.073
85	1.447	0.989	0.994	0.058
90	0.864	0.926	0.962	0.045
95	0.338	0.855	0.925	0.029
100	0	0.768	0.876	0

L.E. radius: 0.411 per cent c

NACA 66-008 Basic Thickness Form

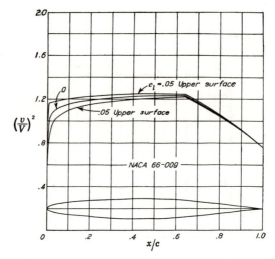

x (per cent c)	y (per cent c)	$(v/V)^2$	v/V	$\Delta v_a/V$
0	0	0	0	3.352
0.5	0.687	0.930	0.964	2.100
0.75	0.824	0.999	0.999	1.750
1.25	1.030	1.036	1.018	1.340
2.5	1.368	1.079	1.039	0.940
5.0	1.880	1.119	1.058	0.686
7.5	2.283	1.142	1.069	0.552
10	2.626	1.159	1.077	0.473
15	3.178	1.178	1.085	0.379
20	3.601	1.190	1.091	0.323
25	3.927	1.201	1.096	0.280
30	4.173	1.210	1.100	0.246
35	4.348	1.217	1.103	0.220
40	4.457	1.221	1.105	0.197
45	4.499	1.228	1.108	0.178
50	4.475	1.232	1.110	0.161
55	4.381	1.237	1.112	0.145
60	4.204	1.240	1.114	0.130
65	3.882	1.230	1.109	0.116
70	3.428	1.172	1.083	0.100
75	2.877	1.113	1.055	0.085
80	2.263	1.050	1.025	0.071
85	1.611	0.985	0.992	0.057
90	0.961	0.915	0.957	0.043
95	0.374	0.839	0.916	0.028
100	0	0.747	0.864	0

L.E. radius: 0.530 per cent c

NACA 66-009 Basic Thickness Form

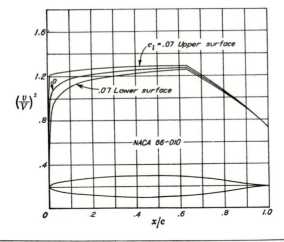

x (per cent c)	y (per cent c)	$(v/V)^2$	v/V	$\Delta v_a/V$
0	0	0	0	3.002
0.5	0.759	0.896	0.947	2.012
0.75	0.913	0.972	0.986	1.686
1.25	1.141	1.023	1.011	1.296
2.5	1.516	1.078	1.038	0.931
5.0	2.087	1.125	1.061	0.682
7.5	2.536	1.154	1.074	0.551
10	2.917	1.174	1.084	0.473
15	3.530	1.198	1.095	0.379
20	4.001	1.215	1.102	0.322
25	4.363	1.226	1.107	0.279
30	4.636	1.236	1.112	0.246
35	4.832	1.243	1.115	0.220
40	4.953	1.249	1.118	0.198
45	5.000	1.255	1.120	0.178
50	4.971	1.261	1.123	0.161
55	4.865	1.265	1.125	0.146
60	4.665	1.270	1.127	0.130
65	4.302	1.250	1.118	0.114
70	3.787	1.190	1.091	0.099
75	3.176	1.121	1.059	0.085
80	2.494	1.052	1.026	0.070
85	1.773	0.979	0.989	0.056
90	1.054	0.904	0.951	0.043
95	0.408	0.821	0.906	0.027
100	0	0.729	0.854	0

L.E. radius: 0.662 per cent c

NACA 66-010 Basic Thickness Form

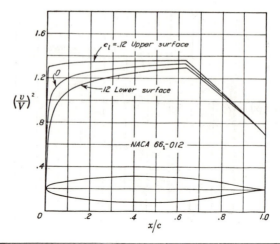

x (per cent c)	y (per cent c)	$(v/V)^2$	v/V	$\Delta v_a/V$
0	0	0	0	2.569
0.5	0.906	0.800	0.894	1.847
0.75	1.087	0.915	0.957	1.575
1.25	1.358	0.980	0.990	1.237
2.5	1.808	1.073	1.036	0.913
5.0	2.496	1.138	1.067	0.674
7.5	3.037	1.177	1.085	0.549
10	3.496	1.204	1.097	0.473
15	4.234	1.237	1.112	0.380
20	4.801	1.259	1.122	0.323
25	5.238	1.275	1.129	0.280
30	5.568	1.287	1.134	0.246
35	5.803	1.297	1.139	0.221
40	5.947	1.303	1.142	0.197
45	6.000	1.311	1.145	0.176
50	5.965	1.318	1.148	0.162
55	5.836	1.323	1.150	0.147
60	5.588	1.331	1.154	0.132
65	5.139	1.302	1.141	0.113
70	4.515	1.221	1.105	0.098
75	3.767	1.139	1.067	0.084
80	2.944	1.053	1.026	0.069
85	2.083	0.968	0.984	0.053
90	1.234	0.879	0.938	0.040
95	0.474	0.788	0.888	0.031
100	0	0.687	0.829	0

L.E. radius: 0.952 per cent c

NACA 66_1-012 Basic Thickness Form

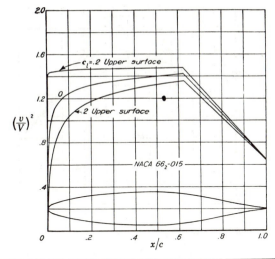

x (per cent c)	y (per cent c)	$(v/V)^2$	v/V	$\Delta v_a/V$
0	0	0	0	2.139
0.5	1.122	0.760	0.872	1.652
0.75	1.343	0.840	0.916	1.431
1.25	1.675	0.929	0.964	1.172
2.5	2.235	1.055	1.027	0.895
5.0	3.100	1.163	1.078	0.663
7.5	3.781	1.208	1.099	0.547
10	4.358	1.242	1.114	0.473
15	5.286	1.288	1.134	0.381
20	5.995	1.317	1.148	0.322
25	6.543	1.340	1.158	0.280
30	6.956	1.356	1.164	0.248
35	7.250	1.370	1.170	0.222
40	7.430	1.380	1.175	0.200
45	7.495	1.391	1.179	0.180
50	7.450	1.401	1.184	0.163
55	7.283	1.411	1.188	0.146
60	6.959	1.420	1.192	0.131
65	6.372	1.367	1.169	0.113
70	5.576	1.260	1.122	0.096
75	4.632	1.156	1.075	0.080
80	3.598	1.053	1.026	0.065
85	2.530	0.949	0.974	0.051
90	1.489	0.847	0.920	0.039
95	0.566	0.744	0.863	0.025
100	0	0.639	0.799	0

L.E. radius: 1.435 per cent c

NACA 66₂-015 Basic Thickness Form

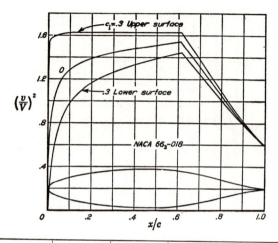

x (per cent c)	y (per cent c)	$(v/V)^2$	v/V	$\Delta v_a/V$
0	0	0	0	1.773
0.5	1.323	0.650	0.806	1.456
0.75	1.571	0.735	0.857	1.312
1.25	1.952	0.850	0.897	1.121
2.5	2.646	1.005	1.002	0.858
5.0	3.690	1.154	1.074	0.649
7.5	4.513	1.234	1.111	0.545
10	5.210	1.285	1.134	0.472
15	6.333	1.350	1.162	0.381
20	7.188	1.393	1.180	0.323
25	7.848	1.423	1.193	0.282
30	8.346	1.445	1.202	0.250
35	8.701	1.464	1.210	0.223
40	8.918	1.481	1.217	0.201
45	8.998	1.496	1.223	0.181
50	8.942	1.509	1.228	0.163
55	8.733	1.522	1.234	0.147
60	8.323	1.534	1.238	0.131
65	7.580	1.438	1.199	0.114
70	6.597	1.302	1.141	0.095
75	5.451	1.172	1.083	0.077
80	4.206	1.045	1.022	0.061
85	2.934	0.922	0.950	0.048
90	1.714	0.803	0.896	0.037
95	0.646	0.692	0.832	0.022
100	0	0.587	0.766	0

L.E. radius: 1.955 per cent c

NACA 66_3-018 Basic Thickness Form

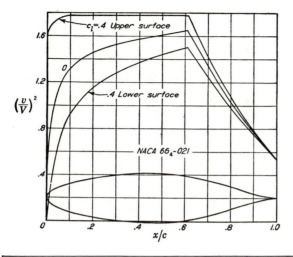

x (per cent c)	y (per cent c)	$(v/V)^2$	v/V	$\Delta v_a/V$
0	0	0	0	1.547
0.5	1.525	0.580	0.761	1.314
0.75	1.804	0.635	0.797	1.218
1.25	2.240	0.755	0.869	1.054
2.5	3.045	0.952	0.976	0.828
5.0	4.269	1.143	1.069	0.635
7.5	5.233	1.246	1.116	0.542
10	6.052	1.318	1.148	0.472
15	7.369	1.405	1.185	0.381
20	8.376	1.459	1.208	0.324
25	9.153	1.499	1.224	0.283
30	9.738	1.528	1.236	0.251
35	10.154	1.551	1.245	0.224
40	10.407	1.574	1.255	0.202
45	10.500	1.594	1.263	0.183
50	10.434	1.611	1.269	0.165
55	10.186	1.629	1.276	0.148
60	9.692	1.648	1.284	0.132
65	8.793	1.508	1.228	0.114
70	7.610	1.335	1.155	0.093
75	6.251	1.176	1.084	0.073
80	4.796	1.031	1.015	0.058
85	3.324	0.891	0.944	0.046
90	1.924	0.763	0.873	0.034
95	0.717	0.648	0.805	0.020
100	0	0.539	0.734	0

L.E. radius: 2.550 per cent c

NACA 66$_4$-021 Basic Thickness Form

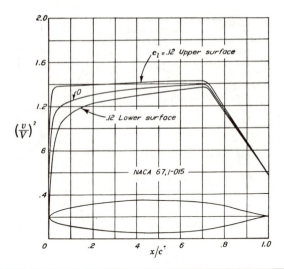

x (per cent c)	y (per cent c)	$(v/V)^2$	v/V	$\Delta v_a/V$
0	0	0	0	2.042
0.5	1.167	0.650	0.806	1.560
0.75	1.394	0.970	0.985	1.370
1.25	1.764	1.059	1.029	1.152
2.5	2.395	1.140	1.068	0.906
5.0	3.245	1.209	1.100	0.667
7.5	3.900	1.239	1.113	0.548
10	4.433	1.259	1.122	0.470
15	5.283	1.285	1.134	0.370
20	5.940	1.304	1.142	0.312
25	6.454	1.318	1.148	0.276
30	6.854	1.330	1.153	0.248
35	7.155	1.341	1.158	0.221
40	7.359	1.351	1.162	0.201
45	7.475	1.360	1.166	0.180
50	7.497	1.368	1.170	0.160
55	7.421	1.375	1.173	0.142
60	7.231	1.381	1.175	0.124
65	6.905	1.388	1.178	0.111
70	6.402	1.390	1.179	0.108
75	5.621	1.321	1.149	0.094
80	4.540	1.176	1.084	0.071
85	3.327	1.018	1.009	0.060
90	2.021	0.864	0.930	0.045
95	0.788	0.712	0.844	0.025
100	0	0.570	0.755	0

L.E. radius: 1.523 per cent c

NACA 67,1-015 Basic Thickness Form

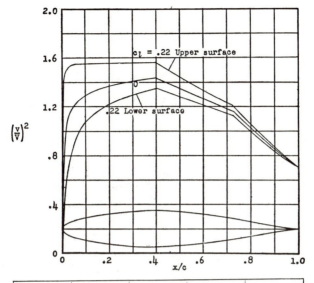

x (per cent c)	y (per cent c)	$(v/V)^2$	v/V	$\Delta v_a/V$
0	0	0	0	2.028
0.5	1.199	0.660	0.812	1.680
0.75	1.435	0.799	0.894	1.560
1.25	1.801	0.942	0.971	1.325
2.5	2.462	1.100	1.049	0.990
5.0	3.419	1.201	1.096	0.695
7.5	4.143	1.259	1.122	0.551
10	4.743	1.295	1.138	0.465
15	5.684	1.339	1.156	0.383
20	6.384	1.369	1.170	0.324
25	6.898	1.390	1.179	0.283
30	7.253	1.409	1.187	0.252
35	7.454	1.423	1.193	0.224
40	7.494	1.435	1.198	0.199
45	7.316	1.391	1.179	0.176
50	7.003	1.348	1.161	0.156
55	6.584	1.306	1.143	0.138
60	6.064	1.265	1.125	0.122
65	5.449	1.221	1.105	0.108
70	4.738	1.178	1.085	0.093
75	3.921	1.115	1.056	0.079
80	3.020	1.027	1.013	0.065
85	2.086	0.938	0.969	0.052
90	1.193	0.852	0.923	0.040
95	0.443	0.774	0.880	0.028
100	0	0.703	0.838	0.018

L.E. radius: 1.544 per cent c

NACA 747A015 Basic Thickness Form

APPENDIX II

MEAN LINES

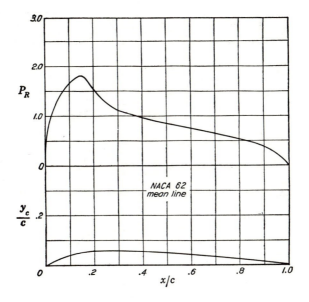

$c_{l_i} = 0.90$	$\alpha_i = 2.81°$	$c_{m_{c/4}} = -0.113$		
x (per cent c)	y_c (per cent c)	dy_c/dx	P_R	$\Delta v/V = P_R/4$
0	0	0.60000	0	0
1.25	0.726	0.56250	0.682	0.171
2.5	1.406	0.52500	1.031	0.258
5.0	2.625	0.45000	1.314	0.328
7.5	3.656	0.37500	1.503	0.376
10	4.500	0.30000	1.651	0.413
15	5.625	0.15000	1.802	0.451
20	6.000	0	1.530	0.383
25	5.977	$-$ 0.00938	1.273	0.318
30	5.906	$-$ 0.01875	1.113	0.279
40	5.625	$-$ 0.03750	0.951	0.238
50	5.156	$-$ 0.05625	0.843	0.211
60	4.500	$-$ 0.07500	0.741	0.185
70	3.656	$-$ 0.09375	0.635	0.159
80	2.625	$-$ 0.11250	0.525	0.131
90	1.406	$-$ 0.13125	0.377	0.094
95	0.727	$-$ 0.14062	0.261	0.065
100	0	$-$ 0.15000	0	0

Data for NACA Mean Line 62

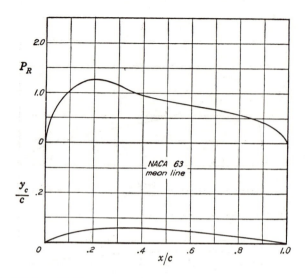

$c_{l_i} = 0.80$	$\alpha_i = 1.60°$	$c_{m_{c/4}} = -0.134$		
x (per cent c)	y_c (per cent c)	dy_c/dx	P_R	$\Delta v/V = P_R/4$
0	0	0.40000	0	0
1.25	0.489	0.38333	0.389	0.097
2.5	0.958	0.36667	0.553	0.138
5.0	1.833	0.33333	0.788	0.197
7.5	2.625	0.30000	0.940	0.235
10	3.333	0.26667	1.066	0.267
15	4.500	0.20000	1.220	0.305
20	5.333	0.13333	1.259	0.315
25	5.833	0.06667	1.233	0.308
30	6.000	0	1.160	0.290
40	5.878	-0.02449	0.949	0.237
50	5.510	-0.04898	0.850	0.213
60	4.898	-0.07347	0.762	0.191
70	4.041	-0.09796	0.673	0.168
80	2.939	-0.12245	0.560	0.140
90	1.592	-0.14694	0.406	0.102
95	0.827	-0.15918	0.291	0.073
100	0	-0.17143	0	0

Data for NACA Mean Line 63

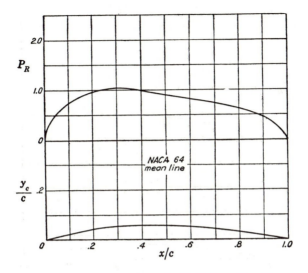

| $c_{l_i} = 0.76$ | $\alpha_i = 0.74°$ | $c_{m_{c/4}} = -0.157$ |

x (per cent c)	y_c (per cent c)	dy_c/dx	P_R	$\Delta v/V = P_R/4$
0	0	0.30000	0	0
1.25	0.369	0.29062	0.257	0.064
2.5	0.726	0.28125	0.391	0.098
5.0	1.406	0.26250	0.546	0.137
7.5	2.039	0.24375	0.668	0.167
10	2.625	0.22500	0.748	0.187
15	3.656	0.18750	0.871	0.218
20	4.500	0.15000	0.966	0.242
25	5.156	0.11250	1.030	0.258
30	5.625	0.07500	1.040	0.260
40	6.000	0	0.999	0.250
50	5.833	− 0.03333	0.910	0.228
60	5.333	− 0.06667	0.827	0.207
70	4.500	− 0.10000	0.750	0.188
80	3.333	− 0.13333	0.635	0.159
90	1.833	− 0.16667	0.466	0.117
95	0.958	− 0.18333	0.334	0.084
100	0	− 0.20000	0	0

Data for NACA Mean Line 64

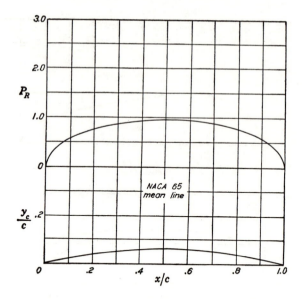

	$c_{l_i} = 0.75$	$\alpha_i = 0°$	$c_{m_{c/4}} = -0.187$	
x (per cent c)	y_c (per cent c)	dy_c/dx	P_R	$\Delta v/V = P_R/4$
0	0	0.24000	0	0
1.25	0.296	0.23400	0.205	0.051
2.5	0.585	0.22800	0.294	0.074
5.0	1.140	0.21600	0.413	0.103
7.5	1.665	0.20400	0.502	0.126
10	2.160	0.19200	0.571	0.143
15	3.060	0.16800	0.679	0.170
20	3.840	0.14400	0.760	0.190
25	4.500	0.12000	0.824	0.206
30	5.040	0.09600	0.872	0.218
40	5.760	0.04800	0.932	0.233
50	6.000	0	0.951	0.238
60	5.760	-0.04800	0.932	0.233
70	5.040	-0.09600	0.872	0.218
80	3.840	-0.14400	0.760	0.190
90	2.160	-0.19200	0.571	0.143
95	1.140	-0.21600	0.413	0.103
100	0	-0.24000	0	0

Data for NACA Mean Line 65

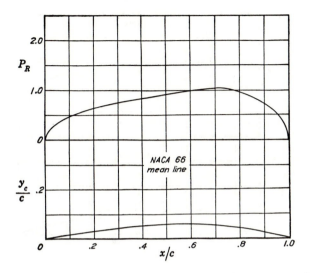

x (per cent c)	y_c (per cent c)	dy_c/dx	P_R	$\Delta v/V = P_R/4$
0	0	0.20000	0	0
1.25	0.247	0.19583	0.135	0.034
2.5	0.490	0.19167	0.244	0.061
5.0	0.958	0.18333	0.334	0.084
7.5	1.406	0.17500	0.408	0.102
10	1.833	0.16667	0.466	0.117
15	2.625	0.15000	0.557	0.139
20	3.333	0.13333	0.635	0.159
25	3.958	0.11667	0.700	0.175
30	4.500	0.10000	0.750	0.188
40	5.333	0.06667	0.827	0.207
50	5.833	0.03333	0.910	0.228
60	6.000	0	0.999	0.250
70	5.625	− 0.07500	1.040	0.260
80	4.500	− 0.15000	0.966	0.242
90	2.625	− 0.22500	0.748	0.187
95	1.406	− 0.26250	0.546	0.137
100	0	− 0.30000	0	0

$$c_{l_i} = 0.76 \qquad \alpha_i = -0.74° \qquad c_{m_{c/4}} = -0.222$$

Data for NACA Mean Line 66

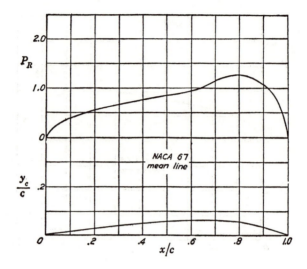

$c_{l_i} = 0.80$		$\alpha_i = -1.60°$	$c_{m_{c/4}} = -0.266$	
x (per cent c)	y_c (per cent c)	dy_c/dx	P_R	$\Delta v/V = P_R/4$
0	0	0.17143	0	0
1.25	0.212	0.16837	0.137	0.034
2.5	0.421	0.16531	0.195	0.049
5.0	0.827	0.15918	0.291	0.073
7.5	1.217	0.15306	0.356	0.089
10	1.592	0.14694	0.406	0.102
15	2.296	0.13469	0.483	0.121
20	2.939	0.12245	0.560	0.140
25	3.520	0.11020	0.616	0.154
30	4.041	0.09796	0.673	0.168
40	4.898	0.07347	0.762	0.191
50	5.510	0.04898	0.850	0.213
60	5.878	0.02449	0.949	0.237
70	6.000	0	1.160	0.290
80	5.333	-0.13333	1.259	0.315
90	3.333	-0.26667	1.066	0.267
95	1.833	-0.33333	0.788	0.197
100	0	-0.40000	0	0

Data for NACA Mean Line 67

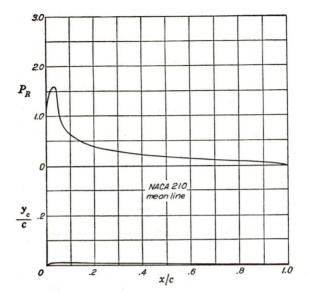

$c_{l_i} = 0.30$	$\alpha_i = 2.09°$	$c_{m_{c/4}} = -0.006$		
x (per cent c)	y_c (per cent c)	dy_c/dx	P_R	$\Delta v/V = P_R/4$
0	0	0.59613	0	0
1.25	0.596	0.36236	1.381	0.345
2.5	0.928	0.18504	1.565	0.391
5.0	1.114	-0.00018	1.221	0.305
7.5	1.087		0.781	0.195
10	1.058		0.626	0.156
15	0.999		0.489	0.122
20	0.940		0.408	0.102
25	0.881		0.348	0.087
30	0.823		0.302	0.075
40	0.705	-0.01175	0.242	0.061
50	0.588		0.198	0.049
60	0.470		0.160	0.040
70	0.353		0.128	0.032
80	0.235		0.098	0.025
90	0.118		0.065	0.016
95	0.059		0.044	0.011
100	0		0	0

Data for NACA Mean Line 210

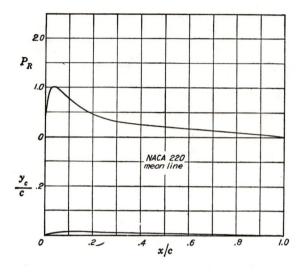

$c_{l_i} = 0.30$	$\alpha_i = 1.86°$	$c_{m_{c/4}} = -0.010$		
x (per cent c)	y_c (per cent c)	dy_c/dx	P_R	$\Delta v/V = P_R/4$
0	0	0.39270	0	0
1.25	0.442	0.31541	0.822	0.206
2.5	0.793	0.24618	1.003	0.251
5.0	1.257	0.13192	0.988	0.247
7.5	1.479	0.04994	0.900	0.225
10	1.535	0.00024	0.801	0.200
15	1.463		0.615	0.154
20	1.377		0.465	0.116
25	1.291		0.378	0.095
30	1.205		0.326	0.082
40	1.033		0.253	0.063
50	0.861		0.205	0.051
60	0.689	-0.01722	0.169	0.042
70	0.516		0.135	0.034
80	0.344		0.100	0.025
90	0.172		0.064	0.016
95	0.086		0.040	0.010
100	0		0	0

Data for NACA Mean Line 220

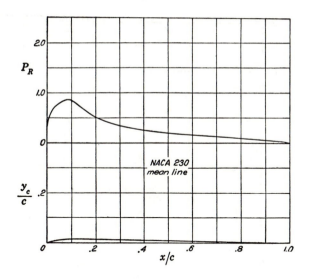

$c_{l_i} = 0.30$ $\quad$ $\alpha_i = 1.65°$ $\quad$ $c_{m_{c/4}} = -0.014$				
x (per cent c)	y_c (per cent c)	dy_c/dx	P_R	$\Delta v/V = P_R/4$
0	0	0.30508	0	0
1.25	0.357	0.26594	0.528	0.132
2.5	0.666	0.22929	0.673	0.168
5.0	1.155	0.16347	0.791	0.198
7.5	1.492	0.10762	0.853	0.213
10	1.701	0.06174	0.859	0.215
15	1.838	$-$ 0.00009	0.678	0.170
20	1.767	$-$ 0.02203	0.519	0.130
25	1.656		0.419	0.105
30	1.546		0.361	0.090
40	1.325		0.274	0.069
50	1.104		0.217	0.054
60	0.883	$-$ 0.02208	0.177	0.044
70	0.662		0.144	0.036
80	0.442		0.105	0.026
90	0.221		0.069	0.017
95	0.110		0.042	0.011
100	0		0	0

Data for NACA Mean Line 230

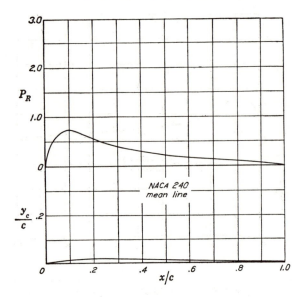

$c_{l_i} = 0.30$		$\alpha_i = 1.45°$	$c_{m_{c/4}} = -0.019$	
x (per cent c)	y_c (per cent c)	dy_c/dx	P_R	$\Delta v/V = P_R/4$
0	0	0.25233	0	0
1.25	0.301	0.22877	0.377	0.094
2.5	0.572	0.20625	0.491	0.123
5.0	1.035	0.16432	0.625	0.156
7.5	1.397	0.12653	0.718	0.180
10	1.671	0.09290	0.750	0.188
15	1.991	0.03810	0.677	0.169
20	2.079	− 0.00010	0.566	0.142
25	2.018	− 0.02169	0.477	0.119
30	1.890		0.410	0.103
40	1.620		0.304	0.076
50	1.350		0.234	0.059
60	1.080		0.186	0.047
70	0.810	− 0.02700	0.150	0.038
80	0.540		0.110	0.028
90	0.270		0.071	0.018
95	0.135		0.047	0.012
100	0		0	0

Data for NACA Mean Line 240

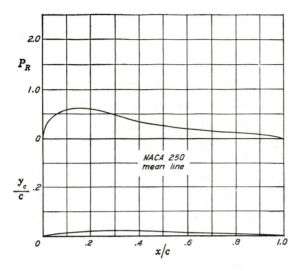

$c_{l_i} = 0.30$		$\alpha_i = 1.26°$	$c_{m_{c/4}} = -0.026$	
x (per cent c)	y_c (per cent c)	dy_c/dx	P_R	$\Delta v/V = P_R/4$
0	0	0.21472	0	0
1.25	0.258	0.19920	0.281	0.070
2.5	0.498	0.18416	0.369	0.092
5.0	0.922	0.15562	0.477	0.119
7.5	1.277	0.12909	0.552	0.138
10	1.570	0.10458	0.592	0.148
15	1.982	0.06162	0.624	0.156
20	2.199	0.02674	0.610	0.153
25	2.263	− 0.00007	0.547	0.137
30	2.212	− 0.01880	0.470	0.117
40	1.931		0.346	0.087
50	1.609		0.255	0.064
60	1.287		0.197	0.049
70	0.965	− 0.03218	0.154	0.038
80	0.644		0.119	0.030
90	0.322		0.076	0.019
95	0.161		0.051	0.013
100	0		0	0

Data for NACA Mean Line 250

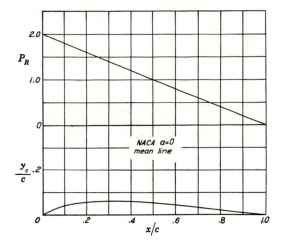

$c_{l_i} = 1.0$	$\alpha_i = 4.56°$	$c_{m_{c/4}} = -0.083$		
x (per cent c)	y_c (per cent c)	dy_c/dx	P_R	$\Delta v/V = P_R/4$
0	0			
0.5	0.460	0.75867	1.990	0.498
0.75	0.641	0.69212	1.985	0.496
1.25	0.964	0.60715	1.975	0.494
2.5	1.641	0.48892	1.950	0.488
5.0	2.693	0.36561	1.900	0.475
7.5	3.507	0.29028	1.850	0.463
10	4.161	0.23515	1.800	0.450
15	5.124	0.15508	1.700	0.425
20	5.747	0.09693	1.600	0.400
25	6.114	0.05156	1.500	0.375
30	6.277	0.01482	1.400	0.350
35	6.273	$-$ 0.01554	1.300	0.325
40	6.130	$-$ 0.04086	1.200	0.300
45	5.871	$-$ 0.06201	1.100	0.275
50	5.516	$-$ 0.07958	1.000	0.250
55	5.081	$-$ 0.09395	0.900	0.225
60	4.581	$-$ 0.10539	0.800	0.200
65	4.032	$-$ 0.11406	0.700	0.175
70	3.445	$-$ 0.12003	0.600	0.150
75	2.836	$-$ 0.12329	0.500	0.125
80	2.217	$-$ 0.12371	0.400	0.100
85	1.604	$-$ 0.12099	0.300	0.075
90	1.013	$-$ 0.11455	0.200	0.050
95	0.467	$-$ 0.10301	0.100	0.025
100	0	$-$ 0.07958	0	0

Data for NACA Mean Line $a = 0$

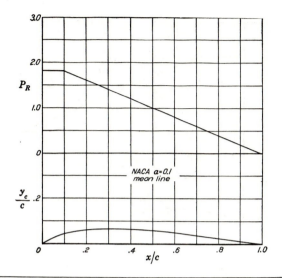

NACA a=0.1
mean line

$c_{l_i} = 1.0$		$\alpha_i = 4.43°$	$c_{m_{c/4}} = -0.086$	
x (per cent c)	y_c (per cent c)	dy_c/dx	P_R	$\Delta v/V = P_R/4$
0	0			
0.5	0.440	0.73441		
0.75	0.616	0.67479		
1.25	0.933	0.59896		
2.5	1.608	0.49366	1.818	0.455
5.0	2.689	0.38235		
7.5	3.551	0.31067		
10	4.253	0.25057		
15	5.261	0.16087	1.717	0.429
20	5.905	0.09981	1.616	0.404
25	6.282	0.05281	1.515	0.379
30	6.449	0.01498	1.414	0.354
35	6.443	-0.01617	1.313	0.328
40	6.296	-0.04210	1.212	0.303
45	6.029	-0.06373	1.111	0.278
50	5.664	-0.08168	1.010	0.253
55	5.218	-0.09637	0.909	0.227
60	4.706	-0.10806	0.808	0.202
65	4.142	-0.11694	0.707	0.177
70	3.541	-0.12307	0.606	0.152
75	2.916	-0.12644	0.505	0.126
80	2.281	-0.12693	0.404	0.101
85	1.652	-0.12425	0.303	0.076
90	1.045	-0.11781	0.202	0.050
95	0.482	-0.10620	0.101	0.025
100	0	-0.08258	0	0

Data for NACA Mean Line $a = 0.1$

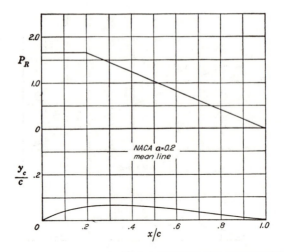

NACA a=0.2
mean line

$c_{l_i} = 1.0$		$\alpha_i = 4.17°$		$c_{m_{c/4}} = -0.094$
x (per cent c)	y_c (per cent c)	dy_c/dx	P_R	$\Delta v/V = P_R/4$
0	0			
0.5	0.414	0.69492		
0.75	0.581	0.64047		
1.25	0.882	0.57135		
2.5	1.530	0.47592		
5.0	2.583	0.37661	1.667	0.417
7.5	3.443	0.31487		
10	4.169	0.26803		
15	5.317	0.19373		
20	6.117	0.12405		
25	6.572	0.06345	1.563	0.391
30	6.777	0.02030	1.459	0.365
35	6.789	-0.01418	1.355	0.339
40	6.646	-0.04246	1.250	0.313
45	6.373	-0.06588	1.146	0.287
50	5.994	-0.08522	1.042	0.260
55	5.527	-0.10101	0.938	0.234
60	4.989	-0.11359	0.834	0.208
65	4.396	-0.12317	0.729	0.182
70	3.762	-0.12985	0.625	0.156
75	3.102	-0.13363	0.521	0.130
80	2.431	-0.13440	0.417	0.104
85	1.764	-0.13186	0.313	0.078
90	1.119	-0.12541	0.208	0.052
95	0.518	-0.11361	0.104	0.026
100	0	-0.08941	0	0

Data for NACA Mean Line $a = 0.2$

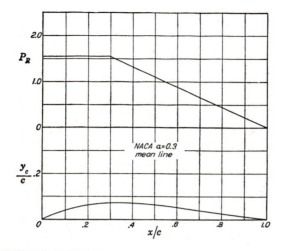

$c_{l_i} = 1.0$		$\alpha_i = 3.84°$	$c_{m_{c/4}} = -0.106$		
x (per cent c)	y_c (per cent c)	dy_c/dx	P_R	$\Delta v/V = P_R/4$	
0	0				
0.5	0.389	0.65536			
0.75	0.546	0.60524			
1.25	0.832	0.54158			
2.5	1.448	0.45399			
5.0	2.458	0.36344			
7.5	3.293	0.30780	1.538	0.385	
10	4.008	0.26621			
15	5.172	0.20246			
20	6.052	0.15068			
25	6.685	0.10278			
30	7.072	0.04833			
35	7.175	− 0.00205	1.429	0.357	
40	7.074	− 0.03710	1.319	0.330	
45	6.816	− 0.06492	1.209	0.302	
50	6.433	− 0.08746	1.099	0.275	
55	5.949	− 0.10567	0.989	0.247	
60	5.383	− 0.12014	0.879	0.220	
65	4.753	− 0.13119	0.769	0.192	
70	4.076	− 0.13901	0.659	0.165	
75	3.368	− 0.14365	0.549	0.137	
80	2.645	− 0.14500	0.440	0.110	
85	1.924	− 0.14279	0.330	0.082	
90	1.224	− 0.13638	0.220	0.055	
95	0.570	− 0.12430	0.110	0.028	
100	0	− 0.09907	0	0	

Data for NACA Mean Line $a = 0.3$

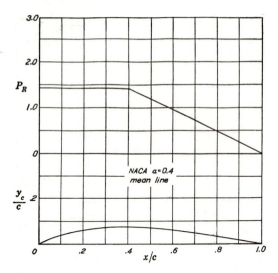

$c_{l_i} = 1.0$		$\alpha_i = 3.46°$	$c_{m_{c/4}} = -0.121$	
x (per cent c)	y_c (per cent c)	dy_c/dx	P_R	$\Delta v/V = P_R/4$
0	0			
0.5	0.366	0.61759		
0.75	0.514	0.57105		
1.25	0.784	0.51210		
2.5	1.367	0.43106		
5.0	2.330	0.34764		
7.5	3.131	0.29671		
10	3.824	0.25892	1.429	0.357
15	4.968	0.20185		
20	5.862	0.15682		
25	6.546	0.11733		
30	7.039	0.07988		
35	7.343	0.04136		
40	7.439	− 0.00721		
45	7.275	− 0.05321	1.310	0.327
50	6.929	− 0.08380	1.190	0.298
55	6.449	− 0.10734	1.071	0.268
60	5.864	− 0.12567	0.952	0.238
65	5.199	− 0.13962	0.833	0.208
70	4.475	− 0.14963	0.714	0.179
75	3.709	− 0.15589	0.595	0.149
80	2.922	− 0.15837	0.476	0.119
85	2.132	− 0.15683	0.357	0.089
90	1.361	− 0.15062	0.238	0.060
95	0.636	− 0.13816	0.119	0.030
100	0	− 0.11138	0	0

Data for NACA Mean Line $a = 0.4$

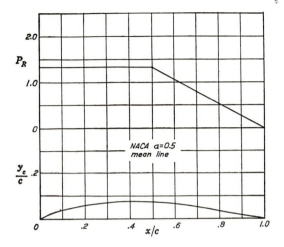

$c_{l_i} = 1.0$		$\alpha_i = 3.04°$		$c_{m_{c/4}} = -0.139$	
x (per cent c)	y_c (per cent c)	dy_c/dx	P_R	$\Delta v/V = P_R/4$
0	0			
0.5	0.345	0.58195		
0.75	0.485	0.53855		
1.25	0.735	0.48360		
2.5	1.295	0.40815		
5.0	2.205	0.33070		
7.5	2.970	0.28365		
10	3.630	0.24890		
15	4.740	0.19690	1.333	0.333
20	5.620	0.15650		
25	6.310	0.12180		
30	6.840	0.09000		
35	7.215	0.05930		
40	7.430	0.02800		
45	7.490	− 0.00630		
50	7.350	− 0.05305		
55	6.965	− 0.09765	1.200	0.300
60	6.405	− 0.12550	1.067	0.267
65	5.725	− 0.14570	0.933	0.233
70	4.955	− 0.16015	0.800	0.200
75	4.130	− 0.16960	0.667	0.167
80	3.265	− 0.17435	0.533	0.133
85	2.395	− 0.17415	0.400	0.100
90	1.535	− 0.16850	0.267	0.067
95	0.720	− 0.15565	0.133	0.033
100	0	− 0.12660	0	0

Data for NACA Mean Line $a = 0.5$

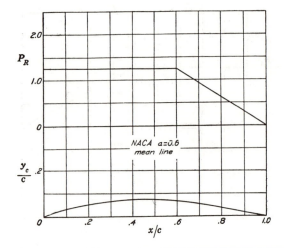

$c_{l_i} = 1.0$	$\alpha_i = 2.58°$	$c_{m_{c/4}} = -0.158$		
x (per cent c)	y_c (per cent c)	dy_c/dx	P_R	$\Delta v/V = P_R/4$
0	0			
0.5	0.325	0.54825		
0.75	0.455	0.50760		
1.25	0.695	0.45615		
2.5	1.220	0.38555		
5.0	2.080	0.31325		
7.5	2.805	0.26950		
10	3.435	0.23730		
15	4.495	0.18935		
20	5.345	0.15250	1.250	0.312
25	6.035	0.12125		
30	6.570	0.09310		
35	6.965	0.06660		
40	7.235	0.04060		
45	7.370	0.01405		
50	7.370	-0.01435		
55	7.220	-0.04700		
60	6.880	-0.09470		
65	6.275	-0.14015	1.094	0.273
70	5.505	-0.16595	0.938	0.234
75	4.630	-0.18270	0.781	0.195
80	3.695	-0.19225	0.625	0.156
85	2.720	-0.19515	0.469	0.117
90	1.755	-0.19095	0.312	0.078
95	0.825	-0.17790	0.156	0.039
100	0	-0.14550	0	0

Data for NACA Mean Line $a = 0.6$

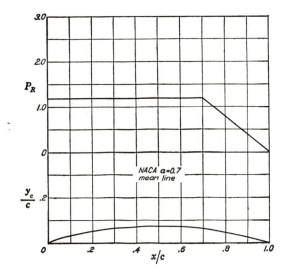

$c_{l_i} = 1.0$		$\alpha_i = 2.09°$	$c_{m_{c/4}} = -0.179$	
x (per cent c)	y_c (per cent c)	dy_c/dx	P_R	$\Delta v/V = P_R/4$
0	0			
0.5	0.305	0.51620		
0.75	0.425	0.47795		
1.25	0.655	0.42960		
2.5	1.160	0.36325		
5.0	1.955	0.29545		
7.5	2.645	0.25450		
10	3.240	0.22445		
15	4.245	0.17995		
20	5.060	0.14595		
25	5.715	0.11740	1.176	0.294
30	6.240	0.09200		
35	6.635	0.06840		
40	6.925	0.04570		
45	7.095	0.02315		
50	7.155	0		
55	7.090	-0.02455		
60	6.900	-0.05185		
65	6.565	-0.08475		
70	6.030	-0.13650		
75	5.205	-0.18510	0.980	0.245
80	4.215	-0.20855	0.784	0.196
85	3.140	-0.21955	0.588	0.147
90	2.035	-0.21960	0.392	0.098
95	0.965	-0.20725	0.196	0.049
100	0	-0.16985	0	0

Data for NACA Mean Line $a = 0.7$

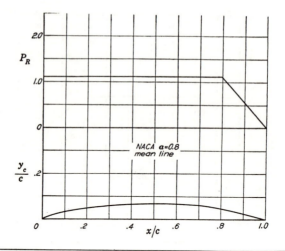

$c_{l_i} = 1.0$		$\alpha_i = 1.54°$	$c_{m_{c/4}} = -0.202$	
x (per cent c)	y_c (per cent c)	dy_c/dx	P_R	$\Delta v/V = P_R/4$
0	0			
0.5	0.287	0.48535		
0.75	0.404	0.44925		
1.25	0.616	0.40359		
2.5	1.077	0.34104		
5.0	1.841	0.27718		
7.5	2.483	0.23868		
10	3.043	0.21050		
15	3.985	0.16892		
20	4.748	0.13734		
25	5.367	0.11101		
30	5.863	0.08775	1.111	0.278
35	6.248	0.06634		
40	6.528	0.04601		
45	6.709	0.02613		
50	6.790	0.00620		
55	6.770	− 0.01433		
60	6.644	− 0.03611		
65	6.405	− 0.06010		
70	6.037	− 0.08790		
75	5.514	− 0.12311		
80	4.771	− 0.18412		
85	3.683	− 0.23921	0.833	0.208
90	2.435	− 0.25583	0.556	0.139
95	1.163	− 0.24904	0.278	0.069
100	0	− 0.20385	0	0

Data for NACA Mean Line $a = 0.8$

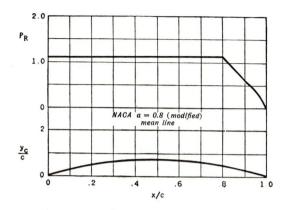

$c_{l_i} = 1.0$	$\alpha_i = 1.40°$	$c_{m_{c/4}} = 0.219$		
x (per cent c)	y_c (per cent c)	dy_c/dx	P_R	$\Delta v/V = P_R/4$
0	0			
0.5	0.281	0.47539		
0.75	0.396	0.44004		
1.25	0.603	0.39531		
2.5	1.055	0.33404	1.092	0.273
5.0	1.803	0.27149		
7.5	2.432	0.23378		
10	2.981	0.20618		
15	3.903	0.16546		
20	4.651	0.13452	1.096	0.274
25	5.257	0.10873		
30	5.742	0.08595		
35	6.120	0.06498		
40	6.394	0.04507	1.100	0.275
45	6.571	0.02559		
50	6.651	0.00607		
55	6.631	− 0.01404	1.104	0.276
60	6.508	− 0.03537		
65	6.274	− 0.05887	1.108	0.277
70	5.913	− 0.08610	1.108	0.277
75	5.401	− 0.12058	1.112	0.278
80	4.673	− 0.18034	1.112	0.278
85	3.607	− 0.23430	0.840	0.210
90	2.452	− 0.24521	0.588	0.147
95	1.226	− 0.24521	0.368	0.092
100	0	− 0.24521	0	0

Data for NACA Mean Line $a = 0.8$ (modified)

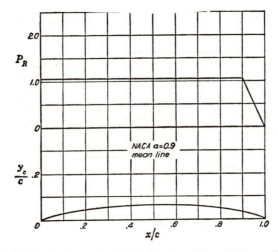

$c_{l_i} = 1.0$		$\alpha_i = 0.90°$	$c_{m_{c/4}} = -0.225$	
x (per cent c)	y_c (per cent c)	dy_c/dx	P_R	$\Delta v/V = P_R/4$
0	0			
0.5	0.269	0.45482		
0.75	0.379	0.42064		
1.25	0.577	0.37740		
2.5	1.008	0.31821		
5.0	1.720	0.25786		
7.5	2.316	0.22153		
10	2.835	0.19500		
15	3.707	0.15595		
20	4.410	0.12644		
25	4.980	0.10196		
30	5.435	0.08047		
35	5.787	0.06084	1.053	0.263
40	6.045	0.04234		
45	6.212	0.02447		
50	6.290	0.00678		
55	6.279	-0.01111		
60	6.178	-0.02965		
65	5.981	-0.04938		
70	5.681	-0.07103		
75	5.265	-0.09583		
80	4.714	-0.12605		
85	3.987	-0.16727		
90	2.984	-0.25204		
95	1.503	-0.31463	0.526	0.132
100	0	-0.26086	0	0

Data for NACA Mean Line $a = 0.9$

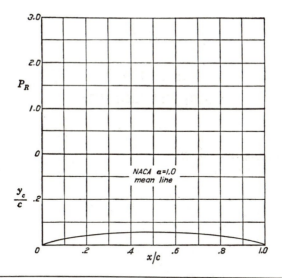

$c_{l_i} = 1.0$	$\alpha_i = 0°$	$c_{m_{c/4}} = -0.250$			
x (per cent c)	y_c (per cent c)	dy_c/dx	P_R	$\Delta v/V = P_R/4$
0	0			
0.5	0.250	0.42120		
0.75	0.350	0.38875		
1.25	0.535	0.34770		
2.5	0.930	0.29155		
5.0	1.580	0.23430		
7.5	2.120	0.19995		
10	2.585	0.17485		
15	3.365	0.13805		
20	3.980	0.11030		
25	4.475	0.08745		
30	4.860	0.06745		
35	5.150	0.04925		
40	5.355	0.03225	1.000	0.250
45	5.475	0.01595		
50	5.515	0		
55	5.475	-0.01595		
60	5.355	-0.03225		
65	5.150	-0.04925		
70	4.860	-0.06745		
75	4.475	-0.08745		
80	3.980	-0.11030		
85	3.365	-0.13805		
90	2.585	-0.17485		
95	1.580	-0.23430		
100	0			

Data for NACA Mean Line $a = 1.0$

APPENDIX III

AIRFOIL ORDINATES

Contents

NACA 0010-34

a = 0.8 (modified)　　c_{l_i} = 0.2

(Stations and ordinates given in per cent of airfoil chord)

Upper surface		Lower surface	
Station	Ordinate	Station	Ordinate
0	0	0	0
0.687	0.790	0.813	− 0.632
1.176	1.062	1.324	− 0.820
2.407	1.608	2.593	− 1.186
4.887	2.436	5.113	− 1.714
7.378	3.094	7.622	− 2.122
9.875	3.637	10.125	− 2.445
14.876	4.523	15.124	− 2.961
19.886	5.172	20.114	− 3.312
29.917	5.980	30.083	− 3.684
39.955	6.279	40.045	− 3.721
49.994	6.186	50.006	− 3.526
60.031	5.735	59.969	− 3.131
70.064	4.915	69.936	− 2.549
80.100	3.700	79.900	− 1.830
90.076	2.044	89.924	− 1.064
95.042	1.100	94.958	− 0.610
100.000	0.100	100.000	− 0.100

L.E. radius: 0.272
Slope of radius through L.E.: 0.095

NACA 0012-64

a = 0.8 (modified)　　c_{l_i} = 0.2

(Stations and ordinates given in per cent of airfoil chord)

Upper surface		Lower surface	
Station	Ordinate	Station	Ordinate
0	0	0	0
1.107	1.928	1.393	− 1.686
2.336	2.659	2.664	− 2.237
4.823	3.623	5.177	− 2.901
7.322	4.295	7.678	− 3.323
9.825	4.832	10.175	− 3.640
14.839	5.645	15.161	− 4.083
19.858	6.221	20.142	− 4.361
29.900	6.974	30.100	− 4.678
39.946	7.279	40.054	− 4.721
49.993	7.157	50.007	− 4.497
60.038	6.622	59.962	− 4.018
70.077	5.662	69.923	− 3.296
80.120	4.253	79.880	− 2.383
90.091	2.355	89.909	− 1.375
95.050	1.271	94.950	− 0.781
100.000	0.120	100.000	− 0.120

L.E. radius: 1.582
Slope of radius through L.E.: 0.095

NACA 1408

(Stations and ordinates given in per cent of airfoil chord)

Upper surface		Lower surface	
Station	Ordinate	Station	Ordinate
0	0	0	0
1.189	1.324	1.311	− 1.200
2.418	1.862	2.582	− 1.620
4.896	2.602	5.104	− 2.134
7.386	3.138	7.614	− 2.458
9.883	3.558	10.117	− 2.682
14.889	4.171	15.111	− 2.953
19.904	4.574	20.096	− 3.074
24.926	4.819	25.074	− 3.101
29.950	4.939	30.050	− 3.063
40.000	4.869	40.000	− 2.869
50.020	4.502	49.980	− 2.556
60.034	3.931	59.966	− 2.153
70.041	3.193	69.959	− 1.693
80.039	2.305	79.961	− 1.193
90.027	1.271	89.973	− 0.659
95.016	0.698	94.984	− 0.378
100.000	0.084	100.000	− 0.084

L.E. radius: 0.70
Slope of radius through L.E.: 0.05

NACA 1410

(Stations and ordinates given in per cent of airfoil chord)

Upper surface		Lower surface	
Station	Ordinate	Station	Ordinate
0	0	0	0
1.174	1.639	1.326	− 1.515
2.398	2.297	2.602	− 2.055
4.870	3.194	5.130	− 2.726
7.358	3.837	7.642	− 3.157
9.854	4.338	10.146	− 3.462
14.861	5.062	15.139	− 3.844
19.880	5.531	20.120	− 4.031
24.907	5.809	25.093	− 4.091
29.937	5.940	30.063	− 4.064
40.000	5.836	40.000	− 3.836
50.025	5.385	49.975	− 3.439
60.042	4.692	59.958	− 2.914
70.051	3.804	69.949	− 2.304
80.049	2.741	79.951	− 1.629
90.034	1.513	89.966	− 0.901
95.021	0.832	94.979	− 0.512
100.000	0.105	100.000	− 0.105

L.E. radius: 1.10
Slope of radius through L.E.: 0.05

NACA 1412

(Stations and ordinates given in per cent of airfoil chord)

Upper surface		Lower surface	
Station	Ordinate	Station	Ordinate
0	0	0	0
1.158	1.954	1.342	− 1.830
2.378	2.733	2.622	− 2.491
4.845	3.786	5.155	− 3.318
7.330	4.537	7.670	− 3.857
9.824	5.118	10.176	− 4.242
14.833	5.951	15.167	− 4.733
19.857	6.486	20.143	− 4.986
24.889	6.799	25.111	− 5.081
29.925	6.940	30.075	− 5.064
40.000	6.803	40.000	− 4.803
50.029	6.267	49.971	− 4.321
60.051	5.453	59.949	− 3.675
70.061	4.413	69.939	− 2.913
80.058	3.178	79.942	− 2.066
90.040	1.753	89.960	− 1.141
95.025	0.966	94.975	− 0.646
100.000	0.126	100.000	− 0.126

L.E. radius: 1.58
Slope of radius through L.E.: 0.05

NACA 2408

(Stations and ordinates given in per cent of airfoil chord)

Upper surface		Lower surface	
Station	Ordinate	Station	Ordinate
0	0	0	0
1.128	1.380	1.372	− 1.134
2.337	1.977	2.663	− 1.493
4.794	2.829	5.206	− 1.891
7.273	3.471	7.727	− 2.111
9.768	3.987	10.232	− 2.237
14.778	4.776	15.222	− 2.338
19.809	5.320	20.191	− 2.320
24.852	5.677	25.148	− 2.239
29.900	5.875	30.100	− 2.125
40.000	5.869	40.000	− 1.869
50.039	5.473	49.961	− 1.585
60.068	4.820	59.932	− 1.264
70.081	3.942	69.919	− 0.942
80.078	2.858	79.922	− 0.636
90.054	1.575	89.946	− 0.353
95.033	0.855	94.967	− 0.217
100.000	0.084	100.000	− 0.084

L.E. radius: 0.70
Slope of radius through L.E.: 0.1

NACA 2410
(Stations and ordinates given in per cent of airfoil chord)

Upper surface		Lower surface	
Station	Ordinate	Station	Ordinate
0	0	0	0
1.098	1.694	1.402	− 1.448
2.297	2.411	2.703	− 1.927
4.742	3.420	5.258	− 2.482
7.217	4.169	7.783	− 2.809
9.710	4.766	10.290	− 3.016
14.722	5.665	15.278	− 3.227
19.761	6.276	20.239	− 3.276
24.814	6.668	25.186	− 3.230
29.875	6.875	30.125	− 3.125
40.000	6.837	40.000	− 2.837
50.049	6.356	49.951	− 2.468
60.085	5.580	59.915	− 2.024
70.102	4.551	69.898	− 1.551
80.097	3.296	79.903	− 1.074
90.067	1.816	89.933	− 0.594
95.041	0.990	94.959	− 0.352
100.000	0.105	100.000	− 0.105

L.E. radius: 1.10
Slope of radius through L.E.: 0.1

NACA 2412
(Stations and ordinates given in per cent of airfoil chord)

Upper surface		Lower surface	
Station	Ordinate	Station	Ordinate
0	. . . . .	0	0
1.25	2.15	1.25	− 1.65
2.5	2.99	2.5	− 2.27
5.0	4.13	5.0	− 3.01
7.5	4.96	7.5	− 3.46
10	5.63	10	− 3.75
15	6.61	15	− 4.10
20	7.26	20	− 4.23
25	7.67	25	− 4.22
30	7.88	30	− 4.12
40	7.80	40	− 3.80
50	7.24	50	− 3.34
60	6.36	60	− 2.76
70	5.18	70	− 2.14
80	3.75	80	− 1.50
90	2.08	90	− 0.82
95	1.14	95	− 0.48
100	(0.13)	100	(− 0.13)
100	. . . . .	100	0

L.E. radius: 1.58
Slope of radius through L.E.: 0.10

NACA 2415
(Stations and ordinates given in per cent of airfoil chord)

Upper surface		Lower surface	
Station	Ordinate	Station	Ordinate
0	. . . . .	0	0
1.25	2.71	1.25	− 2.06
2.5	3.71	2.5	− 2.86
5.0	5.07	5.0	− 3.84
7.5	6.06	7.5	− 4.47
10	6.83	10	− 4.90
15	7.97	15	− 5.42
20	8.70	20	− 5.66
25	9.17	25	− 5.70
30	9.38	30	− 5.62
40	9.25	40	− 5.25
50	8.57	50	− 4.67
60	7.50	60	− 3.90
70	6.10	70	− 3.05
80	4.41	80	− 2.15
90	2.45	90	− 1.17
95	1.34	95	− 0.68
100	(0.16)	100	(− 0.16)
100	. . . . .	100	0

L.E. radius: 2.48
Slope of radius through L.E.: 0.10

NACA 2418

(Stations and ordinates given in per cent of airfoil chord)

Upper surface		Lower surface	
Station	Ordinate	Station	Ordinate
0	......	0	0
1.25	3.28	1.25	− 2.45
2.5	4.45	2.5	− 3.44
5.0	6.03	5.0	− 4.68
7.5	7.17	7.5	− 5.48
10	8.05	10	− 6.03
15	9.34	15	− 6.74
20	10.15	20	− 7.09
25	10.65	25	− 7.18
30	10.88	30	− 7.12
40	10.71	40	− 6.71
50	9.89	50	− 5.99
60	8.65	60	− 5.04
70	7.02	70	− 3.97
80	5.08	80	− 2.80
90	2.81	90	− 1.53
95	1.55	95	− 0.87
100	(0.19)	100	(− 0.19)
100	......	100	0

L.E. radius: 3.56
Slope of radius through L.E.: 0.10

NACA 2421

(Stations and ordinates given in per cent of airfoil chord)

Upper surface		Lower surface	
Station	Ordinate	Station	Ordinate
0	......	0	0
1.25	3.87	1.25	− 2.82
2.5	5.21	2.5	− 4.02
5.0	7.00	5.0	− 5.51
7.5	8.29	7.5	− 6.48
10	9.28	10	− 7.18
15	10.70	15	− 8.05
20	11.59	20	− 8.52
25	12.15	25	− 8.67
30	12.38	30	− 8.62
40	12.16	40	− 8.16
50	11.22	50	− 7.31
60	9.79	60	− 6.17
70	7.94	70	− 4.87
80	5.74	80	− 3.44
90	3.18	90	− 1.88
95	1.76	95	− 1.06
100	(0.22)	100	(− 0.22)
100	......	100	0

L.E. radius: 4.85
Slope of radius through L.E.: 0.10

NACA 2424

(Stations and ordinates given in per cent of airfoil chord)

Upper surface		Lower surface	
Station	Ordinate	Station	Ordinate
0	0	0	0
0.885	3.892	1.615	− 3.646
2.012	5.449	2.988	− 4.965
4.380	7.552	5.620	− 6.614
6.820	9.052	8.180	− 7.692
9.300	10.215	10.700	− 8.465
14.333	11.888	15.667	− 9.450
19.427	12.959	20.573	− 9.959
24.555	13.593	25.445	− 10.155
29.700	13.874	30.300	− 10.124
40.000	13.606	40.000	− 9.606
50.118	12.532	49.882	− 8.644
60.203	10.903	59.797	− 7.347
70.244	8.824	69.756	− 5.824
80.233	6.352	79.767	− 4.130
90.161	3.502	89.839	− 2.280
95.098	1.930	94.902	− 1.292
100.000	......	100.000	0

L.E. radius: 6.33
Slope of radius through L.E.: 0.10

NACA 4412
(Stations and ordinates given in per cent of airfoil chord)

Upper surface		Lower surface	
Station	Ordinate	Station	Ordinate
0	0	0	0
1.25	2.44	1.25	− 1.43
2.5	3.39	2.5	− 1.95
5.0	4.73	5.0	− 2.49
7.5	5.76	7.5	− 2.74
10	6.59	10	− 2.86
15	7.89	15	− 2.88
20	8.80	20	− 2.74
25	9.41	25	− 2.50
30	9.76	30	− 2.26
40	9.80	40	− 1.80
50	9.19	50	− 1.40
60	8.14	60	− 1.00
70	6.69	70	− 0.65
80	4.89	80	− 0.39
90	2.71	90	− 0.22
95	1.47	95	− 0.16
100	(0.13)	100	(− 0.13)
100	......	100	0

L.E. radius: 1.58
Slope of radius through L.E.: 0.20

NACA 4415
(Stations and ordinates given in per cent of airfoil chord)

Upper surface		Lower surface	
Station	Ordinate	Station	Ordinate
0	......	0	0
1.25	3.07	1.25	− 1.79
2.5	4.17	2.5	− 2.48
5.0	5.74	5.0	− 3.27
7.5	6.91	7.5	− 3.71
10	7.84	10	− 3.98
15	9.27	15	− 4.18
20	10.25	20	− 4.15
25	10.92	25	− 3.98
30	11.25	30	− 3.75
40	11.25	40	− 3.25
50	10.53	50	− 2.72
60	9.30	60	− 2.14
70	7.63	70	− 1.55
80	5.55	80	− 1.03
90	3.08	90	− 0.57
95	1.67	95	− 0.36
100	(0.16)	100	(− 0.16)
100	......	100	0

L.E. radius: 2.48
Slope of radius through L.E.: 0.20

NACA 4418
(Stations and ordinates given in per cent of airfoil chord)

Upper surface		Lower surface	
Station	Ordinate	Station	Ordinate
0	......	0	0
1.25	3.76	1.25	− 2.11
2.5	5.00	2.5	− 2.99
5.0	6.75	5.0	− 4.06
7.5	8.06	7.5	− 4.67
10	9.11	10	− 5.06
15	10.66	15	− 5.49
20	11.72	20	− 5.56
25	12.40	25	− 5.49
30	12.76	30	− 5.26
40	12.70	40	− 4.70
50	11.85	50	− 4.02
60	10.44	60	− 3.24
70	8.55	70	− 2.45
80	6.22	80	− 1.67
90	3.46	90	− 0.93
95	1.89	95	− 0.55
100	(0.19)	100	(− 0.19)
100	......	100	0

L.E. radius: 3.56
Slope of radius through L.E.: 0.20

NACA 23012
(Stations and ordinates given in per cent of airfoil chord)

Upper surface		Lower surface	
Station	Ordinate	Station	Ordinate
0	......	0	0
1.25	2.67	1.25	− 1.23
2.5	3.61	2.5	− 1.71
5.0	4.91	5.0	− 2.26
7.5	5.80	7.5	− 2.61
10	6.43	10	− 2.92
15	7.19	15	− 3.50
20	7.50	20	− 3.97
25	7.60	25	− 4.28
30	7.55	30	− 4.46
40	7.14	40	− 4.48
50	6.41	50	− 4.17
60	5.47	60	− 3.67
70	4.36	70	− 3.00
80	3.08	80	− 2.16
90	1.68	90	− 1.23
95	0.92	95	− 0.70
100	(0.13)	100	(− 0.13)
100	......	100	0

L.E. radius: 1.58
Slope of radius through L.E.: 0.305

NACA 4424
(Stations and ordinates given in per cent of airfoil chord)

Upper surface		Lower surface	
Station	Ordinate	Station	Ordinate
0	0	0	0
0.530	3.964	1.970	− 3.472
1.536	5.624	3.464	− 4.656
3.775	7.942	6.225	− 6.066
6.153	9.651	8.847	− 6.931
8.611	11.012	11.389	− 7.512
13.674	13.045	16.326	− 8.169
18.858	14.416	21.142	− 8.416
24.111	15.287	25.889	− 8.411
29.401	15.738	30.599	− 8.238
40.000	15.606	40.000	− 7.606
50.235	14.474	49.765	− 6.698
60.405	12.674	59.595	− 5.562
70.487	10.312	69.513	− 4.312
80.464	7.447	79.536	− 3.003
90.320	4.099	89.680	− 1.655
95.196	2.240	94.804	− 0.964
100.000	......	100.000	0

L.E. radius: 6.33
Slope of radius through L.E.: 0.20

NACA 4421
(Stations and ordinates given in per cent of airfoil chord)

Upper surface		Lower surface	
Station	Ordinate	Station	Ordinate
0	......	0	0
1.25	4.45	1.25	− 2.42
2.5	5.84	2.5	− 3.48
5.0	7.82	5.0	− 4.78
7.5	9.24	7.5	− 5.62
10	10.35	10	− 6.15
15	12.04	15	− 6.75
20	13.17	20	− 6.98
25	13.88	25	− 6.92
30	14.27	30	− 6.76
40	14.16	40	− 6.16
50	13.18	50	− 5.34
60	11.60	60	− 4.40
70	9.50	70	− 3.35
80	6.91	80	− 2.31
90	3.85	90	− 1.27
95	2.11	95	− 0.74
100	(0.22)	100	(− 0.22)
100	......	100	0

L.E. radius: 4.85
Slope of radius through L.E.: 0.20

NACA 23015
(Stations and ordinates given in per cent of airfoil chord)

Upper surface		Lower surface	
Station	Ordinate	Station	Ordinate
0	.....	0	0
1.25	3.34	1.25	− 1.54
2.5	4.44	2.5	− 2.25
5.0	5.89	5.0	− 3.04
7.5	6.90	7.5	− 3.61
10	7.64	10	− 4.09
15	8.52	15	− 4.84
20	8.92	20	− 5.41
25	9.08	25	− 5.78
30	9.05	30	− 5.96
40	8.59	40	− 5.92
50	7.74	50	− 5.50
60	6.61	60	− 4.81
70	5.25	70	− 3.91
80	3.73	80	− 2.83
90	2.04	90	− 1.59
95	1.12	95	− 0.90
100	(0.16)	100	(− 0.16)
100	.....	100	0

L.E. radius: 2.48
Slope of radius through L.E.: 0.305

NACA 23018
(Stations and ordinates given in per cent of airfoil chord)

Upper surface		Lower surface	
Station	Ordinate	Station	Ordinate
0	.....	0	0
1.25	4.09	1.25	− 1.83
2.5	5.29	2.5	− 2.71
5.0	6.92	5.0	− 3.80
7.5	8.01	7.5	− 4.60
10	8.83	10	− 5.22
15	9.86	15	− 6.18
20	10.36	20	− 6.86
25	10.56	25	− 7.27
30	10.55	30	− 7.47
40	10.04	40	− 7.37
50	9.05	50	− 6.81
60	7.75	60	− 5.94
70	6.18	70	− 4.82
80	4.40	80	− 3.48
90	2.39	90	− 1.94
95	1.32	95	− 1.09
100	(0.19)	100	(− 0.19)
100	.....	100	0

L.E. radius: 3.56
Slope of radius through L.E.: 0.305

NACA 23021
(Stations and ordinates given in per cent of airfoil chord)

Upper surface		Lower surface	
Station	Ordinate	Station	Ordinate
0	.....	0	0
1.25	4.87	1.25	− 2.08
2.5	6.14	2.5	− 3.14
5.0	7.93	5.0	− 4.52
7.5	9.13	7.5	− 5.55
10	10.03	10	− 6.32
15	11.19	15	− 7.51
20	11.80	20	− 8.30
25	12.05	25	− 8.76
30	12.06	30	− 8.95
40	11.49	40	− 8.83
50	10.40	50	− 8.14
60	8.90	60	− 7.07
70	7.09	70	− 5.72
80	5.05	80	− 4.13
90	2.76	90	− 2.30
95	1.53	95	− 1.30
100	(0.22)	100	(− 0.22)
100	.....	100	0

L.E. radius: 4.85
Slope of radius through L.E.: 0.305

NACA 23024

(Stations and ordinates given in per cent of airfoil chord)

Upper surface		Lower surface	
Station	Ordinate	Station	Ordinate
0	0	0	0
0.277	4.017	2.223	− 3.303
1.331	5.764	3.669	− 4.432
3.853	8.172	6.147	− 5.862
6.601	9.844	8.399	− 6.860
9.423	11.049	10.577	− 7.647
15.001	12.528	14.999	− 8.852
20.253	13.237	19.747	− 9.703
25.262	13.535	24.738	− 10.223
30.265	13.546	29.735	− 10.454
40.256	12.928	39.744	− 10.278
50.235	11.690	49.766	− 9.482
60.202	10.008	59.798	− 8.242
70.162	7.988	69.838	− 6.664
80.116	5.687	79.884	− 4.803
90.064	3.115	89.936	− 2.673
95.036	1.724	94.964	− 1.504
100	. . . . . .	100	0

L.E. radius: 6.33
Slope of radius through L.E.: 0.305

NACA 63-206

(Stations and ordinates given in percent of airfoil chord)

Upper Surface		Lower Surface	
Station	Ordinate	Station	Ordinate
0	0	0	0
.458	.551	.542	− .451
.703	.677	.797	− .537
1.197	.876	1.303	− .662
2.438	1.241	2.562	− .869
4.932	1.776	5.068	− 1.144
7.429	2.189	7.571	− 1.341
9.930	2.526	10.070	− 1.492
14.934	3.058	15.066	− 1.712
19.941	3.451	20.059	− 1.859
24.950	3.736	25.050	− 1.946
29.960	3.926	30.040	− 1.982
34.970	4.030	35.030	− 1.970
39.981	4.042	40.019	− 1.900
44.991	3.972	45.009	− 1.782
50.000	3.826	50.000	− 1.620
55.008	3.612	54.992	− 1.422
60.015	3.338	59.985	− 1.196
65.020	3.012	64.980	− .952
70.023	2.642	69.977	− .698
75.023	2.237	74.927	− .447
80.022	1.804	79.978	− .212
85.019	1.356	84.981	− .010
90.013	.900	89.987	.134
95.006	.454	94.994	.178
100.000	0	100.000	0

L.E. radius: 0.297
Slope of radius through L.E.: 0.0842

NACA 63-209

(Stations and ordinates given in
per cent of airfoil chord)

Upper surface		Lower surface	
Station	Ordinate	Station	Ordinate
0	0	0	0
0.436	0.796	0.563	− 0.696
0.680	0.973	0.820	− 0.833
1.170	1.255	1.330	− 1.041
2.408	1.765	2.592	− 1.393
4.897	2.510	5.103	− 1.878
7.394	3.077	7.606	− 2.229
9.894	3.539	10.106	− 2.505
14.901	4.263	15.099	− 2.917
19.912	4.792	20.088	− 3.200
24.925	5.169	25.075	− 3.379
29.940	5.414	30.060	− 3.470
34.956	5.530	35.044	− 3.470
39.971	5.518	40.029	− 3.376
44.986	5.391	45.014	− 3.201
50.000	5.159	50.000	− 2.953
55.012	4.834	54.988	− 2.644
60.022	4.429	59.978	− 2.287
65.029	3.958	64.971	− 1.898
70.033	3.430	69.967	− 1.486
75.034	2.861	74.966	− 1.071
80.032	2.267	79.968	− 0.675
85.027	1.663	84.973	− 0.317
90.019	1.067	89.981	− 0.033
95.009	0.512	94.991	0.120
100.000	0	100.000	0

L.E. radius: 0.631
Slope of radius through L.E.: 0.0842

NACA 63-210

(Stations and ordinates given in
per cent of airfoil chord)

Upper surface		Lower surface	
Station	Ordinate	Station	Ordinate
0	0	0	0
0.430	0.876	0.570	− 0.776
0.669	1.107	0.831	− 0.967
1.162	1.379	1.338	− 1.165
2.398	1.939	2.602	− 1.567
4.886	2.753	5.114	− 2.121
7.382	3.372	7.618	− 2.524
9.882	3.877	10.118	− 2.843
14.890	4.665	15.110	− 3.319
19.902	5.240	20.098	− 3.648
24.917	5.647	25.083	− 3.857
29.933	5.910	30.067	− 3.966
34.951	6.030	35.049	− 3.970
39.968	6.009	40.032	− 3.867
44.985	5.861	45.015	− 3.671
50.000	5.599	50.000	− 3.393
55.013	5.235	54.987	− 3.045
60.024	4.786	59.976	− 2.644
65.032	4.264	64.968	− 2.204
70.036	3.684	69.964	− 1.740
75.038	3.061	74.962	− 1.271
80.036	2.414	79.964	− 0.822
85.030	1.761	84.970	− 0.415
90.021	1.121	89.979	− 0.087
95.010	0.530	94.990	0.102
100.000	0	100.000	0

L.E. radius: 0.770
Slope of radius through L.E.: 0.0842

NACA 63₁-212

(Stations and ordinates given in per cent of airfoil chord)

Upper surface		Lower surface	
Station	Ordinate	Station	Ordinate
0	0	0	0
0.417	1.032	0.583	− 0.932
0.657	1.260	0.843	− 1.120
1.145	1.622	1.355	− 1.408
2.378	2.284	2.622	− 1.912
4.863	3.238	5.137	− 2.606
7.358	3.963	7.642	− 3.115
9.859	4.554	10.141	− 3.520
14.868	5.470	15.132	− 4.124
19.882	6.137	20.118	− 4.545
24.900	6.606	25.100	− 4.816
29.920	6.901	30.080	− 4.957
34.941	7.030	35.059	− 4.970
39.962	6.991	40.038	− 4.849
44.982	6.799	45.018	− 4.609
50.000	6.473	50.000	− 4.267
55.016	6.030	54.984	− 3.840
60.029	5.491	59.971	− 3.349
65.038	4.870	64.962	− 2.810
70.043	4.182	69.957	− 2.238
75.045	3.451	74.955	− 1.661
80.042	2.698	79.958	− 1.106
85.035	1.947	84.965	− 0.601
90.025	1.224	89.975	− 0.190
95.012	0.566	94.988	0.066
100.000	0	100.000	0

L.E. radius: 1.087
Slope of radius through L.E.: 0.0842

NACA 63₁-412

(Stations and ordinates given in per cent of airfoil chord)

Upper surface		Lower surface	
Station	Ordinate	Station	Ordinate
0	0	0	0
0.336	1.071	0.664	− 0.871
0.567	1.320	0.933	− 1.040
1.041	1.719	1.459	− 1.291
2.257	2.460	2.743	− 1.716
4.727	3.544	5.273	− 2.280
7.218	4.379	7.782	− 2.685
9.718	5.063	10.282	− 2.995
14.735	6.138	15.265	− 3.446
19.765	6.929	20.235	− 3.745
24.800	7.499	25.200	− 3.919
29.840	7.872	30.160	− 3.984
34.882	8.059	35.118	− 3.939
39.924	8.062	40.076	− 3.778
44.964	7.894	45.036	− 3.514
50.000	7.576	50.000	− 3.164
55.031	7.125	54.969	− 2.745
60.057	6.562	59.943	− 2.278
65.076	5.899	64.924	− 1.779
70.087	5.153	69.913	− 1.265
75.089	4.344	74.911	− 0.764
80.084	3.492	79.916	− 0.308
85.070	2.618	84.930	0.074
90.049	1.739	89.951	0.329
95.023	0.881	94.977	0.383
100.000	0	100.000	0

L.E. radius: 1.087
Slope of radius through L.E.: 0.1685

NACA 63₂-215

(Stations and ordinates given in
per cent of airfoil chord)

Upper surface		Lower surface	
Station	Ordinate	Station	Ordinate
0	0	0	0
0.399	1.250	0.601	− 1.150
0.637	1.528	0.863	− 1.388
1.120	1.980	1.380	− 1.766
2.348	2.792	2.652	− 2.420
4.829	3.960	5.171	− 3.328
7.323	4.847	7.677	− 3.999
9.823	5.569	10.177	− 4.535
14.834	6.682	15.166	− 5.336
19.852	7.487	20.148	− 5.895
24.875	8.049	25.125	− 6.259
29.900	8.392	30.100	− 6.448
34.926	8.530	35.074	− 6.470
39.952	8.457	40.048	− 6.315
44.977	8.194	45.023	− 6.004
50.000	7.768	50.000	− 5.562
55.019	7.203	54.981	− 5.013
60.035	6.524	59.965	− 4.382
65.047	5.751	64.953	− 3.691
70.053	4.906	69.947	− 2.962
75.055	4.014	74.945	− 2.224
80.051	3.105	79.949	− 1.513
85.043	2.213	84.957	− 0.867
90.030	1.368	89.970	− 0.334
95.014	0.616	94.986	0.016
100.000	0	100.000	0

L.E. radius: 1.594
Slope of radius through L.E.: 0.0842

NACA 63₂-415

(Stations and ordinates given in
per cent of airfoil chord)

Upper surface		Lower surface	
Station	Ordinate	Station	Ordinate
0	0	0	0
0.300	1.287	0.700	− 1.087
0.525	1.585	0.975	− 1.305
0.991	2.074	1.509	− 1.646
2.198	2.964	2.802	− 2.220
4.660	4.264	5.340	− 3.000
7.147	5.261	7.853	− 3.565
9.647	6.077	10.353	− 4.009
14.669	7.348	15.331	− 4.656
19.705	8.279	20.295	− 5.095
24.750	8.941	25.250	− 5.361
29.800	9.362	30.200	− 5.474
34.852	9.559	35.148	− 5.439
39.905	9.527	40.095	− 5.243
44.955	9.289	45.045	− 4.909
50.000	8.871	50.000	− 4.459
55.039	8.298	54.961	− 3.918
60.070	7.595	59.930	− 3.311
65.093	6.780	64.907	− 2.660
70.106	5.877	69.894	− 1.989
75.109	4.907	74.891	− 1.327
80.102	3.900	79.898	− 0.716
85.085	2.885	84.915	− 0.193
90.059	1.884	89.941	0.184
95.028	0.931	94.972	0.333
100.000	0	100.000	0

L.E. radius: 1.594
Slope of radius through L.E.: 0.1685

NACA 63₂-615

(Stations and ordinates given in
per cent of airfoil chord)

Upper surface		Lower surface	
Station	Ordinate	Station	Ordinate
0	0	0	0
0.205	1.317	0.795	− 1.017
0.418	1.634	1.082	− 1.214
0.866	2.159	1.634	− 1.517
2.050	3.129	2.950	− 2.013
4.492	4.560	5.508	− 2.664
6.973	5.667	8.027	− 3.123
9.473	6.578	10.527	− 3.476
14.504	8.010	15.496	− 3.972
19.558	9.066	20.442	− 4.290
24.625	9.830	25.375	− 4.460
29.700	10.331	30.300	− 4.499
34.778	10.587	35.222	− 4.407
39.857	10.598	40.143	− 4.172
44.932	10.384	45.068	− 3.814
50.000	9.974	50.000	− 3.356
55.058	9.393	54.942	− 2.823
60.105	8.665	59.895	− 2.239
65.139	7.809	64.861	− 1.629
70.159	6.847	69.841	− 1.015
75.163	5.800	74.837	− 0.430
80.153	4.693	79.847	0.083
85.127	3.555	84.873	0.483
90.089	2.398	89.911	0.704
95.042	1.245	94.958	0.651
100.000	0	100.000	0

L.E. radius: 1.594
Slope of radius through L.E.: 0.2527

NACA 63₃-218

(Stations and ordinates given in
per cent of airfoil chord)

Upper surface		Lower surface	
Station	Ordinate	Station	Ordinate
0	0	0	0
0.382	1.449	0.618	− 1.349
0.617	1.778	0.883	− 1.638
1.096	2.319	1.404	− 2.105
2.319	3.285	2.681	− 2.913
4.796	4.673	5.204	− 4.041
7.288	5.728	7.712	− 4.880
9.788	6.581	10.212	− 5.547
14.801	7.895	15.199	− 6.549
19.822	8.842	20.178	− 7.250
24.850	9.494	25.150	− 7.704
29.880	9.884	30.120	− 7.940
34.911	10.030	35.089	− 7.970
39.943	9.916	40.057	− 7.774
44.973	9.577	45.027	− 7.387
50.000	9.045	50.000	− 6.839
55.023	8.351	54.977	− 6.161
60.042	7.526	59.958	− 5.384
65.055	6.597	64.945	− 4.537
70.062	5.594	69.938	− 3.650
75.064	4.544	74.936	− 2.754
80.059	3.486	79.941	− 1.894
85.049	2.459	84.951	− 1.113
90.034	1.501	89.966	− 0.467
95.016	0.664	94.984	− 0.032
100.000	0	100.000	0

L.E. radius: 2.120
Slope of radius through L.E.: 0.0842

NACA 63₃-418

(Stations and ordinates given in
per cent of airfoil chord)

Upper surface		Lower surface	
Station	Ordinate	Station	Ordinate
0	0	0	0
0.267	1.484	0.733	− 1.284
0.487	1.833	1.013	− 1.553
0.945	2.410	1.555	− 1.982
2.140	3.455	2.860	− 2.711
4.593	4.975	5.407	− 3.711
7.077	6.139	7.923	− 4.443
9.577	7.087	10.423	− 5.019
14.602	8.560	15.398	− 5.868
19.645	9.632	20.355	− 6.448
24.699	10.385	25.301	− 6.805
29.760	10.854	30.240	− 6.966
34.823	11.058	35.177	− 6.938
39.886	10.986	40.114	− 6.702
44.946	10.672	45.054	− 6.292
50.000	10.148	50.000	− 5.736
55.046	9.446	54.954	− 5.066
60.083	8.596	59.917	− 4.312
65.110	7.626	64.890	− 3.506
70.125	6.564	69.875	− 2.676
75.128	5.438	74.872	− 1.858
80.119	4.280	79.881	− 1.096
85.099	3.130	84.901	− 0.438
90.069	2.017	89.931	0.051
95.032	0.978	94.968	0.286
100.000	0	100.000	0

L.E. radius: 2.120
Slope of radius through L.E.: 0.1685

NACA 63₃-618

(Stations and ordinates given in
per cent of airfoil chord)

Upper surface		Lower surface	
Station	Ordinate	Station	Ordinate
0	0	0	0
0.156	1.511	0.844	− 1.211
0.361	1.878	1.139	− 1.458
0.797	2.491	1.703	− 1.849
1.965	3.616	3.035	− 2.500
4.393	5.268	5.607	− 3.372
6.868	6.542	8.132	− 3.998
9.367	7.586	10.633	− 4.484
14.404	9.219	15.596	− 5.181
19.469	10.418	20.531	− 5.642
24.549	11.273	25.451	− 5.903
29.640	11.822	30.360	− 5.990
34.734	12.086	35.266	− 5.906
39.829	12.056	40.171	− 5.630
44.919	11.767	45.081	− 5.197
50.000	11.251	50.000	− 4.633
55.069	10.541	54.931	− 3.971
60.125	9.667	59.875	− 3.241
65.164	8.655	64.836	− 2.475
70.187	7.534	69.813	− 1.702
75.191	6.330	74.809	− 0.960
80.178	5.073	79.822	− 0.297
85.147	3.800	84.853	0.238
90.103	2.531	89.897	0.571
95.048	1.293	94.952	0.603
100.000	0	100.000	0

L.E. radius: 2.120
Slope of radius through L.E.: 0.2527

NACA 63₄-221

(Stations and ordinates given in per cent of airfoil chord)

Upper surface		Lower surface	
Station	Ordinate	Station	Ordinate
0	0	0	0
0.367	1.627	0.633	− 1.527
0.600	2.001	0.900	− 1.861
1.075	2.628	1.425	− 2.414
2.292	3.757	2.708	− 3.385
4.763	5.375	5.237	− 4.743
7.253	6.601	7.747	− 5.753
9.753	7.593	10.247	− 6.559
14.767	9.111	15.233	− 7.765
19.792	10.204	20.208	− 8.612
24.824	10.946	25.176	− 9.156
29.860	11.383	30.140	− 9.439
34.897	11.529	35.103	− 9.469
39.934	11.369	40.066	− 9.227
44.969	10.949	45.031	− 8.759
50.000	10.309	50.000	− 8.103
55.027	9.485	54.973	− 7.295
60.048	8.512	59.952	− 6.370
65.063	7.426	64.937	− 5.366
70.071	6.262	69.929	− 4.318
75.073	5.054	74.927	− 3.264
80.067	3.849	79.933	− 2.257
85.056	2.693	84.944	− 1.347
90.039	1.629	89.961	− 0.595
95.018	0.708	94.982	− 0.076
100.000	0	100.000	0

L.E. radius: 2.650
Slope of radius through L.E.: 0.0842

NACA 63₄-421

(Stations and ordinates given in per cent of airfoil chord)

Upper surface		Lower surface	
Station	Ordinate	Station	Ordinate
0	0	0	0
0.237	1.661	0.763	− 1.461
0.452	2.054	1.048	− 1.774
0.902	2.717	1.598	− 2.289
2.086	3.925	2.914	− 3.181
4.527	5.675	5.473	− 4.411
7.007	7.010	7.993	− 5.314
9.506	8.097	10.494	− 6.029
14.535	9.774	15.465	− 7.082
19.585	10.993	20.415	− 7.809
24.649	11.837	25.351	− 8.257
29.719	12.352	30.281	− 8.464
34.793	12.558	35.207	− 8.438
39.867	12.439	40.133	− 8.155
44.937	12.044	45.063	− 7.664
50.000	11.412	50.000	− 7.000
55.054	10.580	54.946	− 6.200
60.096	9.582	59.904	− 5.298
65.126	8.455	64.874	− 4.335
70.143	7.232	69.857	− 3.344
75.145	5.947	74.855	− 2.367
80.135	4.643	79.865	− 1.459
85.111	3.364	84.889	− 0.672
90.078	2.144	89.922	− 0.076
95.037	1.022	94.963	0.242
100.000	0	100.000	0

L.E. radius: 2.650
Slope of radius through L.E.: 0.1685

NACA 63A210

(Stations and ordinates given in
per cent of airfoil chord)

Upper surface		Lower surface	
Station	Ordinate	Station	Ordinate
0	0	0	0
0.423	0.868	0.577	− 0.756
0.664	1.058	0.836	− 0.900
1.151	1.367	1.349	− 1.125
2.384	1.944	2.616	− 1.522
4.869	2.769	5.131	− 2.047
7.364	3.400	7.636	− 2.428
9.863	3.917	10.137	− 2.725
14.869	4.729	15.131	− 3.167
19.882	5.328	20.118	− 3.468
24.898	5.764	25.102	− 3.662
29.916	6.060	30.084	− 3.764
34.935	6.219	35.065	− 3.771
39.955	6.247	40.045	− 3.689
44.975	6.151	45.025	− 3.523
49.994	5.943	50.006	− 3.283
55.012	5.637	54.988	− 2.985
60.028	5.245	59.972	− 2.641
65.041	4.772	64.959	− 2.262
70.052	4.227	69.948	− 1.861
75.061	3.624	74.939	− 1.464
80.074	2.974	79.926	− 1.104
85.072	2.254	84.928	− 0.812
90.050	1.519	89.950	− 0.539
95.026	0.769	94.974	− 0.279
100.000	0.021	100.000	− 0.021

L.E. radius: 0.742
T.E. radius: 0.023
Slope of radius through L.E.: 0.095

NACA 64-108

(Stations and ordinates given in
per cent of airfoil chord)

Upper surface		Lower surface	
Station	Ordinate	Station	Ordinate
0	0	0	0
0.472	0.682	0.528	− 0.632
0.719	0.828	0.781	− 0.758
1.215	1.058	1.285	− 0.950
2.460	1.457	2.540	− 1.271
4.956	2.032	5.044	− 1.716
7.455	2.471	7.545	− 2.047
9.955	2.832	10.045	− 2.316
14.958	3.405	15.042	− 2.733
19.962	3.835	20.038	− 3.039
24.968	4.152	25.032	− 3.256
29.974	4.370	30.026	− 3.398
34.980	4.494	35.020	− 3.464
39.987	4.528	40.013	− 3.456
44.994	4.431	45.006	− 3.335
50.000	4.236	50.000	− 3.132
55.005	3.959	54.995	− 2.863
60.010	3.617	59.990	− 2.545
65.013	3.219	64.987	− 2.189
70.015	2.777	69.985	− 1.805
75.016	2.302	74.984	− 1.406
80.015	1.802	79.985	− 1.006
85.013	1.297	84.987	− 0.625
90.010	0.808	89.990	− 0.292
95.005	0.364	94.995	− 0.048
100.000	0	100.000	0

L.E. radius: 0.455
Slope of radius through L.E.: 0.042

NACA 64-110

(Stations and ordinates given in per cent of airfoil chord)

Upper surface		Lower surface	
Station	Ordinate	Station	Ordinate
0	0	0	0
0.465	0.844	0.535	− 0.794
0.712	1.023	0.788	− 0.953
1.207	1.303	1.293	− 1.195
2.450	1.793	2.550	− 1.607
4.945	2.500	5.055	− 2.184
7.443	3.037	7.557	− 2.613
9.944	3.479	10.056	− 2.963
14.947	4.178	15.053	− 3.506
19.953	4.700	20.047	− 3.904
24.959	5.087	25.041	− 4.191
29.967	5.350	30.033	− 4.378
34.975	5.495	35.025	− 4.465
39.984	5.524	40.016	− 4.452
44.992	5.391	45.008	− 4.295
50.000	5.138	50.000	− 4.034
55.007	4.786	54.993	− 3.690
60.012	4.356	59.988	− 3.284
65.016	3.860	64.984	− 2.830
70.019	3.313	69.981	− 2.341
75.020	2.729	74.980	− 1.833
80.019	2.120	79.981	− 1.324
85.016	1.512	84.984	− 0.840
90.012	0.929	89.988	− 0.413
95.006	0.406	94.994	− 0.090
100.000	0	100.000	0

L.E. radius: 0.720
Slope of radius through L.E.: 0.042

NACA 64-206

(Stations and ordinates given in per cent of airfoil chord)

Upper surface		Lower surface	
Station	Ordinate	Station	Ordinate
0	0	0	0
0.459	0.542	0.541	− 0.442
0.704	0.664	0.796	− 0.524
1.198	0.859	1.302	− 0.645
2.440	1.208	2.560	− 0.836
4.934	1.719	5.066	− 1.087
7.432	2.115	7.568	− 1.267
9.933	2.444	10.067	− 1.410
14.937	2.970	15.063	− 1.624
19.943	3.367	20.057	− 1.775
24.952	3.667	25.048	− 1.877
29.961	3.879	30.039	− 1.935
34.971	4.011	35.029	− 1.951
39.981	4.066	40.019	− 1.924
44.991	4.014	45.009	− 1.824
50.000	3.878	50.000	− 1.672
55.008	3.670	54.992	− 1.480
60.015	3.402	59.985	− 1.260
65.020	3.080	64.980	− 1.020
70.023	2.712	69.977	− 0.768
75.025	2.307	74.975	− 0.517
80.024	1.868	79.976	− 0.276
85.020	1.410	84.980	− 0.064
90.015	0.940	89.985	0.094
95.007	0.473	94.993	0.159
100.000	0	100.000	0

L.E. radius: 0.256
Slope of radius through L.E.: 0.084

NACA 64-208
(Stations and ordinates given in
per cent of airfoil chord)

Upper surface		Lower surface	
Station	Ordinate	Station	Ordinate
0	0	0	0
0.445	0.706	0.555	− 0.606
0.688	0.862	0.812	− 0.722
1.180	1.110	1.320	− 0.896
2.421	1.549	2.579	− 1.177
4.912	2.189	5.088	− 1.557
7.410	2.681	7.590	− 1.833
9.909	3.089	10.091	− 2.055
14.915	3.741	15.085	− 2.395
19.924	4.232	20.076	− 2.640
24.935	4.598	25.065	− 2.808
29.948	4.856	30.052	− 2.912
34.961	5.009	35.039	− 2.949
39.974	5.063	40.026	− 2.921
44.988	4.978	45.012	− 2.788
50.000	4.787	50.000	− 2.581
55.011	4.506	54.989	− 2.316
60.020	4.152	59.980	− 2.010
65.027	3.733	64.973	− 1.673
70.031	3.263	69.969	− 1.319
75.032	2.749	74.968	− 0.959
80.031	2.200	79.969	− 0.608
85.027	1.634	84.973	− 0.288
90.019	1.067	89.981	− 0.033
95.010	0.522	94.990	0.110
100.000	0	100.000	0

L.E. radius: 0.455
Slope of radius through L.E.: 0.084

NACA 64-209
(Stations and ordinates given in
per cent of airfoil chord)

Upper surface		Lower surface	
Station	Ordinate	Station	Ordinate
0	0	0	0
0.438	0.786	0.562	− 0.686
0.680	0.959	0.820	− 0.819
1.172	1.232	1.328	− 1.018
2.411	1.716	2.589	− 1.344
4.901	2.423	5.099	− 1.791
7.398	2.965	7.602	− 2.117
9.899	3.413	10.101	− 2.379
14.905	4.127	15.095	− 2.781
19.915	4.663	20.085	− 3.071
24.927	5.064	25.073	− 3.274
29.941	5.345	30.059	− 3.401
34.956	5.509	35.044	− 3.449
39.971	5.561	40.029	− 3.419
44.986	5.459	45.014	− 3.269
50.000	5.239	50.000	− 3.033
55.012	4.921	54.988	− 2.731
60.022	4.523	59.978	− 2.381
65.030	4.056	64.970	− 1.996
70.035	3.533	69.965	− 1.589
75.036	2.964	74.964	− 1.174
80.035	2.360	79.965	− 0.768
85.030	1.742	84.970	− 0.396
90.021	1.128	89.979	− 0.094
95.011	0.543	94.989	0.089
100.000	0	100.000	0

L.E. radius: 0.579
Slope of radius through L.E.: 0.084

NACA 64-210

(Stations and ordinates given in
per cent of airfoil chord)

Upper surface		Lower surface	
Station	Ordinate	Station	Ordinate
0	0	0	0
0.431	0.867	0.569	− 0.767
0.673	1.056	0.827	− 0.916
1.163	1.354	1.337	− 1.140
2.401	1.884	2.599	− 1.512
4.890	2.656	5.110	− 2.024
7.387	3.248	7.613	− 2.400
9.887	3.736	10.113	− 2.702
14.894	4.514	15.106	− 3.168
19.905	5.097	20.095	− 3.505
24.919	5.533	25.081	− 3.743
29.934	5.836	30.066	− 3.892
34.951	6.010	35.049	− 3.950
39.968	6.059	40.032	− 3.917
44.985	5.938	45.015	− 3.748
50.000	5.689	50.000	− 3.483
55.014	5.333	54.987	− 3.143
60.025	4.891	59.975	− 2.749
65.033	4.375	64.967	− 2.315
70.038	3.799	69.962	− 1.855
75.040	3.176	74.960	− 1.386
80.038	2.518	79.962	− 0.926
85.033	1.849	84.968	− 0.503
90.024	1.188	89.977	− 0.154
95.012	0.564	94.988	0.068
100.000	0	100.000	0

L.E. radius: 0.720
Slope of radius through L.E.: 0.084

NACA 64₁-112

(Stations and ordinates given in
per cent of airfoil chord)

Upper surface		Lower surface	
Station	Ordinate	Station	Ordinate
0	0	0	
0.459	1.002	0.541	− 0.952
0.704	1.213	0.796	− 1.143
1.198	1.543	1.302	− 1.435
2.441	2.127	2.559	− 1.941
4.934	2.967	5.066	− 2.651
7.432	3.605	7.568	− 3.181
9.932	4.128	10.068	− 3.612
14.936	4.956	15.064	− 4.284
19.943	5.571	20.057	− 4.775
24.951	6.024	25.049	− 5.128
29.961	6.330	30.039	− 5.358
34.971	6.493	35.029	− 5.463
39.981	6.517	40.019	− 5.445
44.991	6.346	45.009	− 5.250
50.000	6.032	50.000	− 4.928
55.008	5.604	54.992	− 4.508
60.015	5.084	59.985	− 4.012
65.020	4.489	64.980	− 3.459
70.023	3.836	69.977	− 2.864
75.024	3.143	74.976	− 2.247
80.022	2.427	79.978	− 1.631
85.019	1.718	84.981	− 1.046
90.014	1.044	89.986	− 0.528
95.007	0.446	94.993	− 0.130
100.000	0	100.000	0

L.E. radius: 1.040
Slope of radius through L.E.: 0.042

NACA 64₁-212

(Stations and ordinates given in
per cent of airfoil chord)

Upper surface		Lower surface	
Station	Ordinate	Station	Ordinate
0	0	0	0
0.418	1.025	0.582	− 0.925
0.659	1.245	0.841	− 1.105
1.147	1.593	1.353	− 1.379
2.382	2.218	2.618	− 1.846
4.868	3.123	5.132	− 2.491
7.364	3.815	7.636	− 2.967
9.865	4.386	10.135	− 3.352
14.872	5.291	15.128	− 3.945
19.886	5.968	20.114	− 4.376
24.903	6.470	25.097	− 4.680
29.921	6.815	30.079	− 4.871
34.941	7.008	35.059	− 4.948
39.961	7.052	40.039	− 4.910
44.982	6.893	45.018	− 4.703
50.000	6.583	50.000	− 4.377
55.016	6.151	54.984	− 3.961
60.029	5.619	59.971	− 3.477
65.039	5.004	64.961	− 2.944
70.045	4.322	69.955	− 2.378
75.047	3.590	74.953	− 1.800
80.045	2.825	79.955	− 1.233
85.038	2.054	84.962	− 0.708
90.027	1.303	89.973	− 0.269
95.013	0.604	94.987	0.028
100.000	0	100.000	0

L.E. radius: 1.040
Slope of radius through L.E.: 0.084

NACA 64₁-412

(Stations and ordinates given in
per cent of airfoil chord)

Upper surface		Lower surface	
Station	Ordinate	Station	Ordinate
0	0	0	0
0.338	1.064	0.662	− 0.864
0.569	1.305	0.931	− 1.025
1.045	1.690	1.455	− 1.262
2.264	2.393	2.736	− 1.649
4.738	3.430	5.262	− 2.166
7.229	4.231	7.771	− 2.535
9.730	4.896	10.270	− 2.828
14.745	5.959	15.255	− 3.267
19.772	6.760	20.228	− 3.576
24.805	7.363	25.195	− 3.783
29.842	7.786	30.158	− 3.898
34.882	8.037	35.118	− 3.917
39.923	8.123	40.077	− 3.839
44.963	7.988	45.037	− 3.608
50.000	7.686	50.000	− 3.274
55.032	7.246	54.968	− 2.866
60.059	6.690	59.941	− 2.406
65.078	6.033	64.922	− 1.913
70.090	5.293	69.910	− 1.405
75.094	4.483	74.906	− 0.903
80.089	3.619	79.911	− 0.435
85.076	2.722	84.924	− 0.038
90.055	1.818	89.945	0.250
95.027	0.919	94.973	0.345
100.000	0	100.000	0

L.E. radius: 1.040
Slope of radius through L.E.: 0.168

NACA 64₂-215

(Stations and ordinates given in per cent of airfoil chord)

Upper surface		Lower surface	
Station	Ordinate	Station	Ordinate
0	0	0	0
0.399	1.254	0.601	− 1.154
0.637	1.522	0.863	− 1.382
1.122	1.945	1.378	− 1.731
2.353	2.710	2.647	− 2.338
4.836	3.816	5.164	− 3.184
7.331	4.661	7.669	− 3.813
9.831	5.356	10.169	− 4.322
14.840	6.456	15.160	− 5.110
19.857	7.274	20.143	− 5.682
24.878	7.879	25.122	− 6.089
29.901	8.290	30.099	− 6.346
34.926	8.512	35.074	− 6.452
39.952	8.544	40.048	− 6.402
44.977	8.319	45.023	− 6.129
50.000	7.913	50.000	− 5.707
55.020	7.361	54.980	− 5.171
60.036	6.691	59.964	− 4.549
65.048	5.925	64.952	− 3.865
70.055	5.085	69.945	− 3.141
75.058	4.191	74.942	− 2.401
80.055	3.267	79.945	− 1.675
85.046	2.349	84.954	− 1.003
90.033	1.466	89.967	− 0.432
95.016	0.662	94.984	− 0.030
100.000	0	100.000	0

L.E. radius: 1.590
Slope of radius through L.E.: 0.084

NACA 64₂-415

(Stations and ordinates given in per cent of airfoil chord)

Upper surface		Lower surface	
Station	Ordinate	Station	Ordinate
0	0	0	0
0.299	1.291	0.701	− 1.091
0.526	1.579	0.974	− 1.299
0.996	2.038	1.504	− 1.610
2.207	2.883	2.793	− 2.139
4.673	4.121	5.327	− 2.857
7.162	5.075	7.838	− 3.379
9.662	5.864	10.338	− 3.796
14.681	7.122	15.319	− 4.430
19.714	8.066	20.286	− 4.882
24.756	8.771	25.244	− 5.191
29.803	9.260	30.197	− 5.372
34.853	9.541	35.147	− 5.421
39.904	9.614	40.096	− 5.330
44.954	9.414	45.046	− 5.034
50.000	9.016	50.000	− 4.604
55.040	8.456	54.960	− 4.076
60.072	7.762	59.928	− 3.478
65.096	6.954	64.904	− 2.834
70.111	6.055	69.889	− 2.167
75.115	5.084	74.885	− 1.504
80.109	4.062	79.891	− 0.878
85.092	3.020	84.908	− 0.328
90.066	1.982	89.934	0.086
95.032	0.976	94.968	0.288
100.000	0	100.000	0

L.E. radius: 1.590
Slope of radius through L.E.: 0.168

NACA 64₃-218

(Stations and ordinates given in
per cent of airfoil chord)

Upper surface		Lower surface	
Station	Ordinate	Station	Ordinate
0	0	0	0
0.380	1.473	0.620	− 1.373
0.617	1.785	0.883	− 1.645
1.099	2.279	1.401	− 2.065
2.325	3.186	2.675	− 2.814
4.804	4.497	5.196	− 3.865
7.297	5.496	7.703	− 4.648
9.797	6.316	10.203	− 5.282
14.808	7.612	15.192	− 6.266
19.828	8.576	20.172	− 6.984
24.853	9.285	25.147	− 7.495
29.881	9.760	30.119	− 7.816
34.912	10.009	35.088	− 7.949
39.942	10.023	40.058	− 7.881
44.972	9.725	45.028	− 7.535
50.000	9.217	50.000	− 7.011
55.024	8.540	54.976	− 6.350
60.043	7.729	59.957	− 5.587
65.057	6.812	64.943	− 4.752
70.065	5.814	69.935	− 3.870
75.068	4.760	74.932	− 2.970
80.064	3.683	79.936	− 2.091
85.054	2.623	84.946	− 1.277
90.038	1.617	89.962	− 0.583
95.019	0.716	94.981	− 0.084
100.000	0	100.000	0

L.E. radius: 2.208
Slope of radius through L.E.: 0.084

NACA 64₃-418

(Stations and ordinates given in
per cent of airfoil chord)

Upper surface		Lower surface	
Station	Ordinate	Station	Ordinate
0	0	0	0
0.263	1.508	0.737	− 1.308
0.486	1.840	1.014	− 1.560
0.950	2.370	1.550	− 1.942
2.152	3.357	2.848	− 2.613
4.609	4.800	5.391	− 3.536
7.095	5.908	7.905	− 4.212
9.595	6.823	10.405	− 4.755
14.617	8.277	15.383	− 5.585
19.657	9.366	20.343	− 6.182
24.707	10.176	25.293	− 6.596
29.763	10.730	30.237	− 6.842
34.823	11.037	35.177	− 6.917
39.885	11.093	40.115	− 6.809
44.945	10.820	45.055	− 6.440
50.000	10.320	50.000	− 5.908
55.047	9.635	54.953	− 5.255
60.086	8.799	59.914	− 4.515
65.114	7.841	64.886	− 3.721
70.131	6.784	69.869	− 2.896
75.135	5.654	74.865	− 2.074
80.127	4.477	79.873	− 1.293
85.108	3.294	84.892	− 0.602
90.077	2.132	89.923	− 0.064
95.037	1.030	94.963	0.234
100.000	0	100.000	0

L.E. radius: 2.208
Slope of radius through L.E.: 0.168

NACA 64₃-618

(Stations and ordinates given in per cent of airfoil chord)

Upper surface		Lower surface	
Station	Ordinate	Station	Ordinate
0	0	0	0
0.150	1.534	0.850	− 1.234
0.359	1.885	1.141	− 1.465
0.805	2.452	1.695	− 1.810
1.982	3.518	3.018	− 2.402
4.417	5.093	5.583	− 3.197
6.895	6.312	8.105	− 3.768
9.395	7.322	10.605	− 4.220
14.427	8.937	15.573	− 4.899
19.486	10.153	20.514	− 5.377
24.560	11.065	25.440	− 5.695
29.645	11.698	30.355	− 5.866
34.735	12.065	35.265	− 5.885
39.827	12.163	40.173	− 5.737
44.917	11.915	45.083	− 5.345
50.000	11.423	50.000	− 4.805
55.071	10.730	54.929	− 4.160
60.129	9.870	59.871	− 3.444
65.171	8.870	64.829	− 2.690
70.196	7.754	69.804	− 1.922
75.203	6.544	74.797	− 1.174
80.191	5.270	79.809	− 0.494
85.161	3.963	84.839	0.075
90.115	2.646	89.885	0.456
95.056	1.344	94.944	0.552
100.000	0	100.000	0

L.E. radius: 2.208
Slope of radius through L.E.: 0.253

NACA 64₄-221

(Stations and ordinates given in per cent of airfoil chord)

Upper surface		Lower surface	
Station	Ordinate	Station	Ordinate
0	0	0	0
0.362	1.690	0.638	− 1.590
0.596	2.049	0.904	− 1.909
1.075	2.618	1.425	− 2.404
2.297	3.665	2.703	− 3.293
4.772	5.182	5.228	− 4.550
7.264	6.334	7.736	− 5.486
9.763	7.282	10.237	− 6.248
14.776	8.778	15.224	− 7.432
19.799	9.889	20.201	− 8.297
24.829	10.701	25.171	− 8.911
29.861	11.240	30.139	− 9.296
34.897	11.510	35.103	− 9.450
39.933	11.502	40.067	− 9.360
44.968	11.125	45.032	− 8.935
50.000	10.507	50.000	− 8.301
55.027	9.702	54.973	− 7.512
60.050	8.749	59.950	− 6.607
65.065	7.679	64.935	− 5.619
70.075	6.521	69.925	− 4.577
75.077	5.310	74.923	− 3.520
80.073	4.082	79.927	− 2.490
85.061	2.885	84.939	− 1.539
90.044	1.761	89.956	− 0.727
95.021	0.765	94.979	− 0.133
100.000	0	100.000	0

L.E. radius: 2.884
Slope of radius through L.E.: 0.084

NACA 64₁-421

(Stations and ordinates given in
per cent of airfoil chord)

Upper surface		Lower surface	
Station	Ordinate	Station	Ordinate
0	0	0	0
0.227	1.723	0.773	− 1.523
0.445	2.101	1.055	− 1.821
0.903	2.707	1.597	− 2.279
2.096	3.834	2.904	− 3.090
4.545	5.482	5.455	− 4.218
7.028	6.744	7.972	− 5.048
9.528	7.786	10.472	− 5.718
14.553	9.442	15.447	− 6.750
19.599	10.678	20.401	− 7.494
24.657	11.591	25.343	− 8.011
29.723	12.209	30.277	− 8.321
34.794	12.539	35.206	− 8.419
39.865	12.572	40.135	− 8.288
44.936	12.220	45.064	− 7.840
50.000	11.610	50.000	− 7.198
55.055	10.797	54.945	− 6.417
60.099	9.819	59.901	− 5.535
65.131	8.708	64.869	− 4.588
70.150	7.491	69.850	− 3.603
75.154	6.203	74.846	− 2.623
80.145	4.876	79.855	− 1.692
85.122	3.556	84.878	− 0.864
90.087	2.276	89.913	− 0.208
95.042	1.079	94.958	0.185
100.000	0	100.000	0

L.E. radius: 2.884
Slope of radius through L.E.: 0.168

NACA 64A210

(Stations and ordinates given in
per cent of airfoil chord)

Upper surface		Lower surface	
Station	Ordinate	Station	Ordinate
0	0	0	0
0.424	0.856	0.576	− 0.744
0.665	1.044	0.835	− 0.886
1.153	1.342	1.347	− 1.100
2.387	1.895	2.613	− 1.473
4.874	2.685	5.126	− 1.963
7.369	3.288	7.631	− 2.316
9.868	3.792	10.132	− 2.600
14.874	4.592	15.126	− 3.030
19.885	5.200	20.115	− 3.340
24.900	5.656	25.100	− 3.554
29.917	5.984	30.083	− 3.688
34.935	6.192	35.065	− 3.744
39.955	6.274	40.045	− 3.716
44.975	6.208	45.025	− 3.580
49.994	6.014	50.006	− 3.354
55.012	5.714	54.988	− 3.062
60.028	5.323	59.972	− 2.719
65.042	4.852	64.958	− 2.342
70.054	4.310	69.946	− 1.944
75.063	3.702	74.937	− 1.542
80.076	3.037	79.924	− 1.167
85.074	2.301	84.926	− 0.859
90.052	1.551	89.948	− 0.571
95.027	0.785	94.974	− 0.295
100.000	0.021	100.000	− 0.021

L.E. radius: 0.687
T.E. radius: 0.023
Slope of radius through L.E.: 0.095

NACA 64A410

(Stations and ordinates given in per cent of airfoil chord)

Upper surface		Lower surface	
Station	Ordinate	Station	Ordinate
0	0	0	0
0.350	0.902	0.650	− 0.678
0.582	1.112	0.918	− 0.796
1.059	1.451	1.441	− 0.969
2.276	2.095	2.724	− 1.251
4.749	3.034	5.251	− 1.592
7.230	3.865	7.770	− 1.919
9.737	4.380	10.263	− 1.996
14.748	5.366	15.252	− 2.244
19.770	6.126	20.230	− 2.406
24.800	6.705	25.200	− 2.499
29.834	7.131	30.166	− 2.537
34.871	7.414	35.129	− 2.518
39.910	7.552	40.090	− 2.436
44.950	7.522	45.050	− 2.266
49.989	7.344	50.011	− 2.024
55.025	7.040	54.975	− 1.736
60.057	6.624	59.943	− 1.418
65.085	6.106	64.915	− 1.086
70.108	5.490	69.892	− 0.760
75.126	4.780	74.874	− 0.460
80.151	3.967	79.849	− 0.229
85.148	3.018	84.852	− 0.132
90.104	2.038	89.896	− 0.076
95.053	1.028	94.947	− 0.048
100.000	0.021	100.000	− 0.021

L.E. radius: 0.687
T.E. radius: 0.023
Slope of radius through L.E.: 0.190

NACA 64₁A212

(Stations and ordinates given in per cent of airfoil chord)

Upper surface		Lower surface	
Station	Ordinate	Station	Ordinate
0	0	0	0
0.409	1.013	0.591	− 0.901
0.648	1.233	0.852	− 1.075
1.135	1.580	1.365	− 1.338
2.365	2.225	2.635	− 1.803
4.849	3.145	5.151	− 2.423
7.343	3.846	7.657	− 2.874
9.842	4.432	10.158	− 3.240
14.849	5.358	15.151	− 3.796
19.862	6.060	20.138	− 4.200
24.880	6.584	25.120	− 4.482
29.900	6.956	30.100	− 4.660
34.922	7.189	35.078	− 4.741
39.946	7.272	40.054	− 4.714
44.970	7.177	45.030	− 4.549
49.993	6.935	50.007	− 4.275
55.015	6.570	54.985	− 3.918
60.034	6.103	59.966	− 3.499
65.050	5.544	64.950	− 3.034
70.064	4.903	69.936	− 2.537
75.075	4.197	74.925	− 2.037
80.090	3.433	79.910	− 1.563
85.088	2.601	84.912	− 1.159
90.062	1.751	89.938	− 0.771
95.032	0.888	94.968	− 0.398
100.000	0.025	100.000	− 0.025

L.E. radius: 0.994
T.E. radius: 0.028
Slope of radius through L.E.: 0.095

432 THEORY OF WING SECTIONS

NACA 64₂A215

(Stations and ordinates given in per cent of airfoil chord)

Upper surface		Lower surface	
Station	Ordinate	Station	Ordinate
0	0	0	0
0.388	1.243	0.612	− 1.131
0.624	1.509	0.876	− 1.351
1.107	1.930	1.393	− 1.688
2.333	2.713	2.667	− 2.291
4.811	3.833	5.189	− 3.111
7.304	4.683	7.696	− 3.711
9.802	5.391	10.198	− 4.199
14.811	6.510	15.189	− 4.948
19.827	7.351	20.173	− 5.491
24.849	7.975	25.151	− 5.873
29.875	8.417	30.125	− 6.121
34.903	8.686	35.097	− 6.238
39.933	8.766	40.067	− 6.208
44.963	8.627	45.037	− 5.999
49.992	8.308	50.008	− 5.648
55.018	7.843	54.982	− 5.191
60.042	7.258	59.958	− 4.654
65.063	6.566	64.937	− 4.056
70.079	5.782	69.921	− 3.416
75.093	4.926	74.907	− 2.766
80.111	4.017	79.889	− 2.147
85.109	3.039	84.891	− 1.597
90.076	2.046	89.924	− 1.066
95.039	1.039	94.961	− 0.549
100.000	0.032	100.000	− 0.032

L.E. radius: 1.561
T.E. radius: 0.037
Slope of radius through L.E.: 0.095

NACA 65-206

(Stations and ordinates given in per cent of airfoil chord)

Upper surface		Lower surface	
Station	Ordinate	Station	Ordinate
0	0	0	0
0.460	0.524	0.540	− 0.424
0.706	0.642	0.794	− 0.502
1.200	0.822	1.300	− 0.608
2.444	1.140	2.556	− 0.768
4.939	1.625	5.061	− 0.993
7.437	2.012	7.563	− 1.164
9.936	2.340	10.064	− 1.306
14.939	2.869	15.061	− 1.523
19.945	3.277	20.055	− 1.685
24.953	3.592	25.047	− 1.802
29.962	3.824	30.038	− 1.880
34.971	3.982	35.029	− 1.922
39.981	4.069	40.019	− 1.927
44.990	4.078	45.010	− 1.888
50.000	4.003	50.000	− 1.797
55.009	3.836	54.991	− 1.646
60.016	3.589	59.984	− 1.447
65.022	3.276	64.978	− 1.216
70.026	2.907	69.974	− 0.963
75.028	2.489	74.972	− 0.699
80.027	2.029	79.973	− 0.437
85.024	1.538	84.976	− 0.192
90.018	1.027	89.982	0.007
95.009	0.511	94.991	0.121
100.000	0	100.000	0

L.E. radius: 0.240
Slope of radius through L.E.: 0.084

NACA 65-209
(Stations and ordinates given in per cent of airfoil chord)

Upper surface		Lower surface	
Station	Ordinate	Station	Ordinate
0	0	0	0
0.441	0.748	0.559	− 0.648
0.684	0.912	0.816	− 0.772
1.177	1.162	1.323	− 0.948
2.417	1.605	2.583	− 1.233
4.908	2.275	5.092	− 1.643
7.405	2.805	7.595	− 1.957
9.904	3.251	10.096	− 2.217
14.909	3.971	15.091	− 2.625
19.918	4.522	20.082	− 2.930
24.929	4.944	25.071	− 3.154
29.942	5.254	30.058	− 3.310
34.956	5.461	35.044	− 3.401
39.971	5.567	40.029	− 3.425
44.986	5.564	45.014	− 3.374
50.000	5.439	50.000	− 3.233
55.013	5.181	54.987	− 2.991
60.024	4.814	59.976	− 2.672
65.033	4.358	64.967	− 2.298
70.039	3.828	69.961	− 1.884
75.041	3.237	74.959	− 1.447
80.040	2.601	79.960	− 1.009
85.035	1.933	84.965	− 0.587
90.026	1.255	89.974	− 0.221
95.013	0.596	94.987	0.036
100.000	0	100.000	0

L.E. radius: 0.552
Slope of radius through L.E.: 0.084

NACA 65-210
(Stations and ordinates given in per cent of airfoil chord)

Upper surface		Lower surface	
Station	Ordinate	Station	Ordinate
0	0	0	0
0.435	0.819	0.565	− 0.719
0.678	0.999	0.822	− 0.859
1.169	1.273	1.331	− 1.059
2.408	1.757	2.592	− 1.385
4.898	2.491	5.102	− 1.859
7.394	3.069	7.606	− 2.221
9.894	3.555	10.106	− 2.521
14.899	4.338	15.101	− 2.992
19.909	4.938	20.091	− 3.346
24.921	5.397	25.079	− 3.607
29.936	5.732	30.064	− 3.788
34.951	5.954	35.049	− 3.894
39.968	6.067	40.032	− 3.925
44.984	6.058	45.016	− 3.868
50.000	5.915	50.000	− 3.709
55.014	5.625	54.986	− 3.435
60.027	5.217	59.973	− 3.075
65.036	4.712	64.964	− 2.652
70.043	4.128	69.957	− 2.184
75.045	3.479	74.955	− 1.689
80.044	2.783	79.956	− 1.191
85.038	2.057	84.962	− 0.711
90.028	1.327	89.972	− 0.293
95.014	0.622	94.986	0.010
100.000	0	100.000	0

L.E. radius: 0.687
Slope of radius through L.E.: 0.084

NACA 65-410

(Stations and ordinates given in per cent of airfoil chord)

Upper surface		Lower surface	
Station	Ordinate	Station	Ordinate
0	0	0	0
0.372	0.861	0.628	− 0.661
0.607	1.061	0.893	− 0.781
1.089	1.372	1.411	− 0.944
2.318	1.935	2.682	− 1.191
4.797	2.800	5.203	− 1.536
7.289	3.487	7.711	− 1.791
9.788	4.067	10.212	− 1.999
14.798	5.006	15.202	− 2.314
19.817	5.731	20.183	− 2.547
24.843	6.290	25.157	− 2.710
29.872	6.702	30.128	− 2.814
34.903	6.983	35.097	− 2.863
39.936	7.138	40.064	− 2.854
44.968	7.153	45.032	− 2.773
50.000	7.018	50.000	− 2.606
55.029	6.720	54.971	− 2.340
60.053	6.288	59.947	− 2.004
65.073	5.741	64.927	− 1.621
70.085	5.099	69.915	− 1.211
75.090	4.372	74.910	− 0.792
80.088	3.577	79.912	− 0.393
85.076	2.729	84.924	− 0.037
90.057	1.842	89.943	0.226
95.029	0.937	94.971	0.327
100.000	0	100.000	0

L.E. radius: 0.687
Slope of radius through L.E.: 0.168

NACA 65₁-212

(Stations and ordinates given in per cent of airfoil chord)

Upper surface		Lower surface	
Station	Ordinate	Station	Ordinate
0	0	0	0
0.423	0.970	0.577	− 0.870
0.664	1.176	0.836	− 1.036
1.154	1.491	1.346	− 1.277
2.391	2.058	2.609	− 1.686
4.878	2.919	5.122	− 2.287
7.373	3.593	7.627	− 2.745
9.873	4.162	10.127	− 3.128
14.879	5.073	15.121	− 3.727
19.890	5.770	20.110	− 4.178
24.906	6.300	25.094	− 4.510
29.923	6.687	30.077	− 4.743
34.942	6.942	35.058	− 4.882
39.961	7.068	40.039	− 4.926
44.981	7.044	45.019	− 4.854
50.000	6.860	50.000	− 4.654
55.017	6.507	54.983	− 4.317
60.032	6.014	59.968	− 3.872
65.043	5.411	64.957	− 3.351
70.050	4.715	69.950	− 2.771
75.053	3.954	74.947	− 2.164
80.052	3.140	79.948	− 1.548
85.045	2.302	84.955	− 0.956
90.033	1.463	89.967	− 0.429
95.017	0.672	94.983	− 0.040
100.000	0	100.000	0

L.E. radius: 1.000
Slope of radius through L.E.: 0.084

NACA 65₁-212
a = 0.6
(Stations and ordinates given in per cent of airfoil chord)

Upper surface		Lower surface	
Station	Ordinate	Station	Ordinate
0	0	0	0
0.399	0.982	0.601	− 0.852
0.638	1.194	0.862	− 1.012
1.124	1.520	1.376	− 1.242
2.356	2.113	2.644	− 1.625
4.837	3.017	5.163	− 2.185
7.329	3.728	7.671	− 2.606
9.827	4.330	10.173	− 2.956
14.833	5.298	15.167	− 3.500
19.848	6.042	20.152	− 3.904
24.869	6.611	25.131	− 4.197
29.894	7.029	30.106	− 4.401
34.921	7.304	35.079	− 4.518
39.951	7.444	40.049	− 4.550
44.983	7.423	45.017	− 4.475
50.017	7.231	49.983	− 4.283
55.051	6.856	54.949	− 3.968
60.094	6.318	59.906	− 3.566
65.123	5.634	64.877	− 3.124
70.124	4.842	69.876	− 2.640
75.112	3.983	74.888	− 2.131
80.090	3.082	79.910	− 1.604
85.064	2.173	84.936	− 1.085
90.036	1.297	89.964	− 0.595
95.013	0.521	94.987	− 0.191
100.000	0	100.000	0

L.E. radius: 1.000
Slope of radius through L.E.: 0.110

NACA 65₁-412
(Stations and ordinates given in per cent of airfoil chord)

Upper surface		Lower surface	
Station	Ordinate	Station	Ordinate
0	0	0	0
0.347	1.010	0.653	− 0.810
0.580	1.236	0.920	− 0.956
1.059	1.588	1.441	− 1.160
2.283	2.234	2.717	− 1.490
4.757	3.227	5.243	− 1.963
7.247	4.010	7.753	− 2.314
9.746	4.672	10.254	− 2.604
14.757	5.741	15.243	− 3.049
19.781	6.562	20.219	− 3.378
24.811	7.193	25.189	− 3.613
29.846	7.658	30.154	− 3.770
34.884	7.971	35.116	− 3.851
39.923	8.139	40.077	− 3.855
44.962	8.139	45.038	− 3.759
50.000	7.963	50.000	− 3.551
55.035	7.602	54.965	− 3.222
60.064	7.085	59.936	− 2.801
65.086	6.440	64.914	− 2.320
70.101	5.686	69.899	− 1.798
75.107	4.847	74.893	− 1.267
80.103	3.935	79.897	− 0.751
85.090	2.974	84.910	− 0.282
90.066	1.979	89.934	0.089
95.033	0.986	94.967	0.278
100.000	0	100.000	0

L.E. radius: 1.000
Slope of radius through L.E.: 0.168

NACA 65₂-215
(Stations and ordinates given in
per cent of airfoil chord)

Upper surface		Lower surface	
Station	Ordinate	Station	Ordinate
0	0	0	0
0.406	1.170	0.594	− 1.070
0.645	1.422	0.855	− 1.282
1.132	1.805	1.368	− 1.591
2.365	2.506	2.635	− 2.134
4.848	3.557	5.152	− 2.925
7.342	4.380	7.658	− 3.532
9.841	5.069	10.159	− 4.035
14.848	6.175	15.152	− 4.829
19.863	7.018	20.137	− 5.426
24.882	7.658	25.118	− 5.868
29.904	8.123	30.096	− 6.179
34.927	8.426	35.073	− 6.366
39.952	8.569	40.048	− 6.427
44.976	8.522	45.024	− 6.332
50.000	8.271	50.000	− 6.065
55.021	7.815	54.979	− 5.625
60.039	7.189	59.961	− 5.047
65.053	6.433	64.947	− 4.373
70.062	5.572	69.938	− 3.628
75.065	4.638	74.935	− 2.848
80.063	3.653	79.937	− 2.061
85.055	2.649	84.945	− 1.303
90.040	1.660	89.960	− 0.626
95.020	0.744	94.980	− 0.112
100.000	0	100.000	0

L.E. radius: 1.505
Slope of radius through L.E.: 0.084

NACA 65₂-415
(Stations and ordinates given in
per cent of airfoil chord)

Upper surface		Lower surface	
Station	Ordinate	Station	Ordinate
0	0	0	0
0.313	1.208	0.687	− 1.008
0.542	1.480	0.958	− 1.200
1.016	1.900	1.484	− 1.472
2.231	2.680	2.769	− 1.936
4.697	3.863	5.303	− 2.599
7.184	4.794	7.816	− 3.098
9.682	5.578	10.318	− 3.510
14.697	6.842	15.303	− 4.150
19.726	7.809	20.274	− 4.625
24.764	8.550	25.236	− 4.970
29.807	9.093	30.193	− 5.205
34.854	9.455	35.146	− 5.335
39.903	9.639	40.097	− 5.355
44.953	9.617	45.047	− 5.237
50.000	9.374	50.000	− 4.962
55.043	8.910	54.957	− 4.530
60.079	8.260	59.921	− 3.976
65.106	7.462	64.894	− 3.342
70.124	6.542	69.876	− 2.654
75.131	5.532	74.869	− 1.952
80.126	4.447	79.874	− 1.263
85.109	3.320	84.891	− 0.628
90.080	2.175	89.920	− 0.107
95.040	1.058	94.960	0.206
100.000	0	100.000	0

L.E. radius: 1.505
Slope of radius through L.E.: 0.168

NACA 65₂-415
a = 0.5
(Stations and ordinates given in
per cent of airfoil chord)

Upper surface		Lower surface	
Station	Ordinate	Station	Ordinate
0	0	0	0
0.245	1.233	0.755	− 0.957
0.464	1.520	1.036	− 1.132
0.927	1.965	1.573	− 1.377
2.126	2.812	2.874	− 1.776
4.574	4.099	5.426	− 2.335
7.054	5.122	7.946	− 2.746
9.549	5.985	10.451	− 3.081
14.568	7.383	15.432	− 3.591
19.611	8.459	20.389	− 3.963
24.671	9.280	25.329	− 4.232
29.743	9.883	30.257	− 4.411
34.825	10.280	35.175	− 4.508
39.916	10.470	40.084	− 4.526
45.019	10.423	44.981	− 4.431
50.152	10.106	49.848	− 4.226
55.262	9.501	54.738	− 3.929
60.307	8.672	59.693	− 3.548
65.314	7.684	64.686	− 3.104
70.294	6.573	69.706	− 2.609
75.253	5.387	74.747	− 2.083
80.199	4.157	79.801	− 1.545
85.137	2.930	84.863	− 1.014
90.077	1.755	89.923	− 0.527
95.027	0.715	94.973	− 0.139
100.000	0	100.000	0

L.E. radius: 1.505
Slope of radius through L.E.: 0.233

NACA 65₃-218
(Stations and ordinates given in
per cent of airfoil chord)

Upper surface		Lower surface	
Station	Ordinate	Station	Ordinate
0	0	0	0
0.388	1.382	0.612	− 1.282
0.625	1.673	0.875	− 1.533
1.110	2.116	1.390	− 1.902
2.340	2.932	2.660	− 2.560
4.819	4.178	5.181	− 3.546
7.311	5.153	7.689	− 4.305
9.809	5.971	10.191	− 4.937
14.818	7.276	15.182	− 5.930
19.835	8.270	20.165	− 6.676
24.858	9.023	25.142	− 7.233
29.884	9.566	30.116	− 7.622
34.912	9.916	35.088	− 7.856
39.942	10.070	40.058	− 7.928
44.972	9.996	45.028	− 7.806
50.000	9.671	50.000	− 7.465
55.026	9.103	54.974	− 6.913
60.047	8.338	59.953	− 6.196
65.063	7.425	64.937	− 5.365
70.073	6.398	69.927	− 4.454
75.077	5.290	74.923	− 3.500
80.074	4.133	79.926	− 2.541
85.063	2.967	84.937	− 1.621
90.046	1.835	89.954	− 0.801
95.023	0.805	94.977	− 0.173
100.000	0	100.000	0

L.E. radius: 1.96
Slope of radius through L.E.: 0.084

NACA 65₃-418

(Stations and ordinates given in
per cent of airfoil chord)

Upper surface		Lower surface	
Station	Ordinate	Station	Ordinate
0	0	0	0
0.278	1.418	0.722	− 1.218
0.503	1.729	0.997	− 1.449
0.973	2.209	1.527	− 1.781
2.181	3.104	2.819	− 2.360
4.639	4.481	5.361	− 3.217
7.123	5.566	7.877	− 3.870
9.619	6.478	10.381	− 4.410
14.636	7.942	15.364	− 5.250
19.671	9.061	20.329	− 5.877
24.716	9.914	25.284	− 6.334
29.768	10.536	30.232	− 6.648
34.825	10.944	35.175	− 6.824
39.884	11.140	40.116	− 6.856
44.943	11.091	45.057	− 6.711
50.000	10.774	50.000	− 6.362
55.051	10.198	54.949	− 5.818
60.094	9.408	59.906	− 5.124
65.126	8.454	64.874	− 4.334
70.146	7.368	69.854	− 3.480
75.154	6.183	74.846	− 2.603
80.147	4.927	79.853	− 1.743
85.127	3.638	84.873	− 0.946
90.092	2.350	89.908	− 0.282
95.046	1.120	94.954	0.144
100.000	0	100.000	0

L.E. radius: 1.96
Slope of radius through L.E.: 0.168

NACA 65₃-418
$a = 0.5$

(Stations and ordinates given in
per cent of airfoil chord)

Upper surface		Lower surface	
Station	Ordinate	Station	Ordinate
0	0	0	0
0.197	1.440	0.803	− 1.164
0.411	1.766	1.089	− 1.378
0.868	2.271	1.632	− 1.683
2.057	3.233	2.943	− 2.197
4.493	4.715	5.507	− 2.951
6.966	5.891	8.034	− 3.515
9.459	6.882	10.541	− 3.978
14.481	8.482	15.519	− 4.690
19.533	9.709	20.467	− 5.213
24.604	10.643	25.396	− 5.595
29.691	11.325	30.309	− 5.853
34.789	11.770	35.211	− 5.998
39.899	11.970	40.101	− 6.026
45.022	11.897	44.978	− 5.905
50.182	11.506	49.818	− 5.626
55.313	10.788	54.687	− 5.216
60.364	9.820	59.636	− 4.696
65.372	8.674	64.628	− 4.094
70.347	7.397	69.653	− 3.433
75.298	6.038	74.702	− 2.734
80.232	4.636	79.768	− 2.024
85.159	3.247	84.841	− 1.331
90.089	1.930	89.911	− 0.702
95.030	0.777	94.970	− 0.201
100.000	0	100.000	0

L.E. radius: 1.96
Slope of radius through L.E.: 0.233

NACA 65₃-618

(Stations and ordinates given in per cent of airfoil chord)

Upper surface		Lower surface	
Station	Ordinate	Station	Ordinate
0	0	0	0
0.172	1.446	0.828	− 1.146
0.385	1.776	1.115	− 1.356
0.839	2.293	1.661	− 1.651
2.026	3.268	2.974	− 2.152
4.462	4.776	5.538	− 2.880
6.936	5.971	8.064	− 3.427
9.431	6.978	10.569	− 3.876
14.455	8.602	15.545	− 4.564
19.506	9.848	20.494	− 5.072
24.574	10.803	25.426	− 5.433
29.652	11.504	30.348	− 5.672
34.738	11.972	35.262	− 5.792
39.826	12.210	40.174	− 5.784
44.915	12.186	45.085	− 5.616
50.000	11.877	50.000	− 5.259
55.077	11.293	54.923	− 4.723
60.141	10.479	59.859	− 4.053
65.189	9.482	64.811	− 3.302
70.219	8.338	69.781	− 2.506
75.230	7.075	74.770	− 1.705
80.220	5.719	79.780	− 0.943
85.189	4.306	84.811	− 0.268
90.138	2.863	89.862	0.239
95.068	1.433	94.932	0.463
100.000	0	100.000	0

L.E. radius: 1.96
Slope of radius through L.E.: 0.253

NACA 65₃-618
$a = 0.5$

(Stations and ordinates given in per cent of airfoil chord)

Upper surface		Lower surface	
Station	Ordinate	Station	Ordinate
0	0	0	0
0.059	1.469	0.941	− 1.055
0.256	1.821	1.244	− 1.239
0.689	2.375	1.811	− 1.493
1.846	3.449	3.154	− 1.895
4.248	5.115	5.752	− 2.469
6.706	6.448	8.294	− 2.884
9.194	7.575	10.806	− 3.219
14.225	9.404	15.775	− 3.716
19.301	10.815	20.699	− 4.071
24.407	11.893	25.593	− 4.321
29.537	12.687	30.463	− 4.479
34.684	13.209	35.316	− 4.551
39.849	13.456	40.151	− 4.540
45.034	13.395	44.966	− 4.407
50.273	12.974	49.727	− 4.154
55.468	12.173	54.532	− 3.815
60.546	11.090	59.454	− 3.404
65.557	9.806	64.443	− 2.936
70.519	8.374	69.481	− 2.428
75.445	6.851	74.555	− 1.895
80.347	5.279	79.653	− 1.361
85.239	3.720	84.761	− 0.846
90.133	2.233	89.867	− 0.391
95.046	0.920	94.954	− 0.055
100.000	0	100.000	0

L.E. radius: 1.96
Slope of radius through L.E.: 0.349

NACA 65₁-221

NACA 65,-221

(Stations and ordinates given in
per cent of airfoil chord)

Upper surface		Lower surface	
Station	Ordinate	Station	Ordinate
0	0	0	0
0.372	1.567	0.628	− 1.467
0.608	1.902	0.892	− 1.762
1.090	2.402	1.410	− 2.188
2.314	3.335	2.684	− 2.963
4.791	4.783	5.209	− 4.151
7.280	5.918	7.720	− 5.070
9.778	6.865	10.222	− 5.831
14.787	8.370	15.213	− 7.024
19.808	9.514	20.192	− 7.922
24.834	10.381	25.166	− 8.591
29.865	11.007	30.135	− 9.063
34.898	11.404	35.102	− 9.344
39.932	11.570	40.068	− 9.428
44.967	11.461	45.033	− 9.271
50.000	11.055	50.000	− 8.849
55.030	10.372	54.970	− 8.182
60.054	9.461	59.946	− 7.319
65.072	8.390	64.928	− 6.330
70.084	7.195	69.916	− 5.251
75.088	5.918	74.912	− 4.128
80.084	4.595	79.916	− 3.003
85.072	3.270	84.928	− 1.924
90.052	2.000	89.948	− 0.966
95.026	0.861	94.974	− 0.229
100.000	0	100.000	0

L.E. radius: 2.50
Slope of radius through L.E.: 0.084

NACA 65₄-421

NACA 65,-421

(Stations and ordinates given in
per cent of airfoil chord)

Upper surface		Lower surface	
Station	Ordinate	Station	Ordinate
0	0	0	0
0.247	1.601	0.753	− 1.401
0.468	1.956	1.032	− 1.676
0.933	2.493	1.567	− 2.065
2.135	3.505	2.865	− 2.761
4.582	5.085	5.417	− 3.821
7.062	6.329	7.938	− 4.633
9.557	7.371	10.443	− 5.303
14.575	9.034	15.425	− 6.342
19.616	10.304	20.384	− 7.120
24.668	11.271	25.332	− 7.691
29.729	11.976	30.271	− 8.088
34.796	12.433	35.204	− 8.313
39.865	12.640	40.135	− 8.356
44.934	12.556	45.066	− 8.176
50.000	12.158	50.000	− 7.746
55.059	11.467	54.941	− 7.087
60.108	10.531	59.892	− 6.247
65.145	9.419	64.855	− 5.299
70.168	8.166	69.832	− 4.278
75.176	6.811	74.824	− 3.231
80.167	5.388	79.833	− 2.204
85.143	3.940	84.857	− 1.248
90.104	2.514	89.896	− 0.446
95.051	1.176	94.949	0.088
100.000	0	100.000	0

L.E. radius: 2.50
Slope of radius through L.E.: 0.168

NACA 65₄-421
$a = 0.5$
(Stations and ordinates given in
per cent of airfoil chord)

Upper surface		Lower surface	
Station	Ordinate	Station	Ordinate
0	0	0	0
0.155	1.620	0.845	− 1.344
0.363	1.991	1.137	− 1.603
0.813	2.553	1.687	− 1.965
1.992	3.631	3.008	− 2.595
4.414	5.315	5.586	− 3.551
6.880	6.651	8.120	− 4.275
9.371	7.773	10.629	− 4.869
14.395	9.572	15.605	− 5.780
19.455	10.951	20.545	− 6.455
24.538	12.000	25.462	− 6.952
29.639	12.765	30.361	− 7.293
34.754	13.258	35.246	− 7.486
39.882	13.470	40.118	− 7.526
45.026	13.362	44.974	− 7.370
50.211	12.890	49.789	− 7.010
55.362	12.056	54.638	− 6.484
60.421	10.942	59.579	− 5.818
65.428	9.637	64.572	− 5.057
70.398	8.193	69.602	− 4.229
75.340	6.664	74.660	− 3.360
80.264	5.097	79.736	− 2.485
85.181	3.550	84.819	− 1.634
90.100	2.095	89.900	− 0.867
95.034	0.833	94.966	− 0.257
100.000	0	100.000	0

L.E. radius: 2.50
Slope of radius through L.E.: 0.233

NACA 66-206
(Stations and ordinates given in
per cent of airfoil chord)

Upper surface		Lower surface	
Station	Ordinate	Station	Ordinate
0	0	0	0
0.461	0.509	0.539	− 0.409
0.707	0.622	0.793	− 0.482
1.202	0.798	1.298	− 0.584
2.447	1.102	2.553	− 0.730
4.941	1.572	5.059	− 0.940
7.439	1.947	7.561	− 1.099
9.939	2.268	10.061	− 1.234
14.942	2.791	15.058	− 1.445
19.947	3.196	20.053	− 1.604
24.954	3.513	25.046	− 1.723
29.962	3.754	30.038	− 1.810
34.971	3.929	35.029	− 1.869
39.981	4.042	40.019	− 1.900
44.990	4.095	45.010	− 1.905
50.000	4.088	50.000	− 1.882
55.009	4.020	54.991	− 1.830
60.018	3.886	59.982	− 1.744
65.026	3.641	64.974	− 1.581
70.031	3.288	69.969	− 1.344
75.034	2.848	74.966	− 1.058
80.034	2.339	79.966	− 0.747
85.031	1.780	84.969	− 0.434
90.023	1.182	89.977	− 0.148
95.012	0.578	94.988	0.054
100.000	0	100.000	0

L.E. radius: 0.223
Slope of radius through L.E.: 0.084

NACA 66-209

(Stations and ordinates given in
per cent of airfoil chord)

Upper surface		Lower surface	
Station	Ordinate	Station	Ordinate
0	0	0	0
0.442	0.735	0.558	− 0.635
0.686	0.892	0.814	− 0.752
1.179	1.135	1.321	− 0.921
2.420	1.552	2.580	− 1.180
4.912	2.194	5.088	− 1.562
7.409	2.705	7.591	− 1.857
9.908	3.141	10.092	− 2.107
14.912	3.850	15.088	− 2.504
19.921	4.396	20.079	− 2.804
24.931	4.821	25.069	− 3.031
29.944	5.145	30.056	− 3.201
34.957	5.378	35.043	− 3.318
39.971	5.528	40.029	− 3.386
44.986	5.594	45.014	− 3.404
50.000	5.578	50.000	− 3.372
55.014	5.476	54.986	− 3.286
60.027	5.275	59.973	− 3.133
65.038	4.912	64.962	− 2.852
70.046	4.400	69.954	− 2.456
75.050	3.772	74.950	− 1.982
80.050	3.058	79.950	− 1.466
85.044	2.283	84.956	− 0.937
90.034	1.477	89.965	− 0.443
95.018	0.690	94.982	− 0.058
100.000	0	100.000	0

L.E. radius: 0.530
Slope of radius through L.E.: 0.084

NACA 66-210

(Stations and ordinates given in
per cent of airfoil chord)

Upper surface		Lower surface	
Station	Ordinate	Station	Ordinate
0	0	0	0
0.436	0.806	0.564	− 0.706
0.679	0.980	0.821	− 0.840
1.171	1.245	1.329	− 1.031
2.412	1.699	2.588	− 1.327
4.902	2.401	5.098	− 1.769
7.399	2.958	7.601	− 2.110
9.898	3.432	10.102	− 2.389
14.903	4.202	15.097	− 2.856
19.912	4.796	20.088	− 3.204
24.924	5.257	25.076	− 3.467
29.937	5.608	30.063	− 3.664
34.952	5.862	35.048	− 3.802
39.968	6.024	40.032	− 3.882
44.984	6.095	45.016	− 3.905
50.000	6.074	50.000	− 3.868
55.016	5.960	54.984	− 3.770
60.030	5.736	59.970	− 3.594
65.042	5.332	64.958	− 3.272
70.051	4.759	69.949	− 2.815
75.056	4.071	74.944	− 2.281
80.055	3.289	79.945	− 1.697
85.049	2.445	84.951	− 1.099
90.037	1.570	89.963	− 0.536
95.019	0.724	94.981	− 0.092
100.000	0	100.000	0

L.E. radius: 0.662
Slope of radius through L.E.: 0.084

NACA 66₁-212

(Stations and ordinates given in per cent of airfoil chord)

Upper surface		Lower surface	
Station	Ordinate	Station	Ordinate
0	0	0	0
0.424	0.953	0.576	− 0.853
0.666	1.154	0.834	− 1.014
1.156	1.462	1.344	− 1.248
2.395	1.991	2.605	− 1.619
4.883	2.809	5.117	− 2.177
7.379	3.459	7.621	− 2.611
9.878	4.011	10.122	− 2.977
14.883	4.905	15.117	− 3.559
19.894	5.596	20.106	− 4.004
24.908	6.132	25.092	− 4.342
29.925	6.539	30.075	− 4.595
34.943	6.833	35.057	− 4.773
39.962	7.018	40.038	− 4.876
44.981	7.095	45.019	− 4.905
50.000	7.068	50.000	− 4.862
55.019	6.931	54.981	− 4.741
60.036	6.659	59.964	− 4.517
65.051	6.169	64.949	− 4.109
70.061	5.487	69.939	− 3.543
75.066	4.661	74.934	− 2.871
80.065	3.739	79.935	− 2.147
85.057	2.755	84.943	− 1.409
90.043	1.750	89.957	− 0.716
95.022	0.789	94.978	− 0.157
100.000	0	100.000	0

L.E. radius: 0.952
Slope of radius through L.E.: 0.084

NACA 66₂-215

(Stations and ordinates given in per cent of airfoil chord)

Upper surface		Lower surface	
Station	Ordinate	Station	Ordinate
0	0	0	0
0.406	1.168	0.594	− 1.068
0.646	1.409	0.854	− 1.269
1.134	1.778	1.366	− 1.564
2.370	2.417	2.630	− 2.045
4.855	3.413	5.145	− 2.781
7.349	4.202	7.651	− 3.354
9.848	4.872	10.152	− 3.838
14.854	5.957	15.146	− 4.611
19.868	6.790	20.132	− 5.198
24.886	7.437	25.114	− 5.647
29.906	7.927	30.094	− 5.983
34.929	8.280	35.071	− 6.220
39.952	8.501	40.048	− 6.359
44.976	8.590	45.024	− 6.400
50.000	8.553	50.000	− 6.347
55.023	8.378	54.977	− 6.188
60.045	8.030	59.955	− 5.888
65.063	7.402	64.937	− 5.342
70.075	6.547	69.925	− 4.603
75.081	5.526	74.919	− 3.736
80.079	4.393	79.921	− 2.801
85.070	3.202	84.930	− 1.856
90.052	2.005	89.948	− 0.971
95.026	0.881	94.974	− 0.249
100.000	0	100.000	0

L.E. radius: 1.435
Slope of radius through L.E.: 0.084

NACA 66₂-415

(Stations and ordinates given in
per cent of airfoil chord)

Upper surface		Lower surface	
Station	Ordinate	Station	Ordinate
0	0	0	0
0.314	1.206	0.686	− 1.006
0.544	1.467	0.956	− 1.187
1.019	1.873	1.481	− 1.445
2.241	2.592	2.759	− 1.848
4.711	3.718	5.289	− 2.454
7.199	4.617	7.801	− 2.921
9.696	5.381	10.304	− 3.313
14.709	6.624	15.291	− 3.932
19.736	7.581	20.264	− 4.397
24.771	8.329	25.229	− 4.749
29.812	8.897	30.188	− 5.009
34.857	9.309	35.143	− 5.189
39.904	9.571	40.096	− 5.287
44.952	9.685	45.048	− 5.305
50.000	9.656	50.000	− 5.244
55.046	9.473	54.954	− 5.093
60.090	9.100	59.910	− 4.816
65.126	8.431	64.874	− 4.311
70.150	7.518	69.850	− 3.630
75.162	6.419	74.838	− 2.839
80.159	5.187	79.841	− 2.003
85.139	3.872	84.861	− 1.180
90.104	2.519	89.896	− 0.451
95.053	1.196	94.947	0.068
100.000	0	100.000	0

L.E. radius: 1.435
Slope of radius through L.E.: 0.168

NACA 66₃-218

(Stations and ordinates given in
per cent of airfoil chord)

Upper surface		Lower surface	
Station	Ordinate	Station	Ordinate
0	0	0	0
0.389	1.368	0.611	− 1.268
0.628	1.636	0.872	− 1.496
1.115	2.054	1.385	− 1.840
2.346	2.828	2.654	− 2.456
4.827	4.002	5.173	− 3.370
7.320	4.933	7.680	− 4.085
9.818	5.724	10.182	− 4.690
14.825	7.004	15.175	− 5.658
19.841	7.982	20.159	− 6.390
24.863	8.742	25.137	− 6.952
29.887	9.317	30.113	− 7.373
34.914	9.731	35.086	− 7.671
39.942	9.989	40.058	− 7.847
44.971	10.093	45.029	− 7.903
50.000	10.045	50.000	− 7.839
55.028	9.828	54.972	− 7.638
60.054	9.394	59.946	− 7.252
65.075	8.610	64.925	− 6.550
70.089	7.568	69.911	− 5.624
75.095	6.345	74.905	− 4.555
80.093	5.001	79.907	− 3.409
85.081	3.606	84.919	− 2.260
90.060	2.230	89.940	− 1.196
95.030	0.961	94.970	− 0.329
100.000	0	100.000	0

L.E. radius: 1.955
Slope of radius through L.E.: 0.084

NACA 66₃-418

(Stations and ordinates given in per cent of airfoil chord)

Upper surface		Lower surface	
Station	Ordinate	Station	Ordinate
0	0	0	0
0.280	1.405	0.720	− 1.205
0.509	1.692	0.991	− 1.412
0.981	2.147	1.519	− 1.719
2.194	3.000	2.806	− 2.256
4.656	4.306	5.344	− 3.042
7.140	5.347	7.860	− 3.651
9.636	6.231	10.364	− 4.163
14.651	7.669	15.349	− 4.977
19.683	8.773	20.317	− 5.589
24.726	9.633	25.274	− 6.053
29.775	10.287	30.225	− 6.399
34.829	10.759	35.171	− 6.639
39.885	11.059	40.115	− 6.775
44.943	11.188	45.057	− 6.808
50.000	11.148	50.000	− 6.736
55.056	10.923	54.944	− 6.543
60.107	10.464	59.893	− 6.180
65.149	9.639	64.851	− 5.519
70.178	8.539	69.822	− 4.651
75.191	7.238	74.809	− 3.658
80.185	5.794	79.815	− 2.610
85.162	4.276	84.838	− 1.584
90.120	2.744	89.880	− 0.676
95.060	1.275	94.940	− 0.011
100.000	0	100.000	0

L.E. radius: 1.955
Slope of radius through L.E.: 0.168

NACA 66₄-221

(Stations and ordinates given in per cent of airfoil chord)

Upper surface		Lower surface	
Station	Ordinate	Station	Ordinate
0	0	0	0
0.372	1.570	0.628	− 1.470
0.610	1.869	0.890	− 1.729
1.095	2.342	1.405	− 2.128
2.323	3.226	2.677	− 2.854
4.800	4.580	5.200	− 3.948
7.291	5.653	7.709	− 4.805
9.788	6.565	10.212	− 5.531
14.797	8.039	15.203	− 6.693
19.815	9.170	20.185	− 7.578
24.840	10.047	25.160	− 8.257
29.869	10.709	30.131	− 8.765
34.900	11.183	35.100	− 9.123
39.933	11.478	40.067	− 9.336
44.967	11.595	45.033	− 9.405
50.000	11.537	50.000	− 9.331
55.032	11.281	54.968	− 9.091
60.063	10.763	59.937	− 8.621
65.087	9.823	64.913	− 7.763
70.103	8.581	69.897	− 6.637
75.109	7.145	74.891	− 5.355
80.106	5.591	79.894	− 3.999
85.092	3.996	84.908	− 2.650
90.067	2.440	89.933	− 1.406
95.034	1.032	94.966	− 0.400
100.000	0	100.000	0

L.E. radius: 2.550
Slope of radius through L.E.: 0.084

NACA 67,1-215

(Stations and ordinates given in per cent of airfoil chord)

Upper surface		Lower surface	
Station	Ordinate	Station	Ordinate
0	0	0	0
0.402	1.213	0.598	− 1.113
0.642	1.460	0.858	− 1.320
1.128	1.867	1.372	− 1.653
2.361	2.577	2.639	− 2.205
4.848	3.557	5.152	− 2.925
7.344	4.321	7.656	− 3.473
9.845	4.947	10.155	− 3.913
14.854	5.954	15.146	− 4.608
19.869	6.735	20.131	− 5.143
24.887	7.348	25.113	− 5.558
29.908	7.825	30.092	− 5.881
34.930	8.185	35.070	− 6.125
39.953	8.430	40.047	− 6.288
44.976	8.570	45.024	− 6.380
50.000	8.600	50.000	− 6.394
55.024	8.516	54.976	− 6.326
60.047	8.302	59.953	− 6.160
65.068	7.935	64.932	− 5.875
70.086	7.373	69.914	− 5.429
75.098	6.515	74.902	− 4.725
80.100	5.335	79.900	− 3.743
85.092	3.999	84.908	− 2.653
90.071	2.537	89.929	− 1.503
95.037	1.103	94.963	− 0.471
100.000	0	100.000	0

L.E. radius: 1.523
Slope of radius through L.E.: 0.084

NACA 747A315

(Stations and ordinates given in per cent of airfoil chord)

Upper surface		Lower surface	
Station	Ordinate	Station	Ordinate
0	0	0	0
0.229	1.305	0.771	− 1.031
0.449	1.599	1.051	− 1.207
0.911	2.065	1.589	− 1.473
2.109	2.935	2.891	− 1.927
4.564	4.264	5.436	− 2.518
7.053	5.286	7.947	− 2.952
9.558	6.140	10.442	− 3.304
14.599	7.497	15.401	− 3.843
19.668	8.503	20.332	− 4.247
24.758	9.242	25.242	− 4.546
29.867	9.731	30.133	− 4.773
35.001	9.982	34.999	− 4.926
40.200	9.962	39.800	− 5.020
45.375	9.572	44.625	− 5.040
50.447	8.964	49.553	− 5.014
55.463	8.206	54.537	− 4.930
60.435	7.324	59.565	− 4.772
65.366	6.365	64.634	− 4.509
70.241	5.354	69.759	− 4.110
75.130	4.336	74.870	− 3.502
80.073	3.295	79.927	− 2.743
85.038	2.257	84.962	− 1.915
90.016	1.289	89.984	− 1.097
95.004	0.481	94.996	− 0.405
100.000	0	100.000	0

L.E. radius: 1.544
Slope of radius through L.E.: 0.232

NACA 747A415

(Stations and ordinates given in
per cent of airfoil chord)

Upper surface		Lower surface	
Station	Ordinate	Station	Ordinate
0	0	0	0
0.183	1.318	0.817	− 0.994
0.398	1.622	1.102	− 1.160
0.852	2.106	1.648	− 1.406
2.041	3.016	2.959	− 1.822
4.487	4.411	5.513	− 2.349
6.972	5.488	8.028	− 2.730
9.476	6.390	10.524	− 3.038
14.521	7.827	15.479	− 3.501
19.598	8.897	20.402	− 3.845
24.698	9.687	25.302	− 4.095
29.818	10.216	30.182	− 4.286
34.964	10.497	35.036	− 4.411
40.176	10.499	39.824	− 4.485
45.364	10.121	44.636	− 4.493
50.447	9.516	49.553	− 4.462
55.474	8.753	54.526	− 4.381
60.454	7.859	59.546	− 4.235
65.393	6.878	64.607	− 3.992
70.273	5.838	69.727	− 3.622
75.164	4.783	74.836	− 3.053
80.107	3.692	79.893	− 2.344
85.066	2.592	84.934	− 1.578
90.037	1.546	89.963	− 0.838
95.015	0.639	94.985	− 0.247
100.000	0	100.000	0

L.E. radius: 1.544
Slope of radius through L.E.: 0.274

APPENDIX IV

AERODYNAMIC CHARACTERISTICS OF WING SECTIONS

Contents

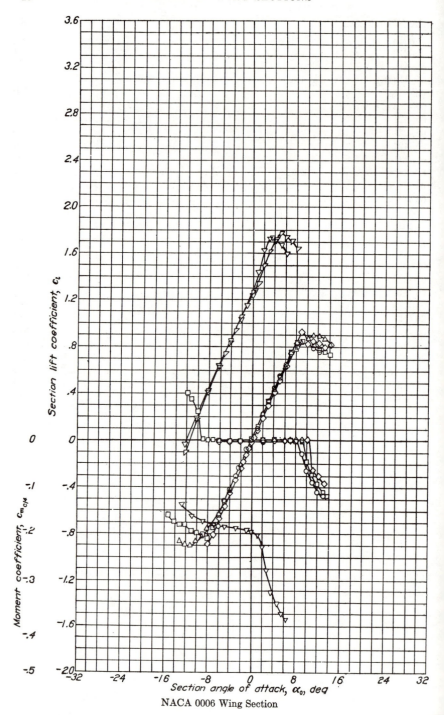

NACA 0006 Wing Section

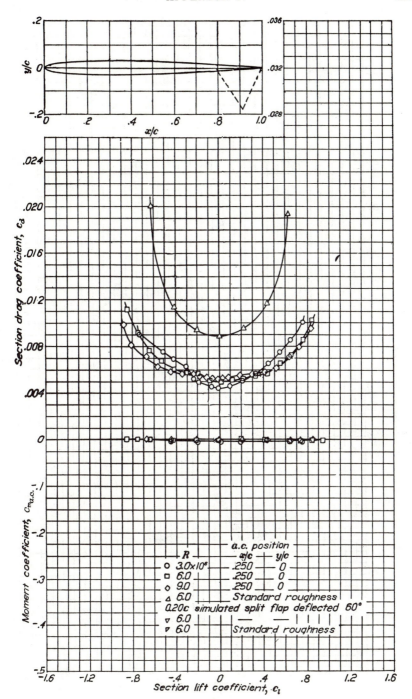

NACA 0006 Wing Section (*Continued*)

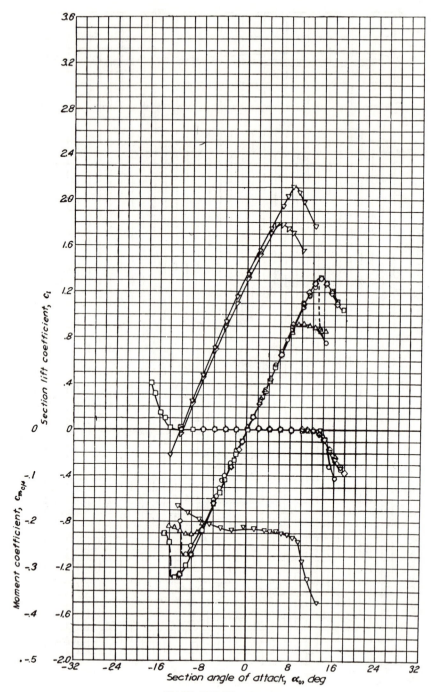

NACA 0009 Wing Section

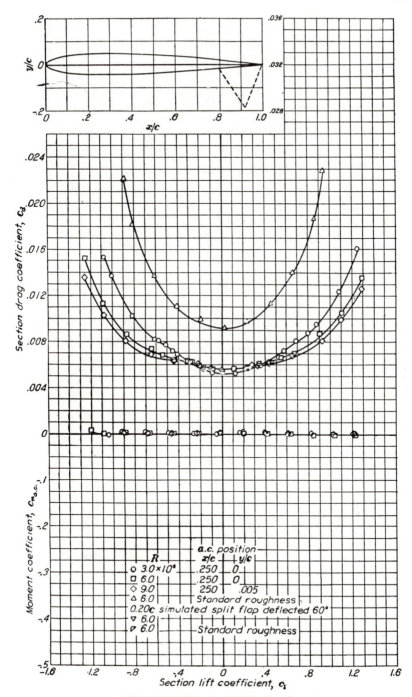

NACA 0009 Wing Section (*Continued*)

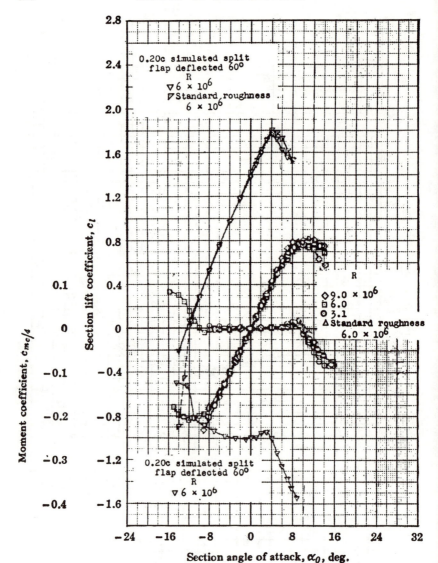

NACA 0010-34 Wing Section

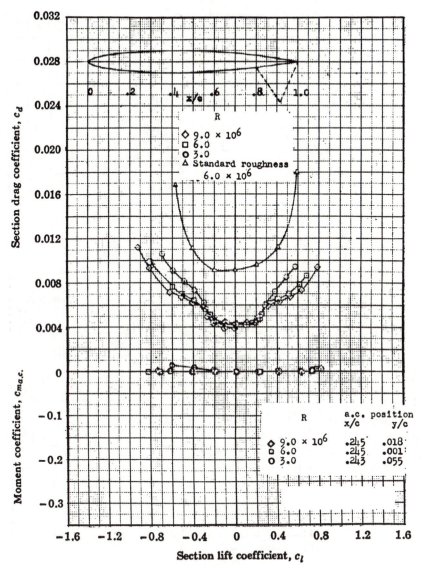

NACA 0010-34 Wing Section (*Continued*)

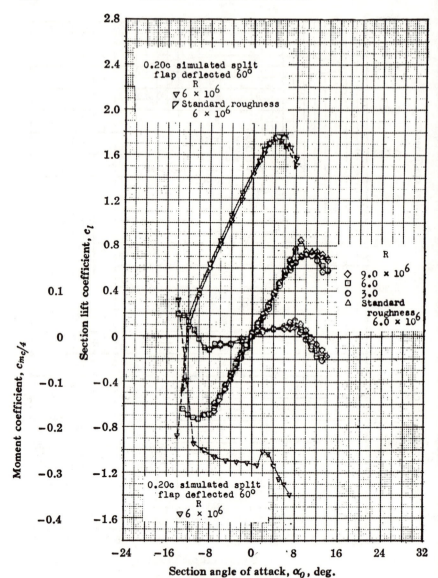

NACA 0010-35 Wing Section

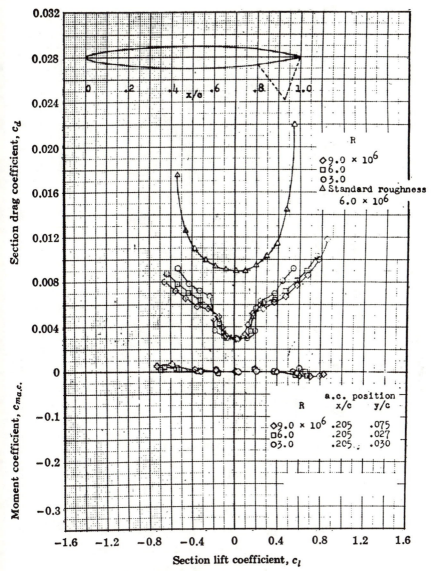

NACA 0010-35 Wing Section (*Continued*)

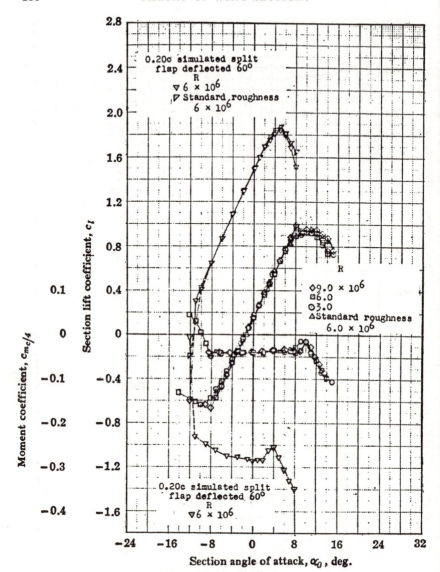

NACA 0010-34 Wing Section
$a = 0.8$ (modified), $c_{l_i} = 0.2$

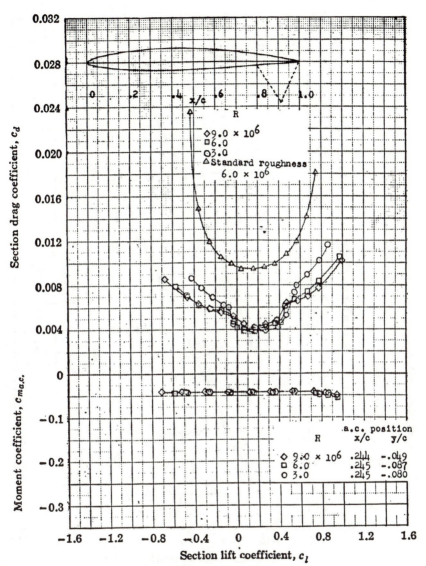

NACA 0010-34 Wing Section
$a = 0.8$ (modified,) $c_{l_i} = 0.2$ (*Continued*)

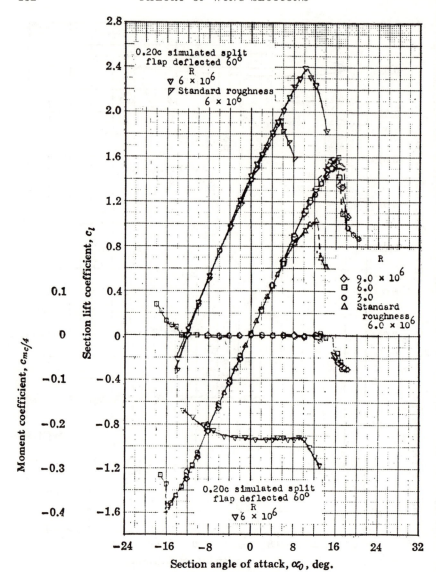

NACA 0012 Wing Section

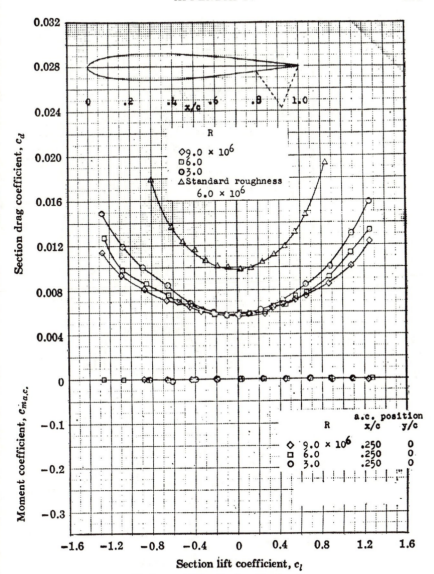

NACA 0012 Wing Section (*Continued*)

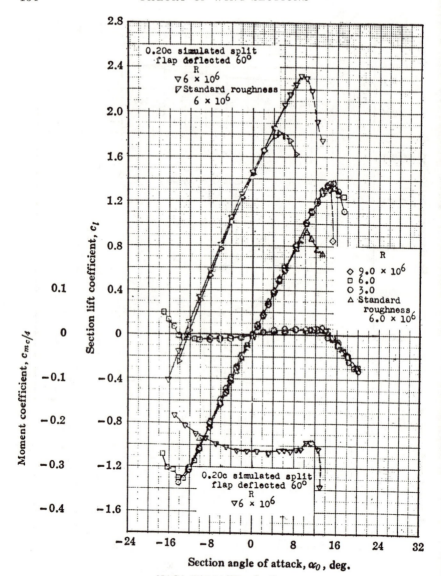

NACA 0012-64 Wing Section

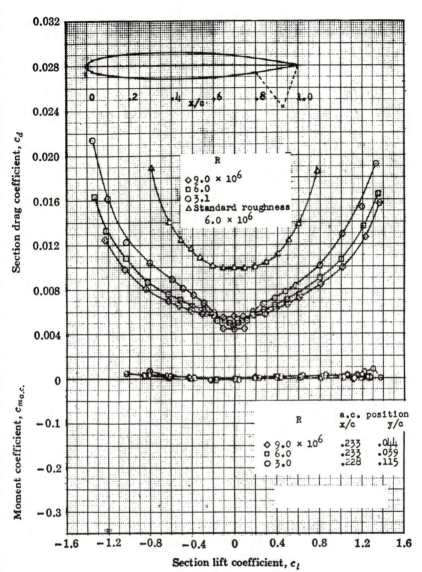

NACA 0012-64 Wing Section (*Continued*)

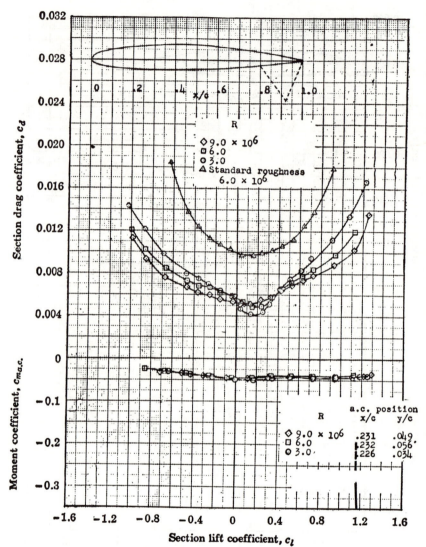

NACA 0012-64 Wing Section,
$a = 0.8$ (modified), $c_{l_i} = 0.2$

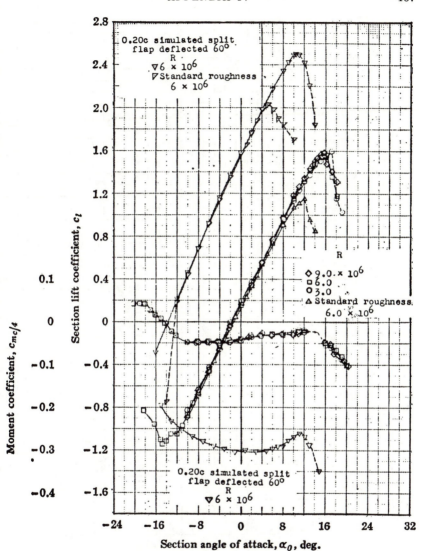

NACA 0012-64 Wing Section,
$a = 0.8$ (modified), $c_{l_i} = 0.2$ (*Continued*)

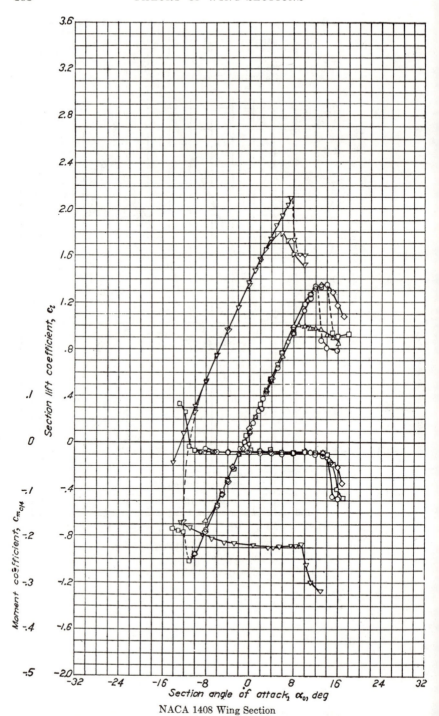

NACA 1408 Wing Section

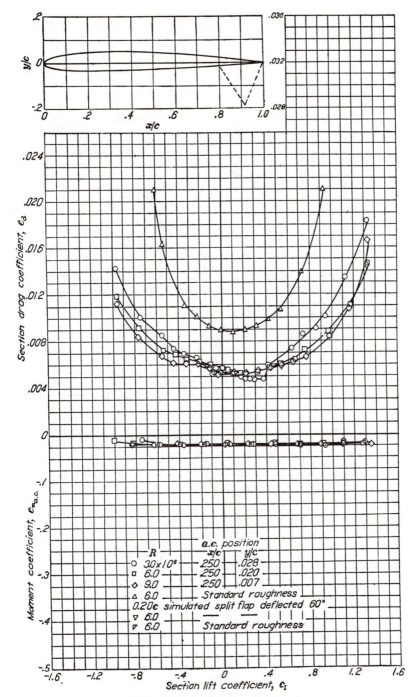

NACA 1408 Wing Section (*Continued*)

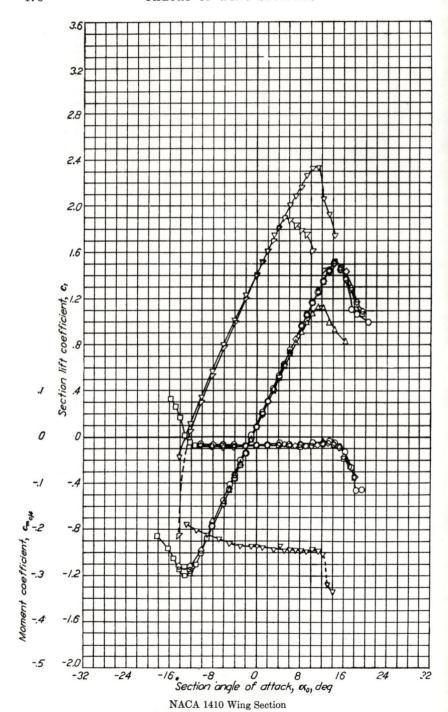

NACA 1410 Wing Section

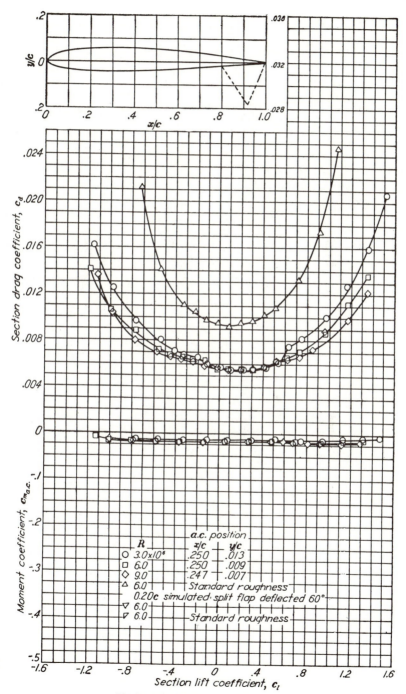

NACA 1410 Wing Section (*Continued*)

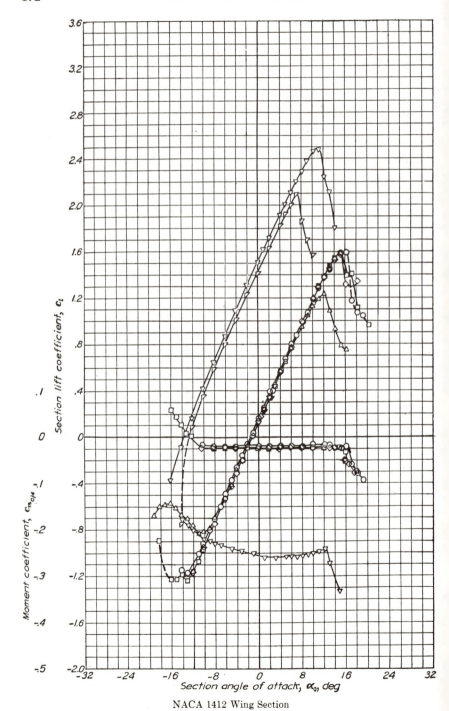

NACA 1412 Wing Section

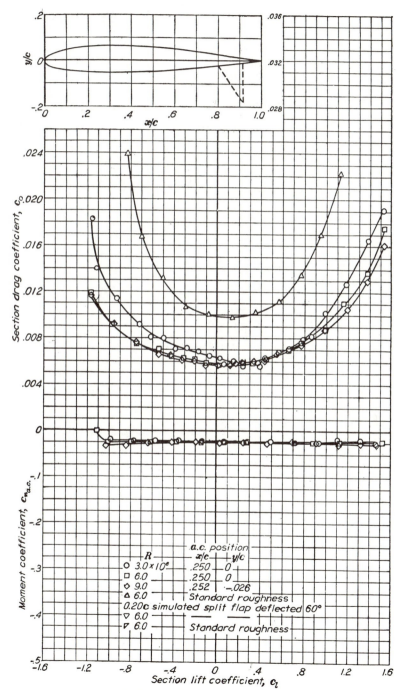

NACA 1412 Wing Section (*Continued*)

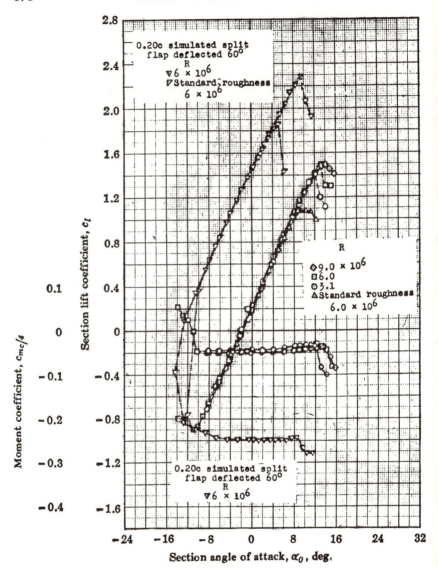

NACA 2408 Wing Section

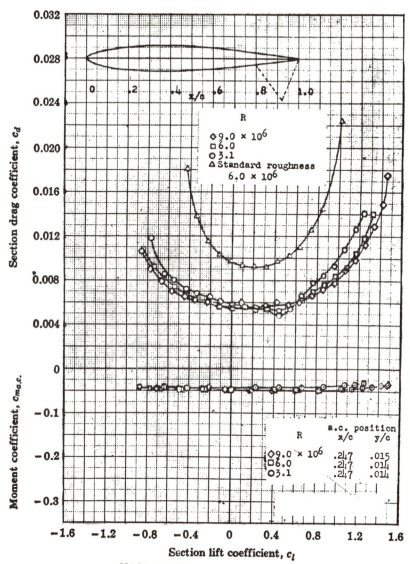

NACA 2408 Wing Section (*Continued*)

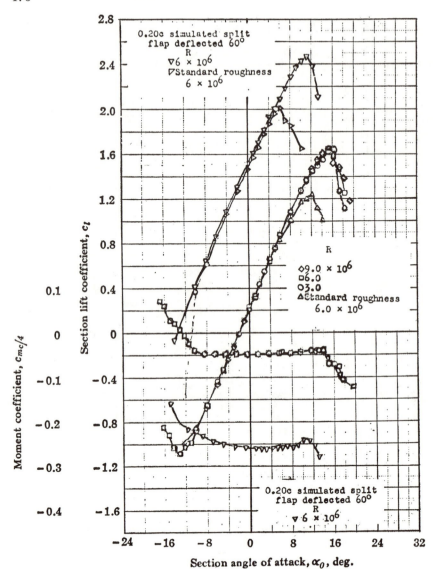

NACA 2410 Wing Section

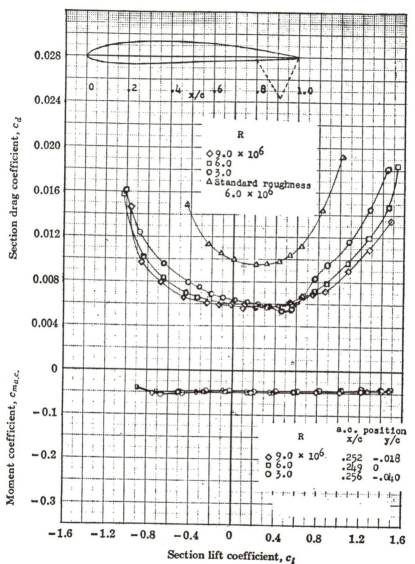

NACA 2410 Wing Section (*Continued*)

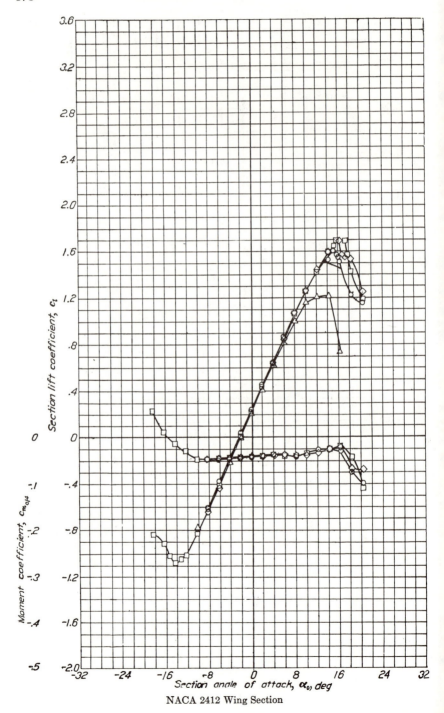

NACA 2412 Wing Section

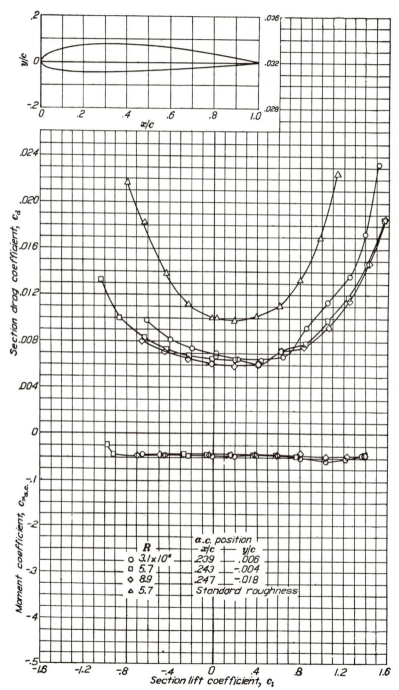

NACA 2412 Wing Section (*Continued*)

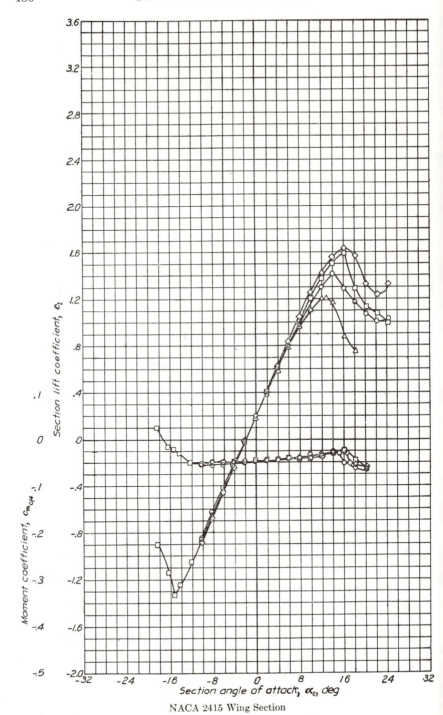

NACA 2415 Wing Section

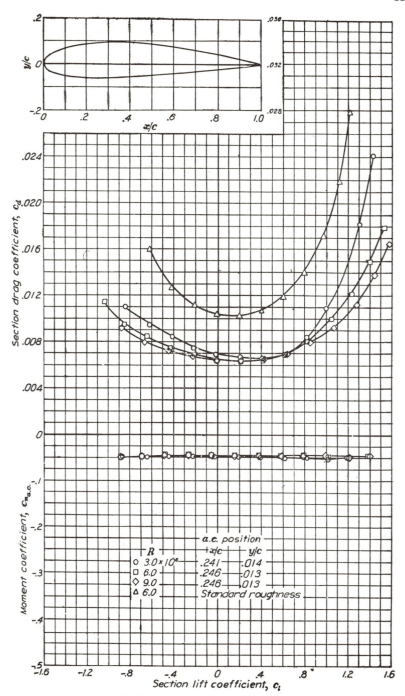

NACA 2415 Wing Section (*Continued*)

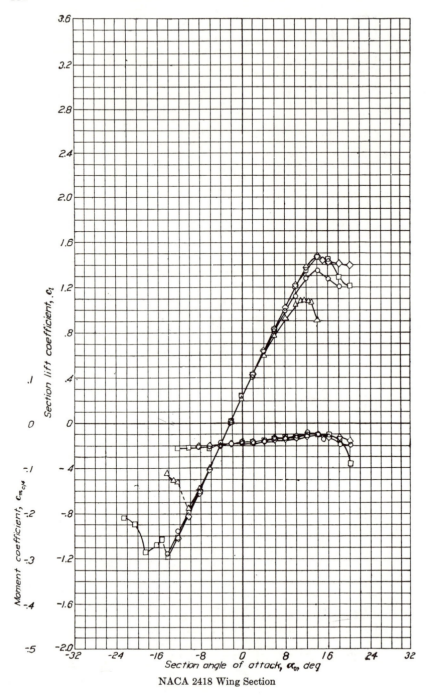

NACA 2418 Wing Section

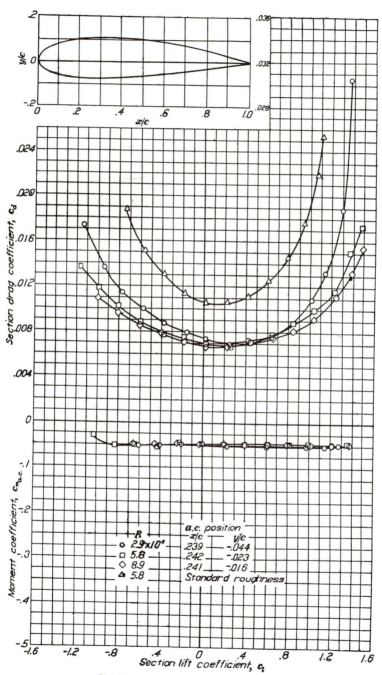

NACA 2418 Wing Section (*Continued*)

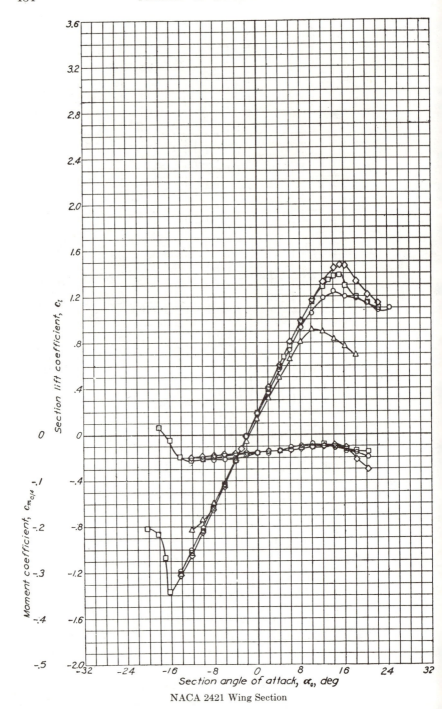

NACA 2421 Wing Section

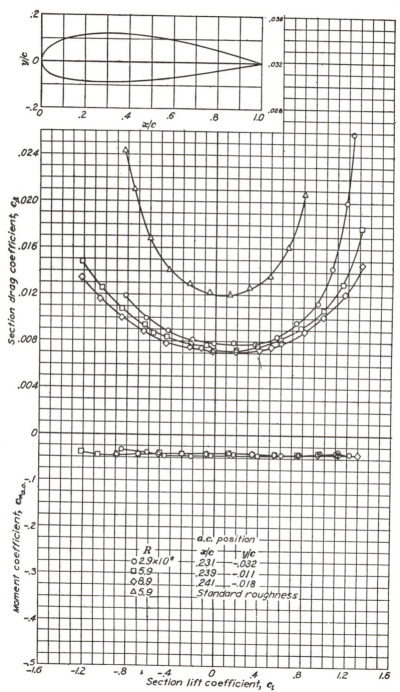

NACA 2421 Wing Section (*Continued*)

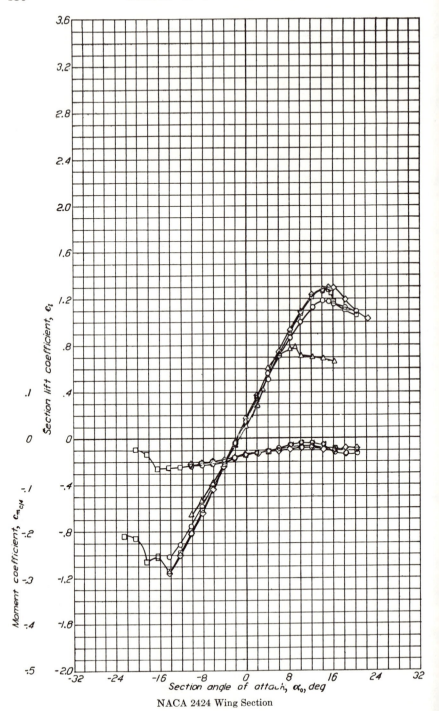

NACA 2424 Wing Section

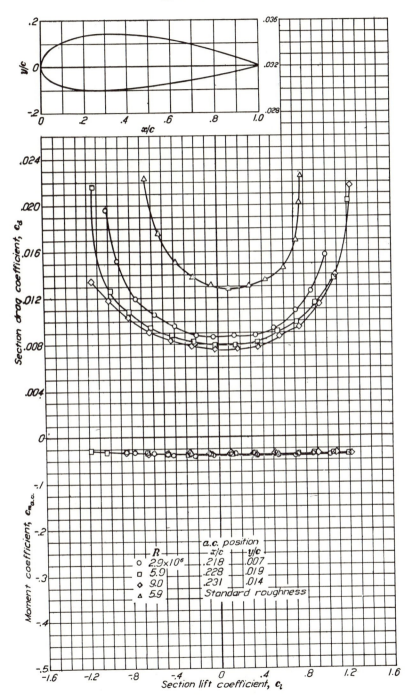

NACA 2424 Wing Section (*Continued*)

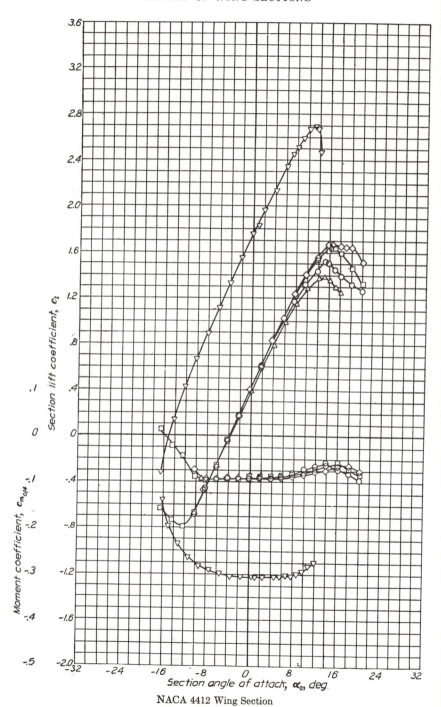

NACA 4412 Wing Section

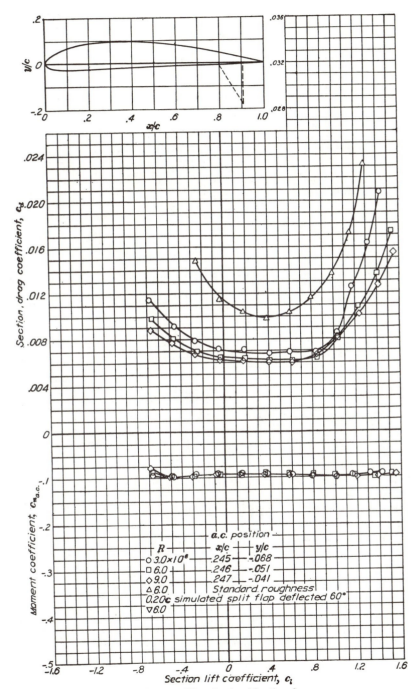

NACA 4412 Wing Section (*Continued*)

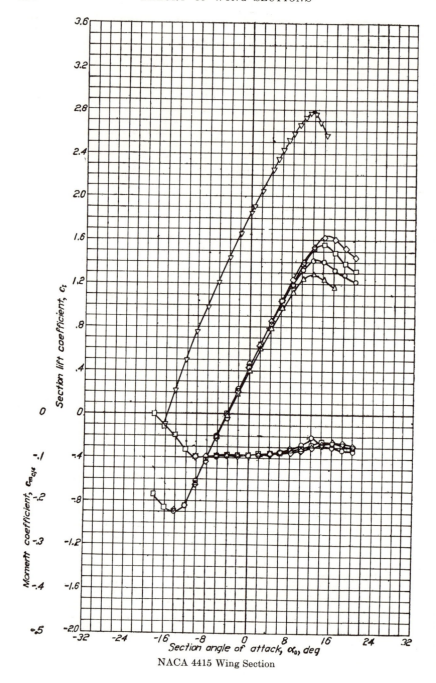

NACA 4415 Wing Section

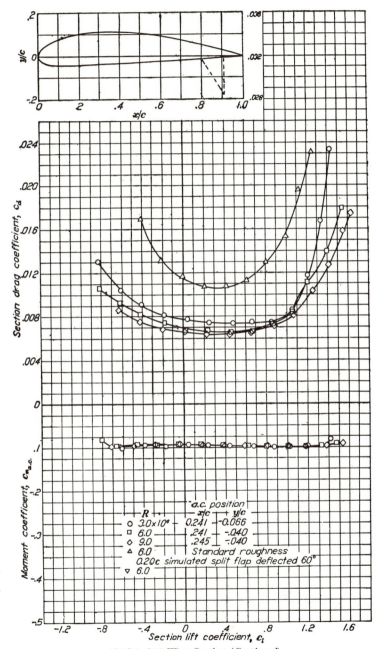

NACA 4415 Wing Section (*Continued*)

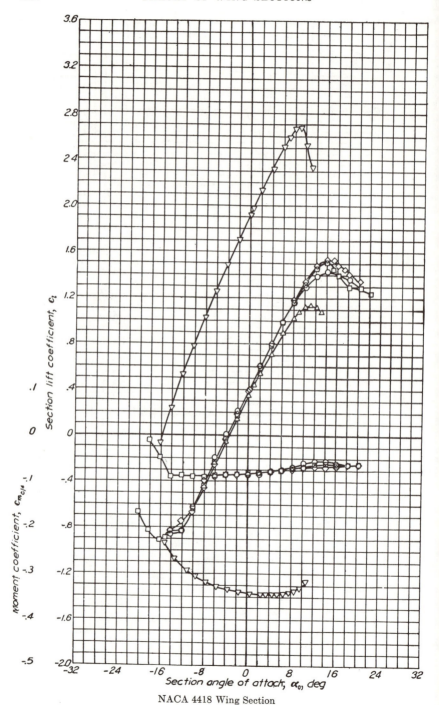

NACA 4418 Wing Section

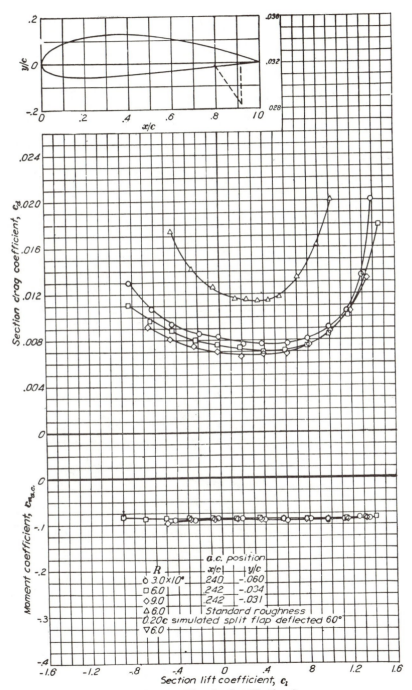

NACA 4418 Wing Section (*Continued*)

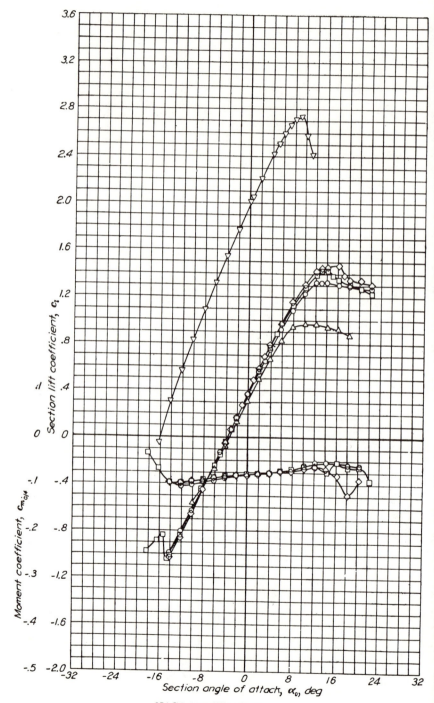

NACA 4421 Wing Section

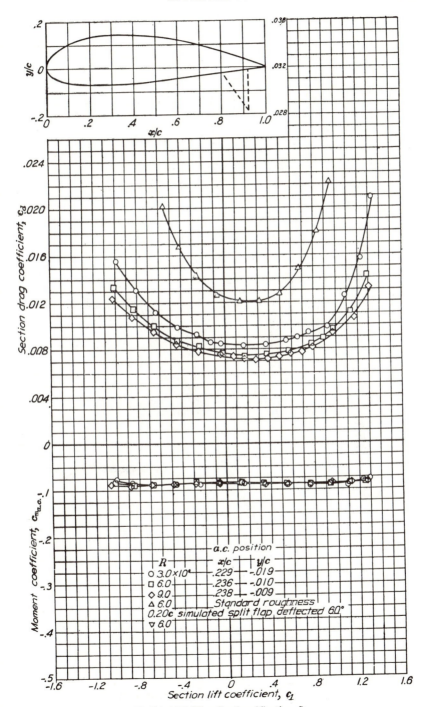

NACA 4421 Wing Section (*Continued*)

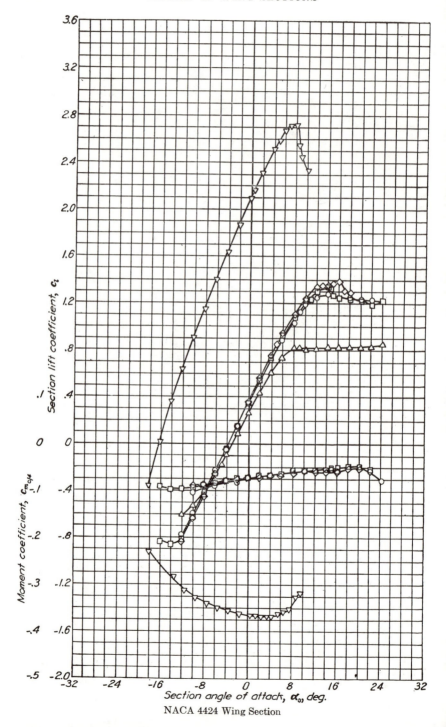

Section lift coefficient, c_l

Moment coefficient, $c_{m_{c/4}}$

Section angle of attack, α_0, deg.

NACA 4424 Wing Section

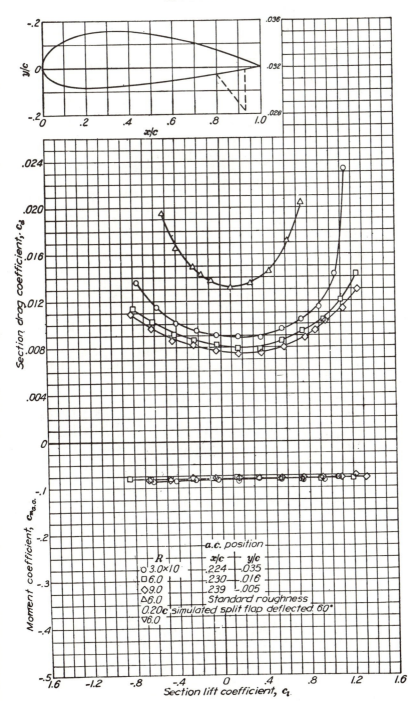

NACA 4424 Wing Section (*Continued*)

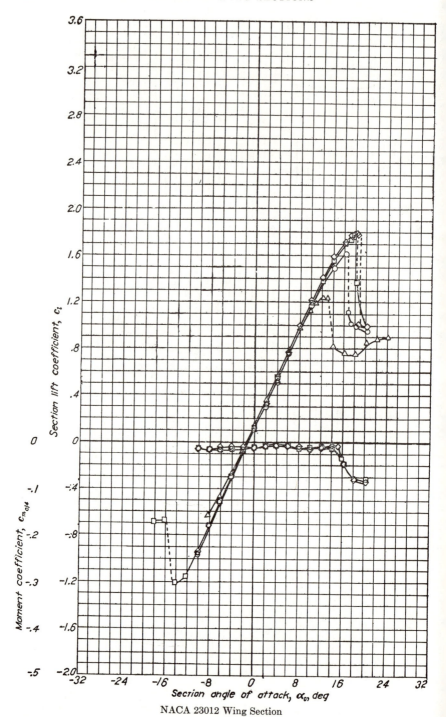

NACA 23012 Wing Section

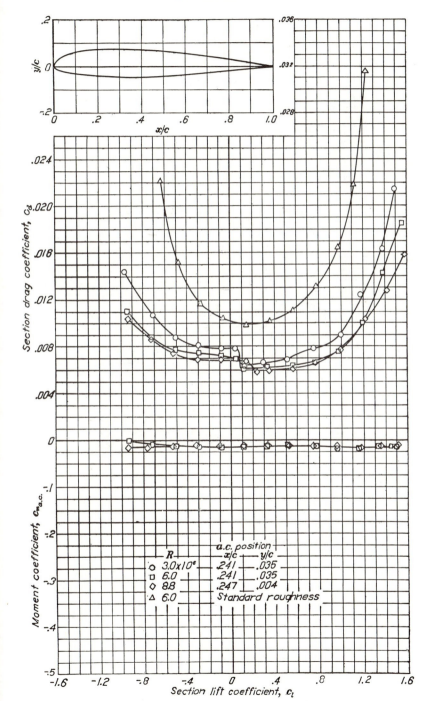

NACA 23012 Wing Section (*Continued*)

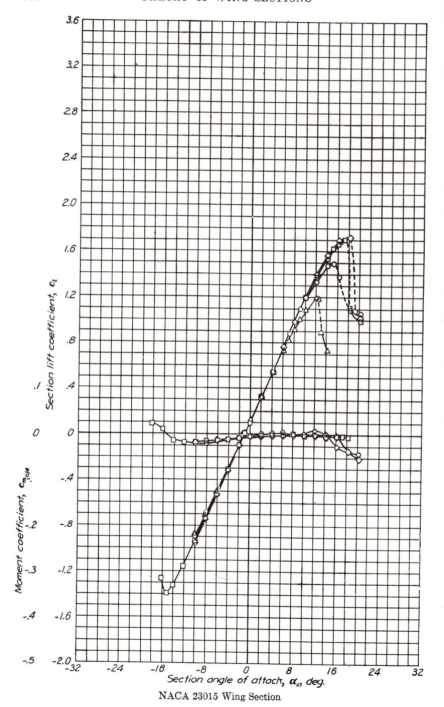

NACA 23015 Wing Section

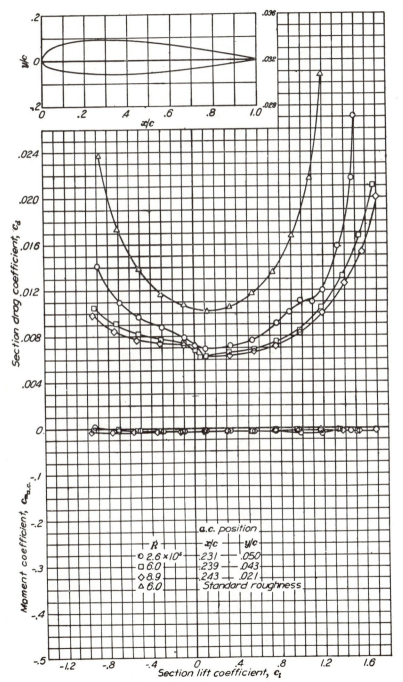

NACA 23015 Wing Section (*Continued*)

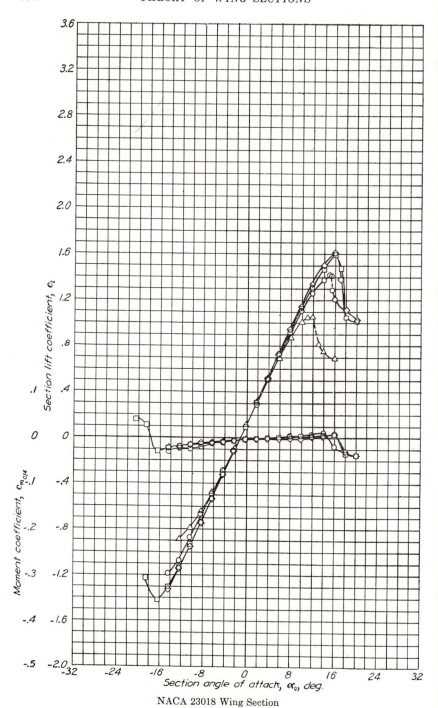

NACA 23018 Wing Section

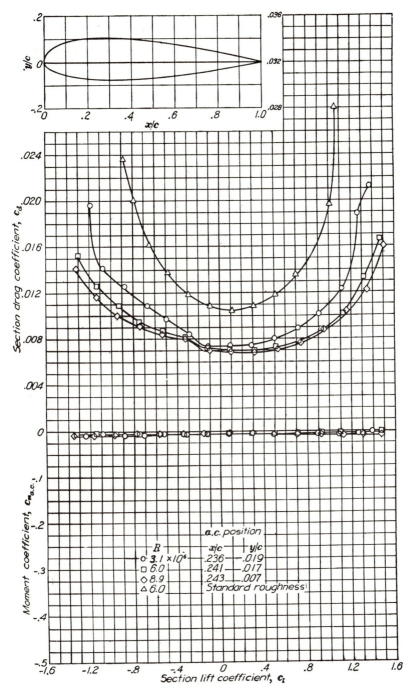

NACA 23018 Wing Section (*Continued*)

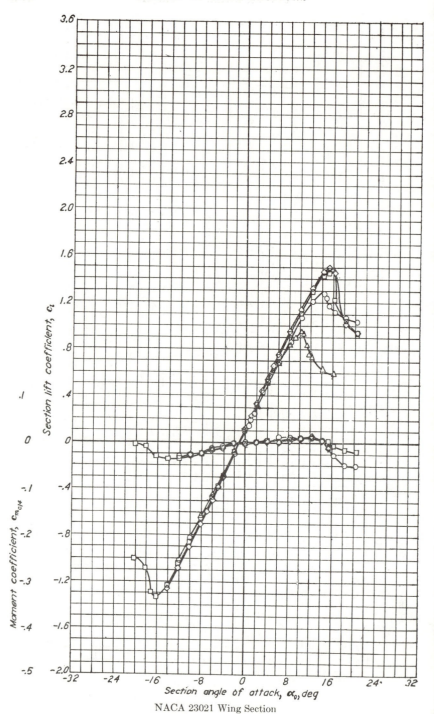

NACA 23021 Wing Section

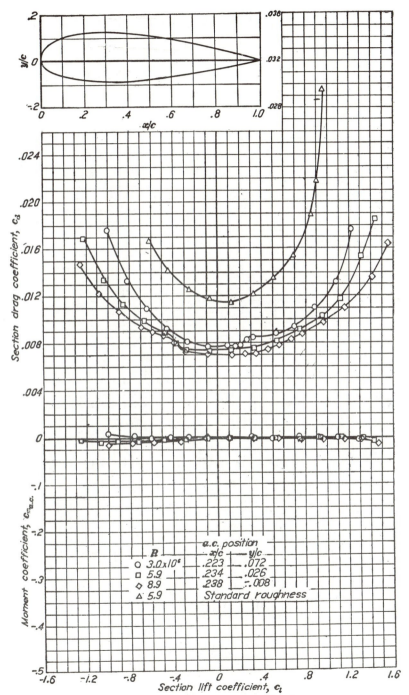

NACA 23021 Wing Section (*Continued*)

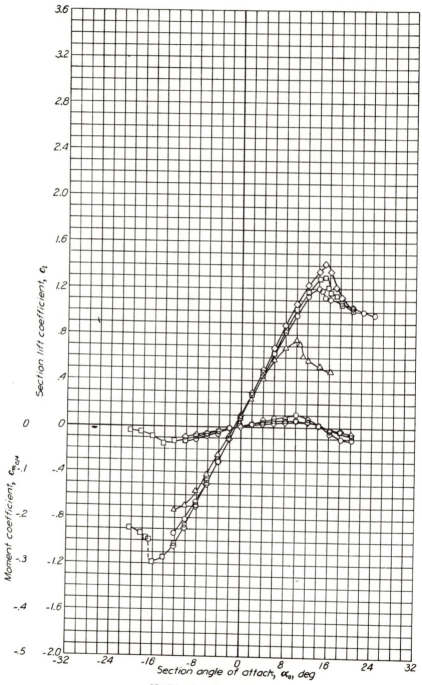

NACA 23024 Wing Section

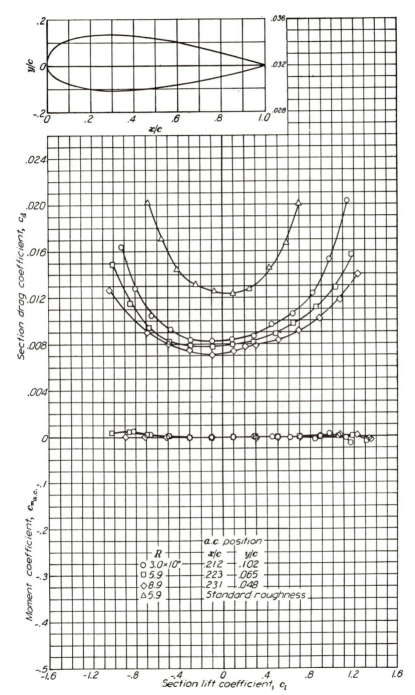

NACA 23024 Wing Section (*Continued*)

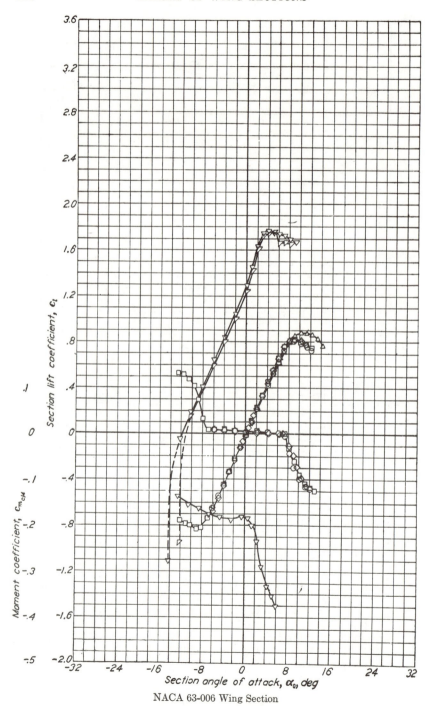

Section lift coefficient, c_l

Moment coefficient, $c_{m_{c/4}}$

Section angle of attack, α_0, deg

NACA 63-006 Wing Section

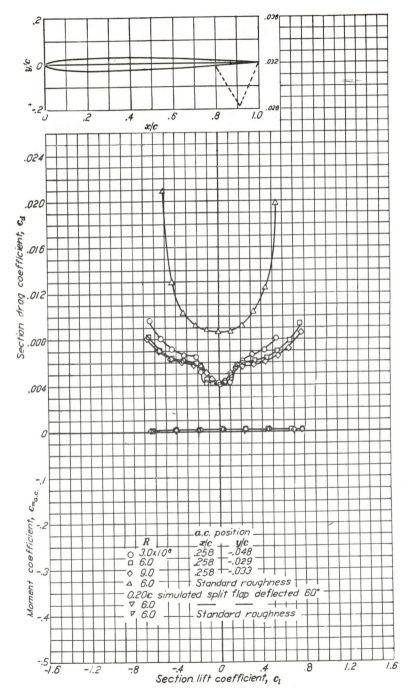

NACA 63-006 Wing Section (*Continued*)

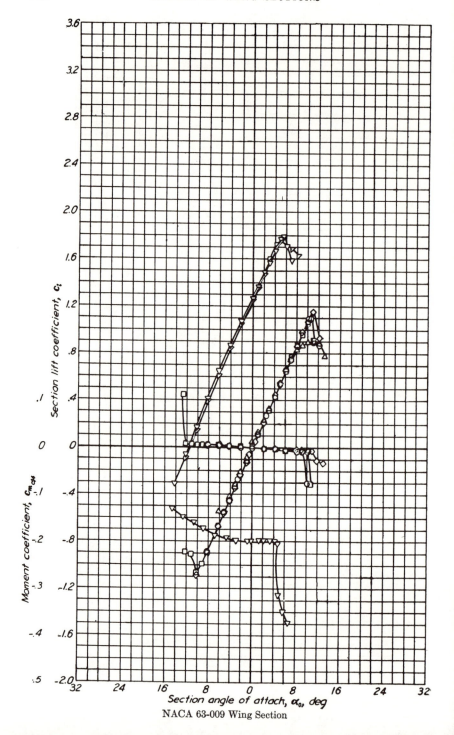

NACA 63-009 Wing Section

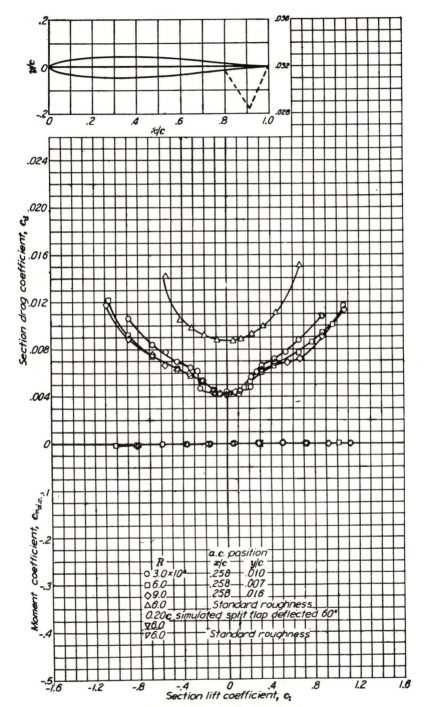

NACA 63-009 Wing Section (*Continued*)

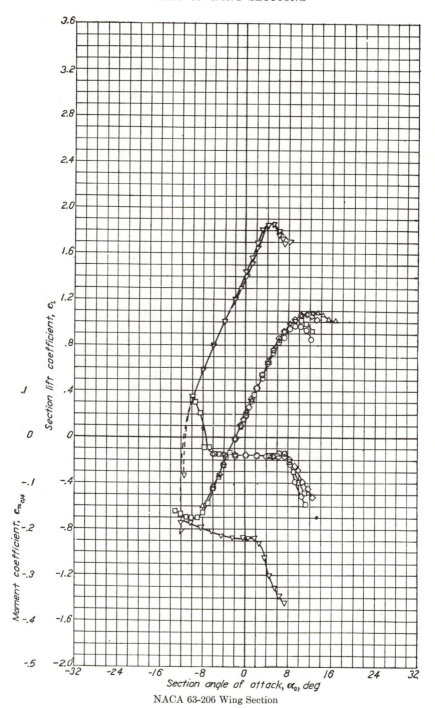

NACA 63-206 Wing Section

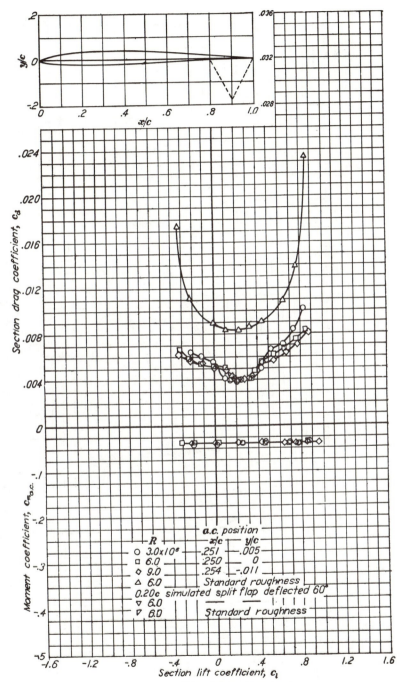

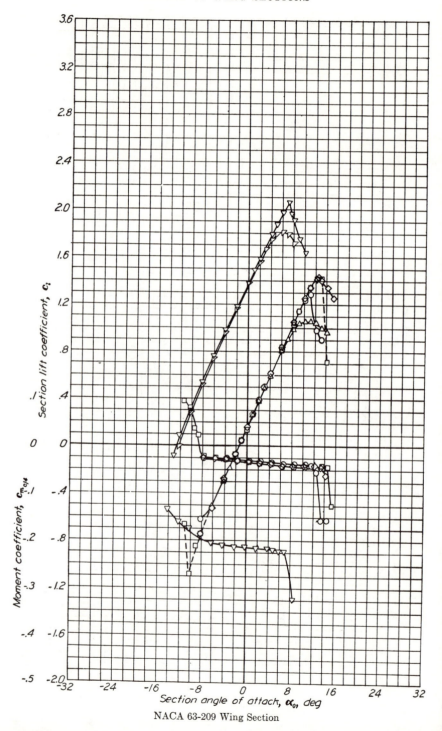

NACA 63-209 Wing Section

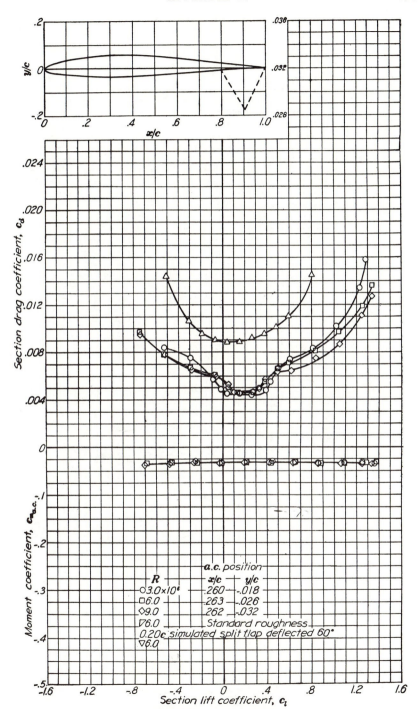

NACA 63-209 Wing Section (*Continued*)

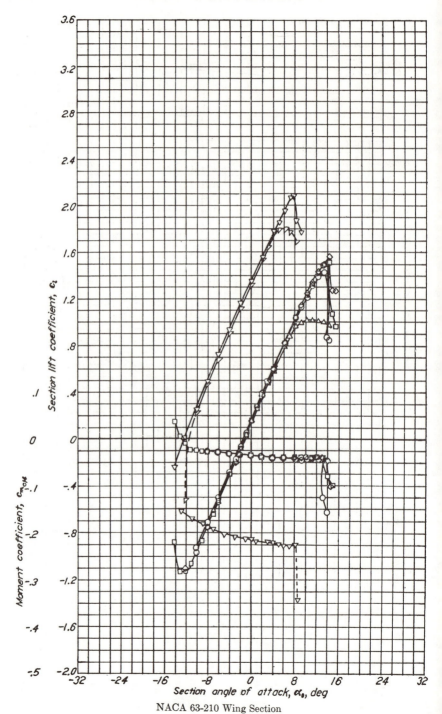

NACA 63-210 Wing Section

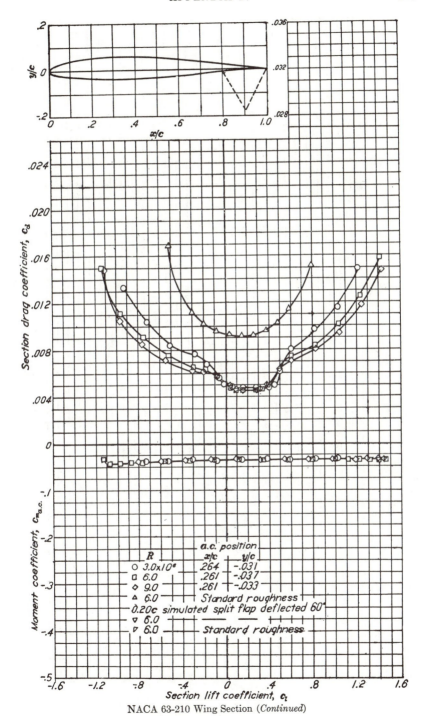

NACA 63-210 Wing Section (*Continued*)

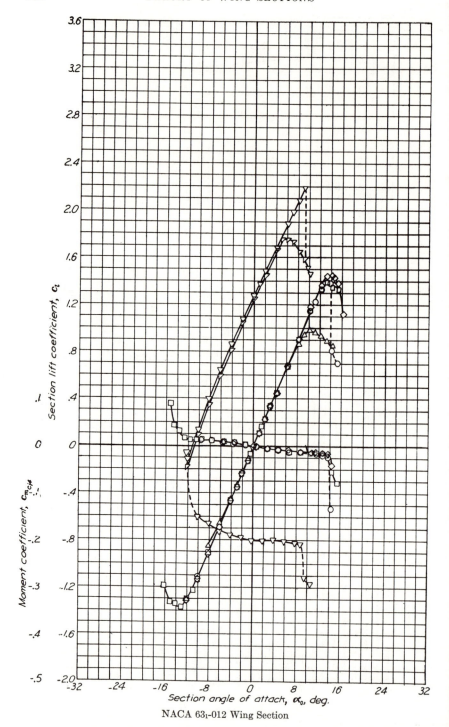

NACA 63₁-012 Wing Section

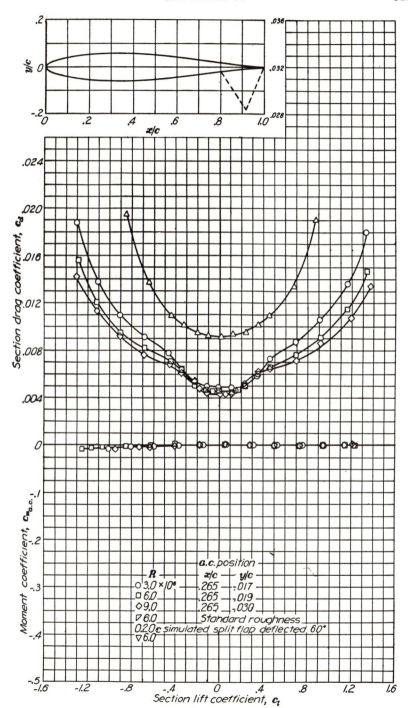

NACA 63₁-012 Wing Section (*Continued*)

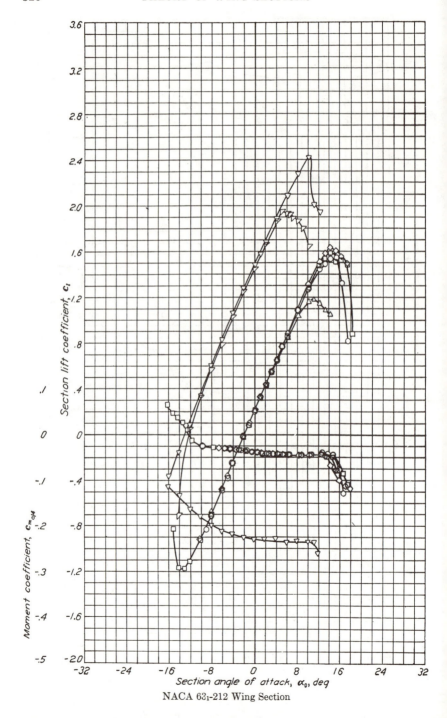

NACA 63_1-212 Wing Section

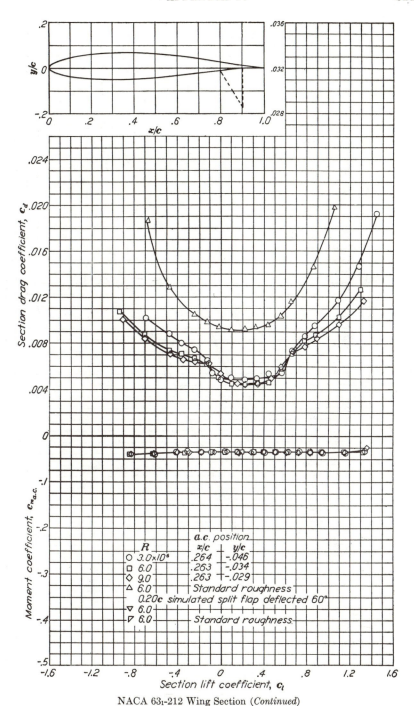

NACA 63₁-212 Wing Section (*Continued*)

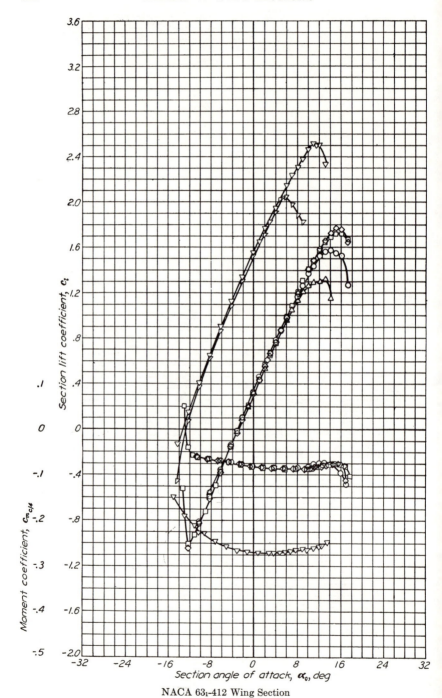

NACA 63₁-412 Wing Section

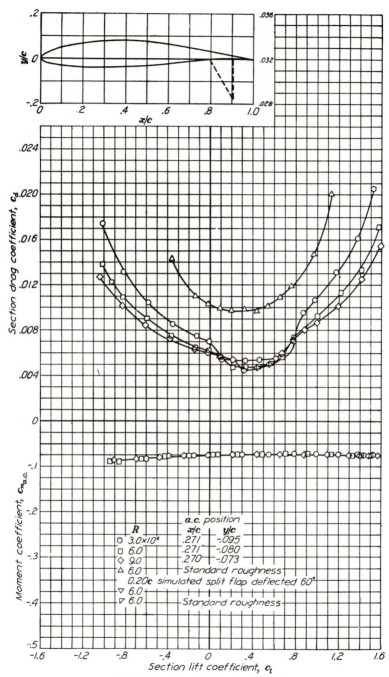

NACA 63₁-412 Wing Section (*Continued*)

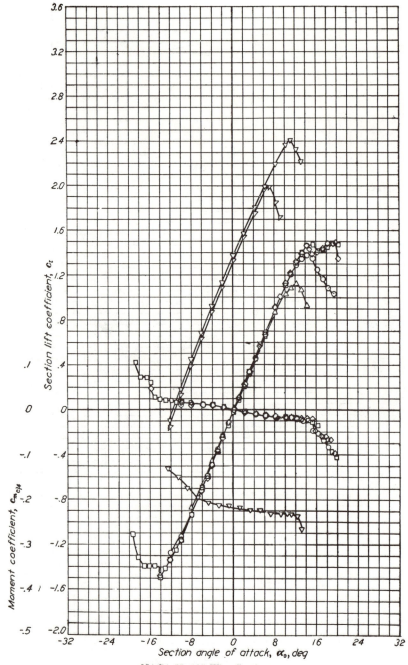

NACA 63₂-015 Wing Section

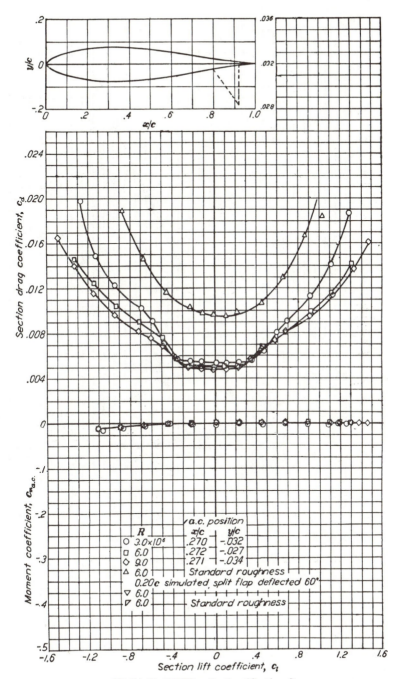

NACA 63₂-015 Wing Section (*Continued*)

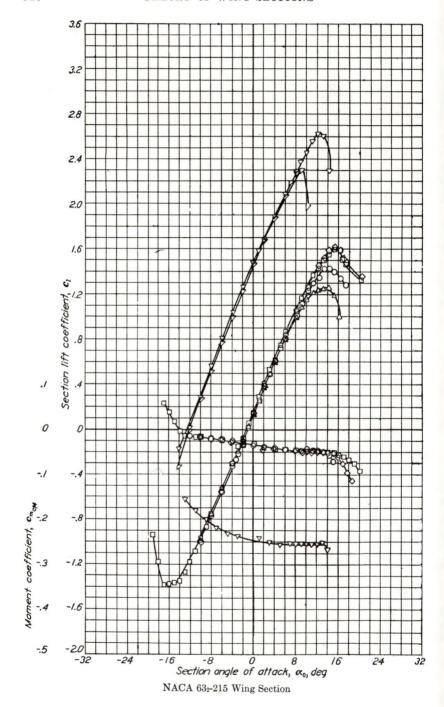

NACA 63₂-215 Wing Section

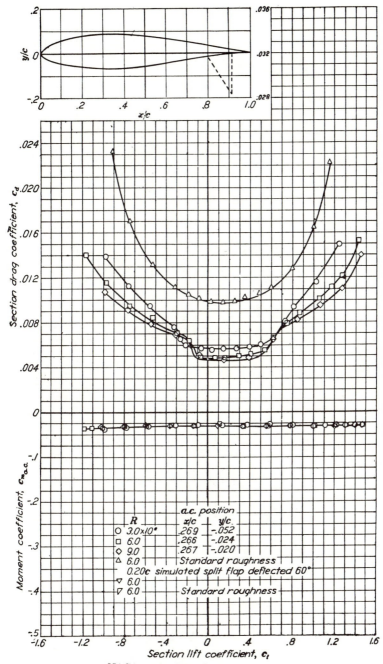

NACA 63$_2$-215 Wing Section (*Continued*)

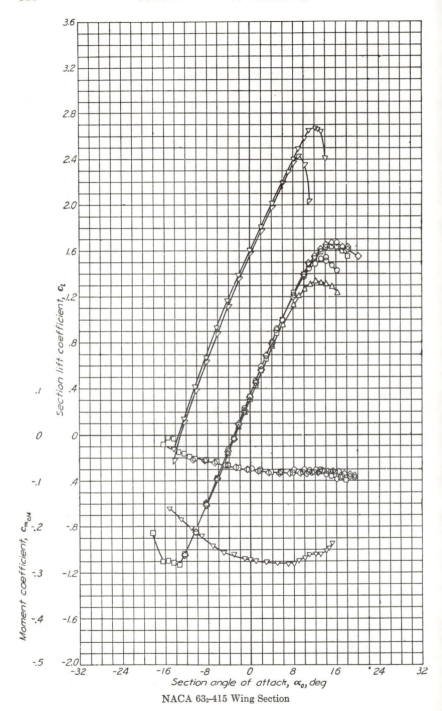

NACA 63₂-415 Wing Section

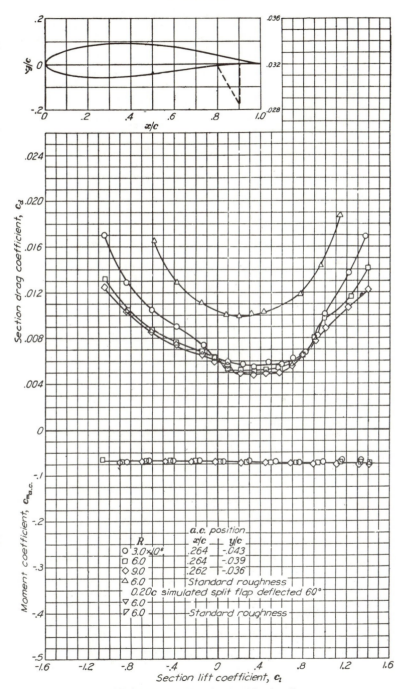

NACA 63₂-415 Wing Section (*Continued*)

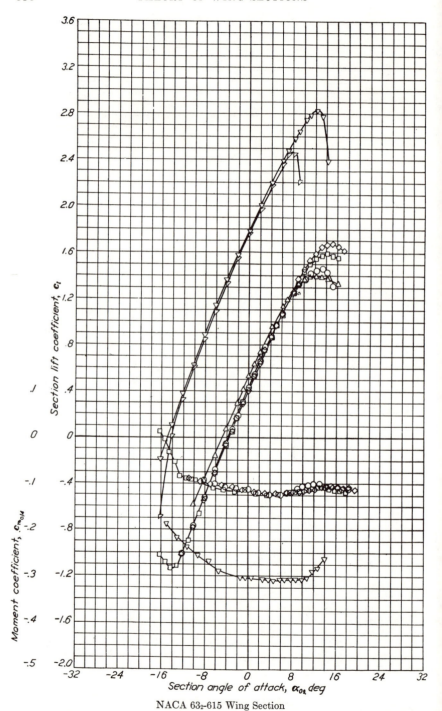

NACA 63₂-615 Wing Section

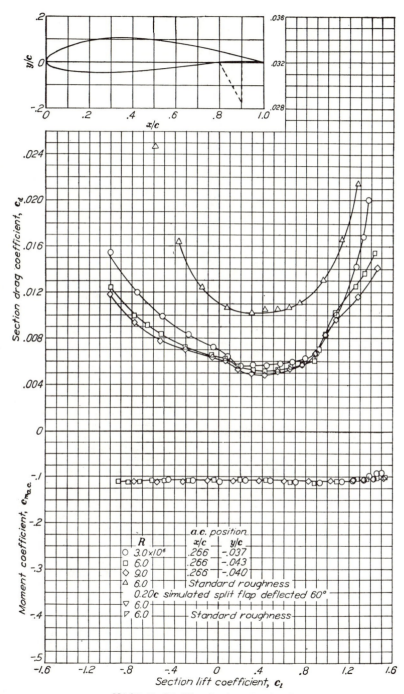

NACA 63₂-615 Wing Section (*Continued*)

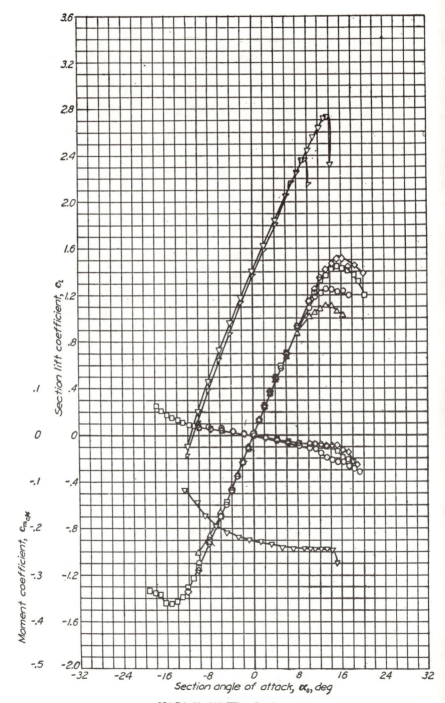

NACA 63₃-018 Wing Section

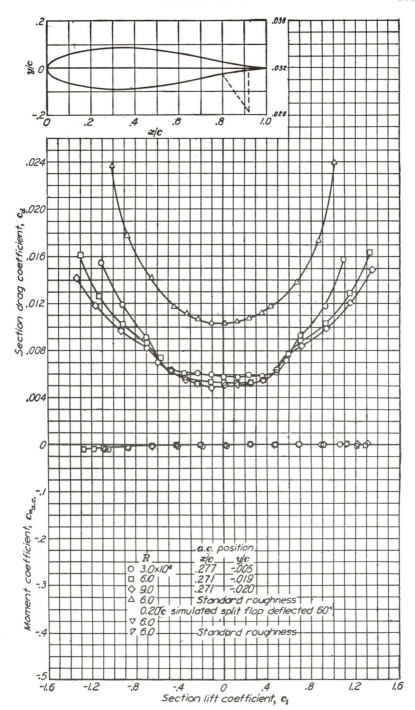

NACA 63₃-018 Wing Section (*Continued*)

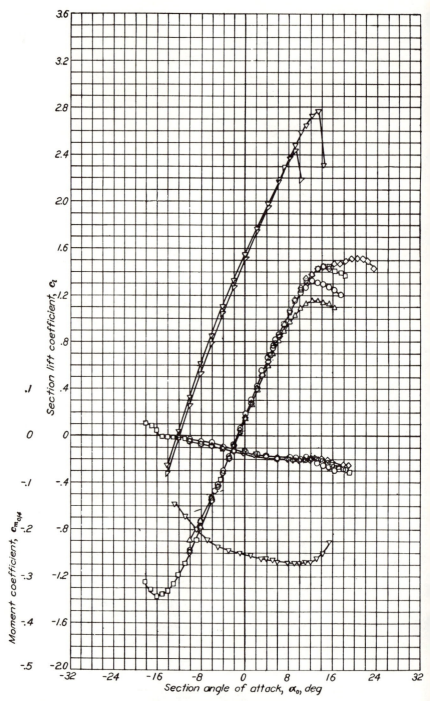

NACA 63₃-218 Wing Section

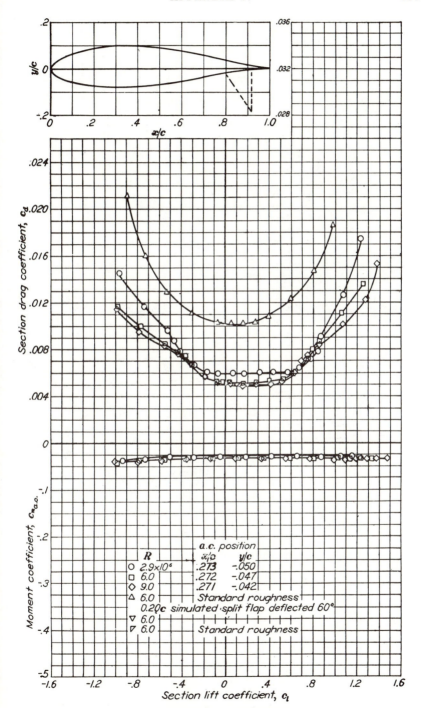

NACA 63₃-218 Wing Section (*Continued*)

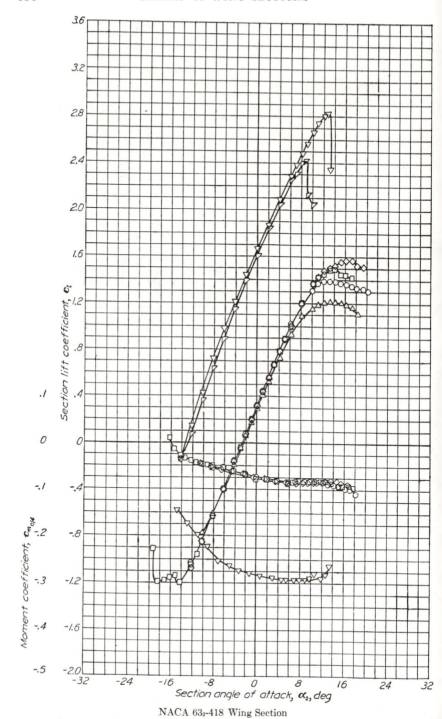

NACA 63₃-418 Wing Section

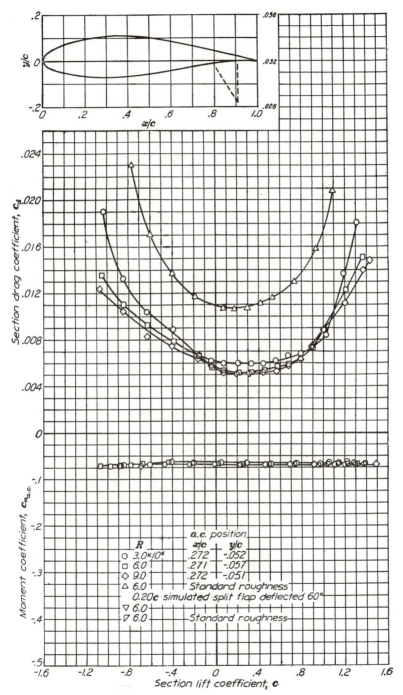

NACA 63₃–418 Wing Section (*Continued*)

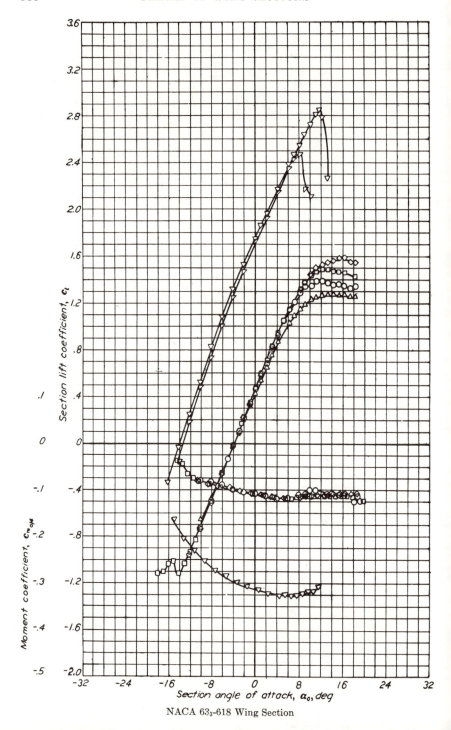

NACA 63₃-618 Wing Section

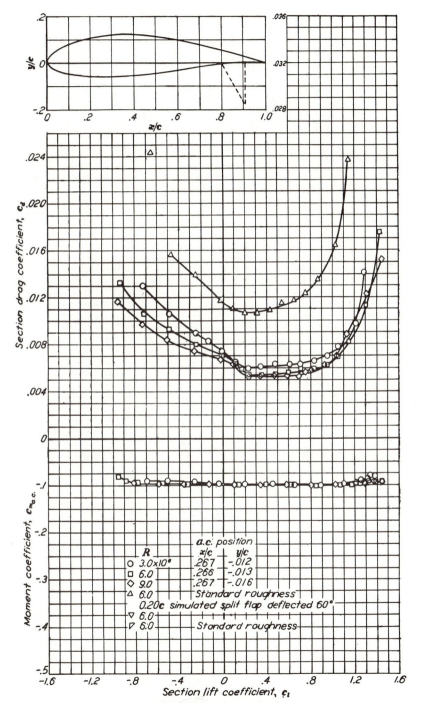

NACA 63₃-618 Wing Section (*Continued*)

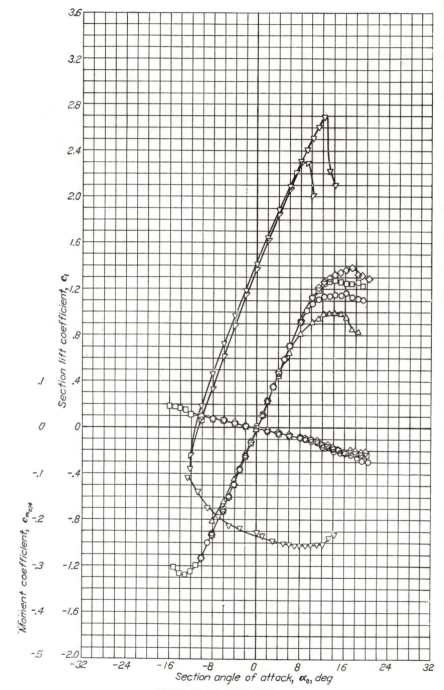

NACA 63₄-021 Wing Section

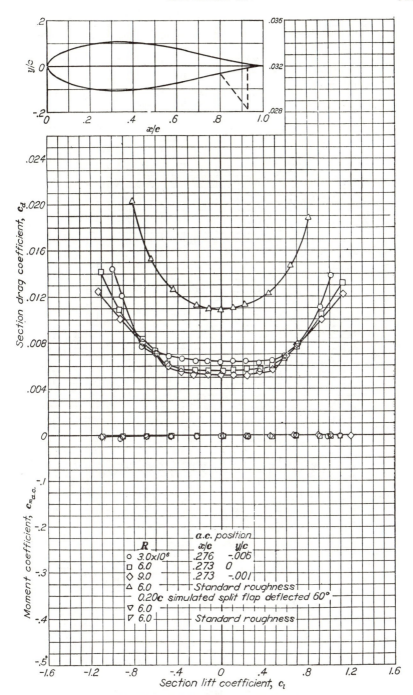

NACA 63₄-021 Wing Section (*Continued*)

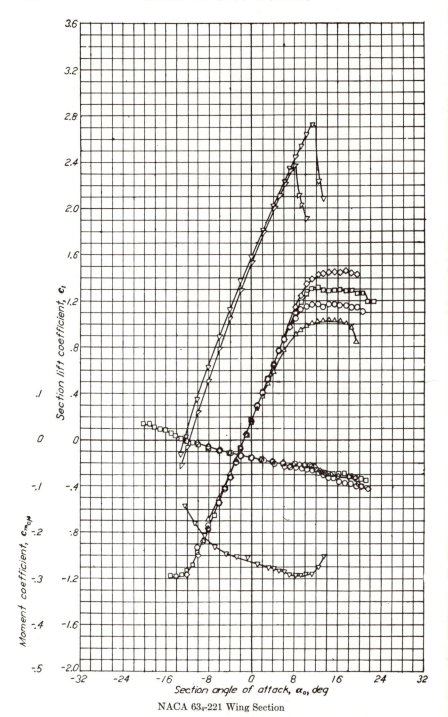

NACA 63₄-221 Wing Section

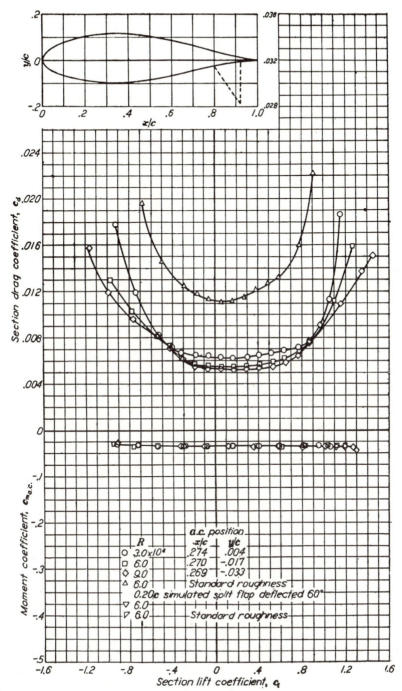

NACA 63₄-221 Wing Section (*Continued*)

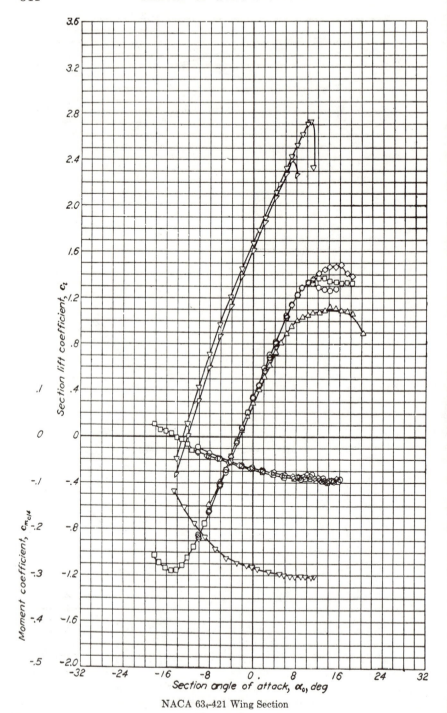

NACA 63₄-421 Wing Section

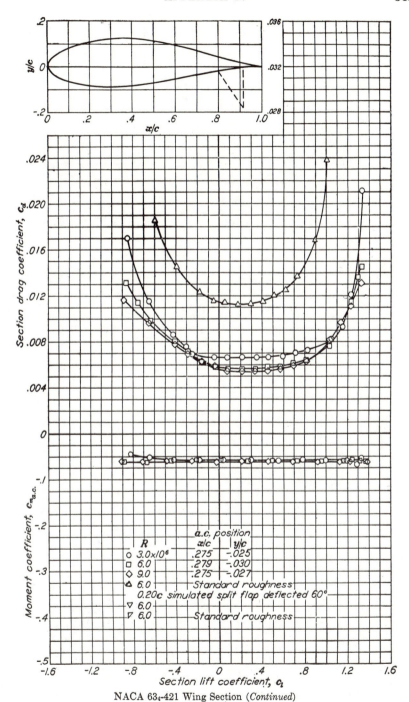

NACA 63₄-421 Wing Section (*Continued*)

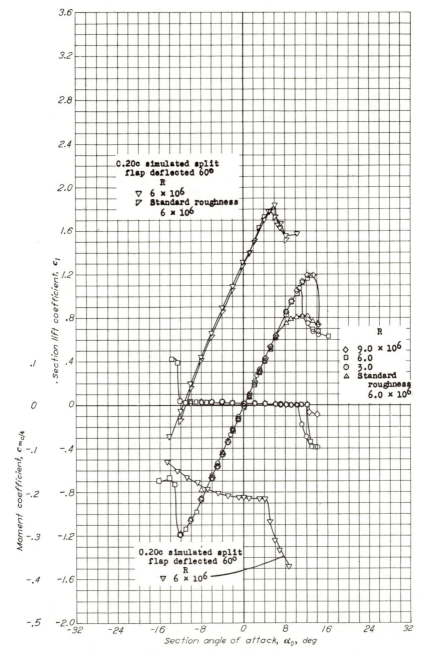

NACA 63A010 Wing Section

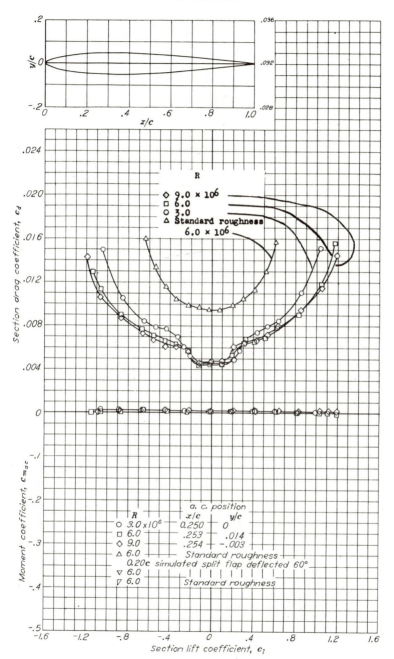

NACA 63A010 Wing Section (*Continued*)

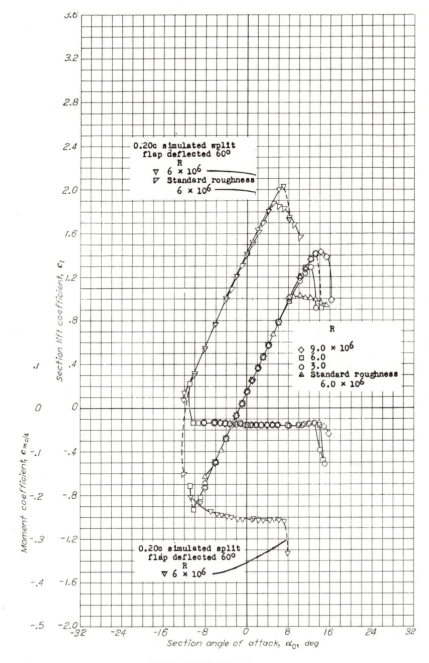

NACA 63A210 Wing Section

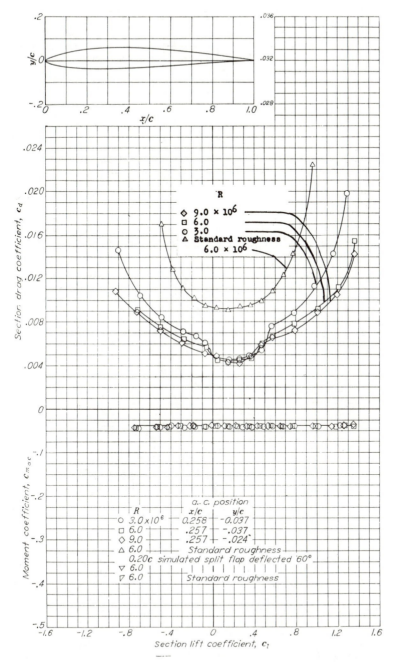

NACA 63A210 Wing Section (*Continued*)

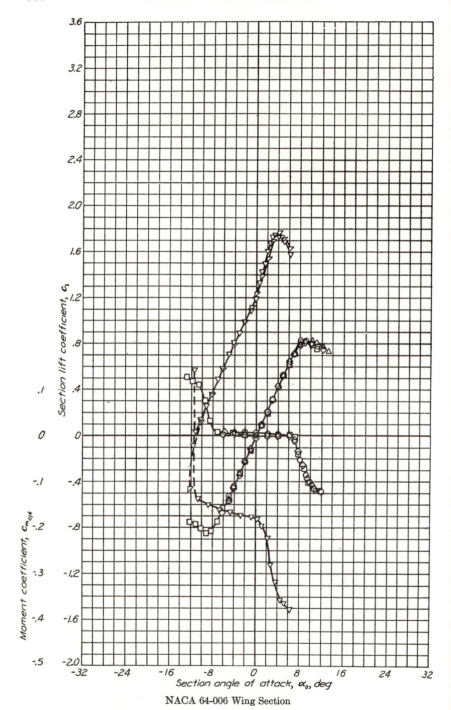

NACA 64-006 Wing Section

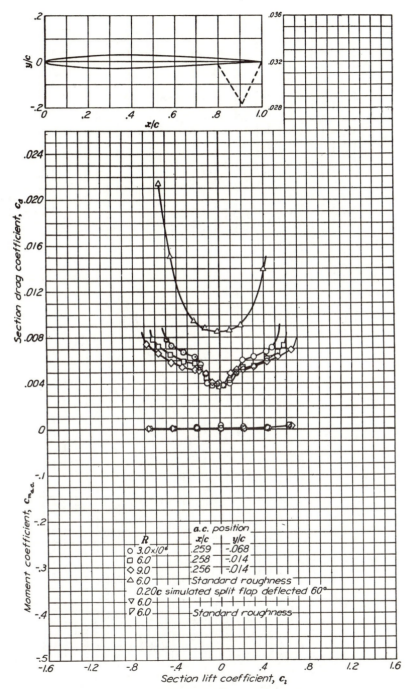

	a.c. position		
R	x/c	y/c	
○ 3.0×10^6	.259	-.068	
□ 6.0	.258	-.014	
◇ 9.0	.256	-.014	
△ 6.0		Standard roughness	
0.20c simulated split flap deflected 60°			
▽ 6.0			
▽ 6.0		Standard roughness	

NACA 64-006 Wing Section (*Continued*)

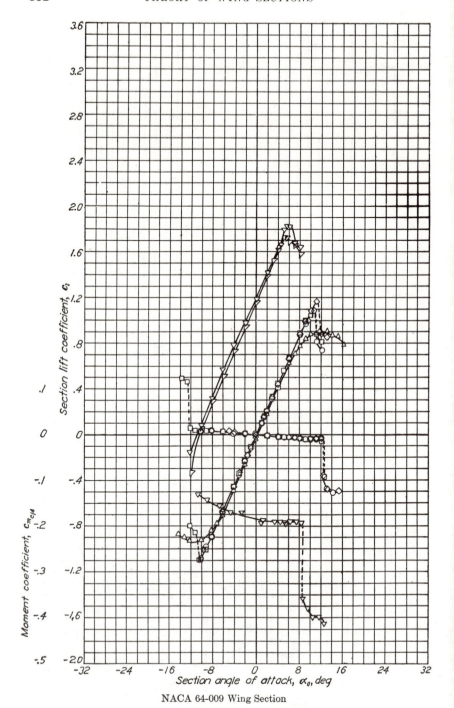

NACA 64-009 Wing Section

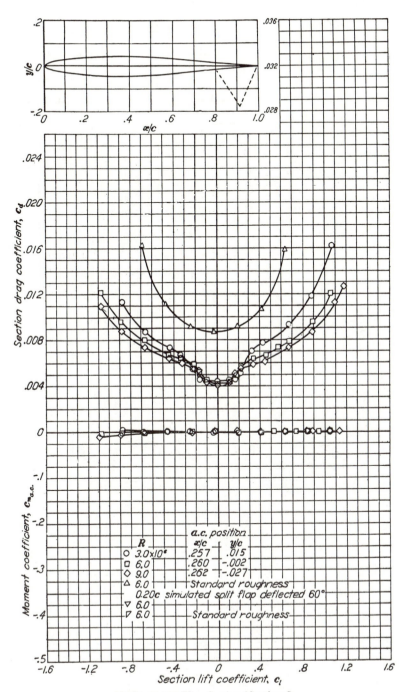

NACA 64-009 Wing Section (*Continued*)

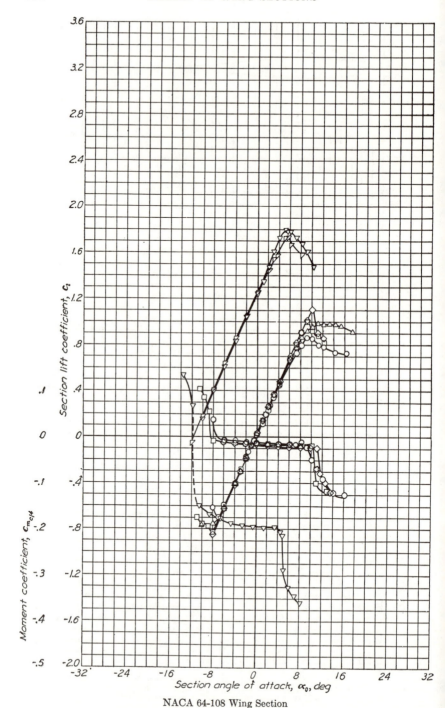

NACA 64-108 Wing Section

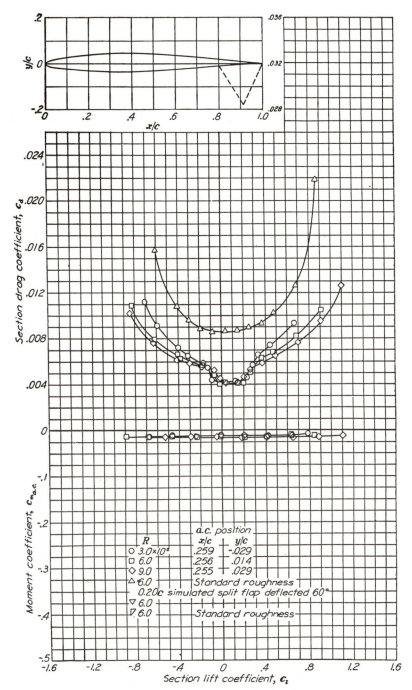

NACA 64-108 Wing Section (*Continued*)

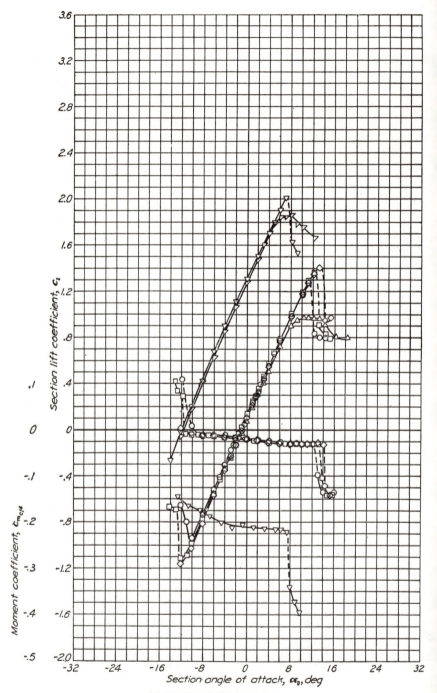

NACA 64-110 Wing Section

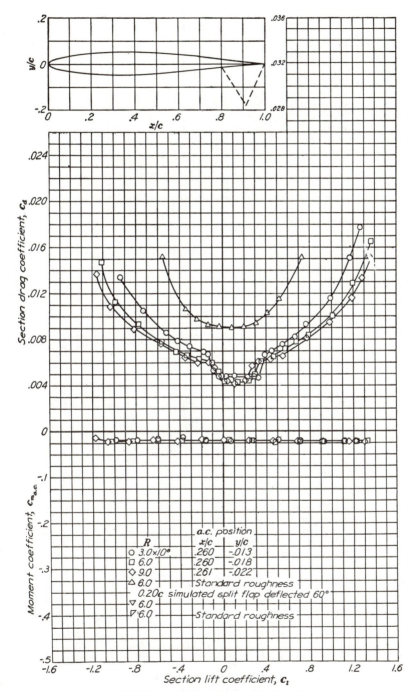

NACA 64-110 Wing Section (*Continued*)

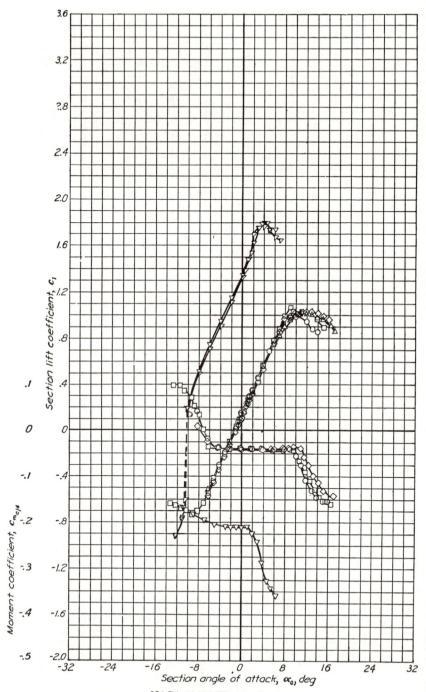

NACA 64-206 Wing Section

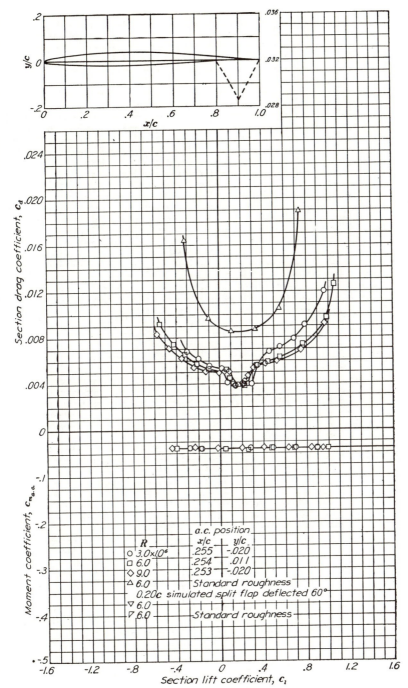

NACA 64-206 Wing Section (*Continued*)

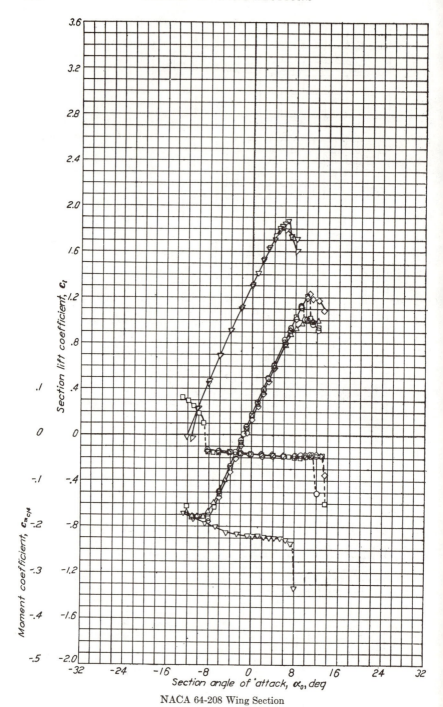

NACA 64-208 Wing Section

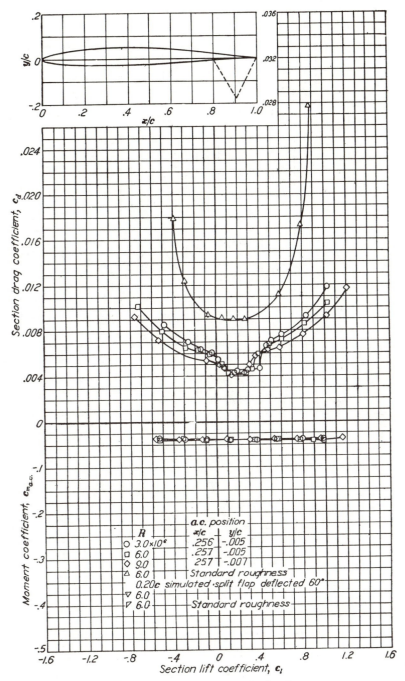

NACA 64-208 Wing Section (*Continued*)

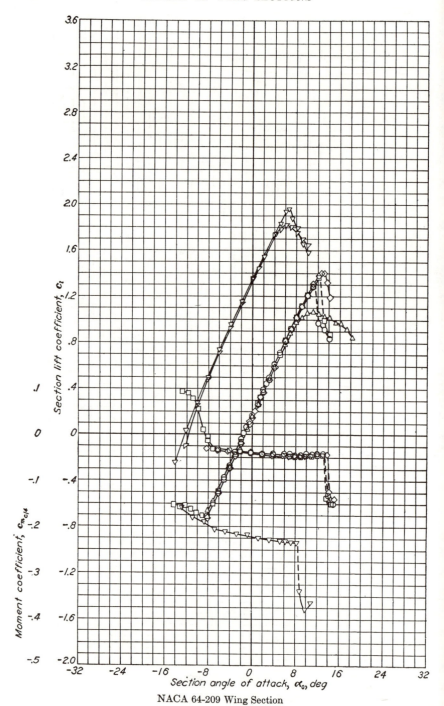

NACA 64-209 Wing Section

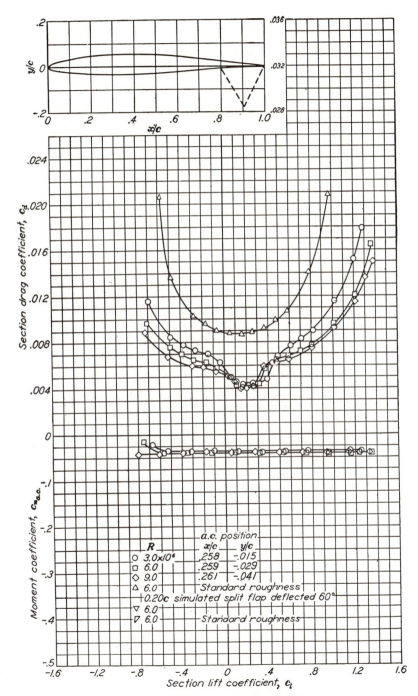

NACA 64-209 Wing Section (*Continued*)

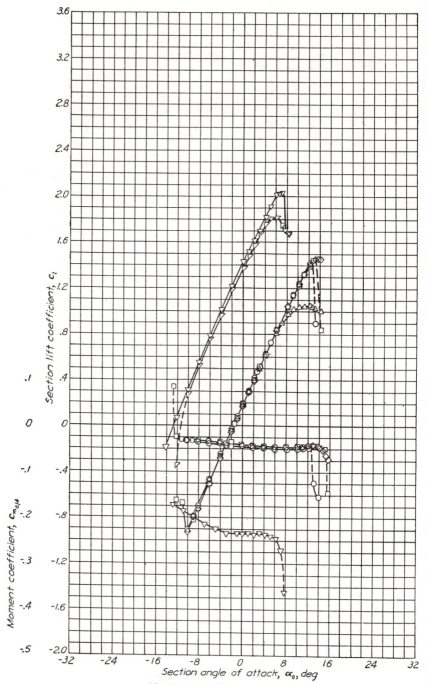

NACA 64-210 Wing Section

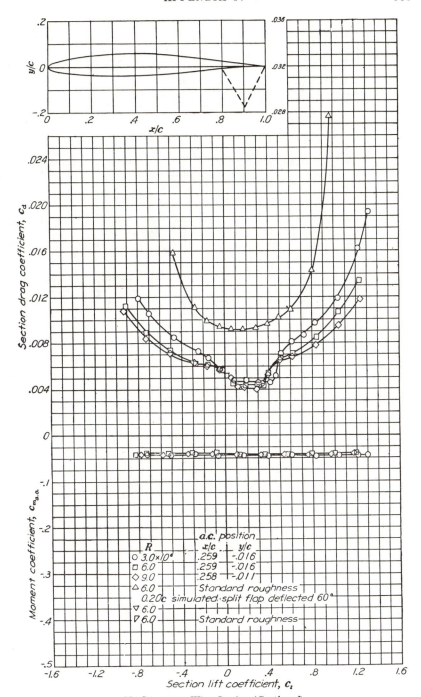

NACA 64-210 Wing Section (*Continued*)

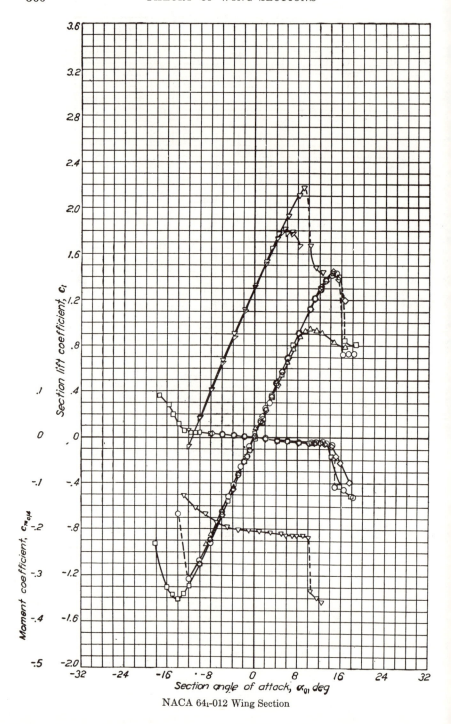

NACA 64₁-012 Wing Section

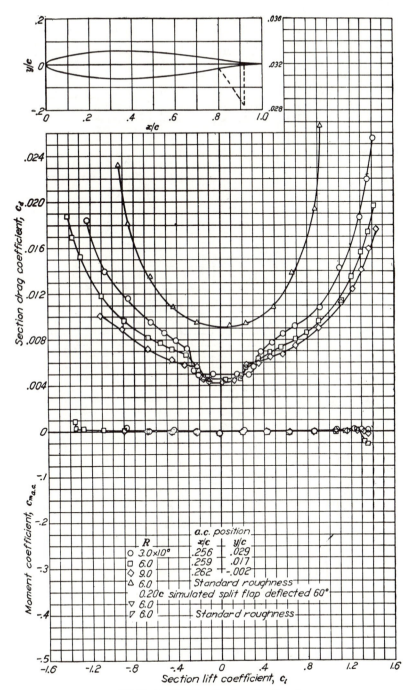

NACA 64₁-012 Wing Section (*Continued*)

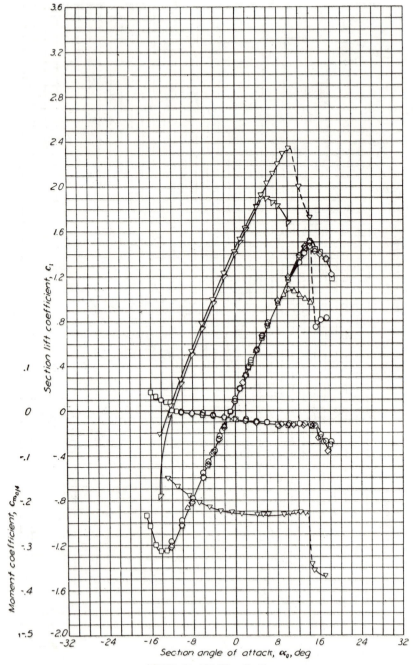

NACA 64₁-112 Wing Section

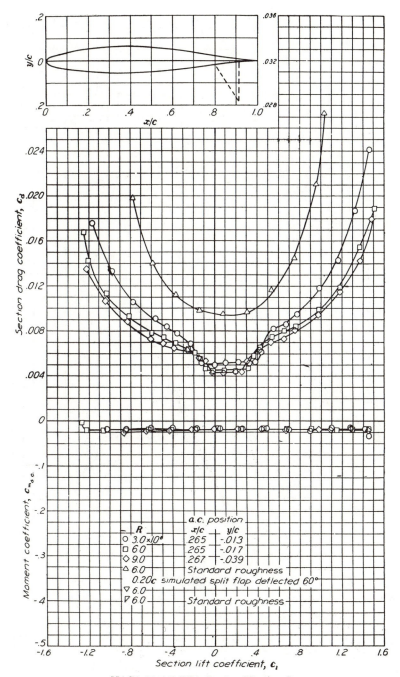

a.c. position

R	x/c	y/c
○ 3.0×10⁶	.265	-.013
□ 6.0	.265	-.017
◇ 9.0	.267	-.039
△ 6.0	Standard roughness	
0.20c simulated split flap deflected 60°		
▽ 6.0		
▽ 6.0	Standard roughness	

NACA 64₁-112 Wing Section (*Continued*)

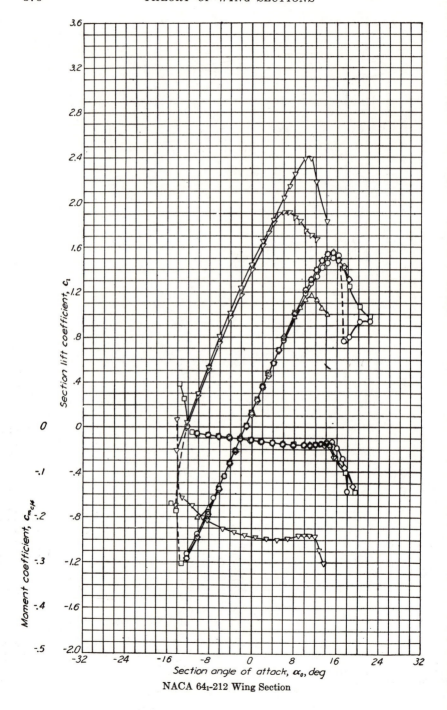

NACA 64₁-212 Wing Section

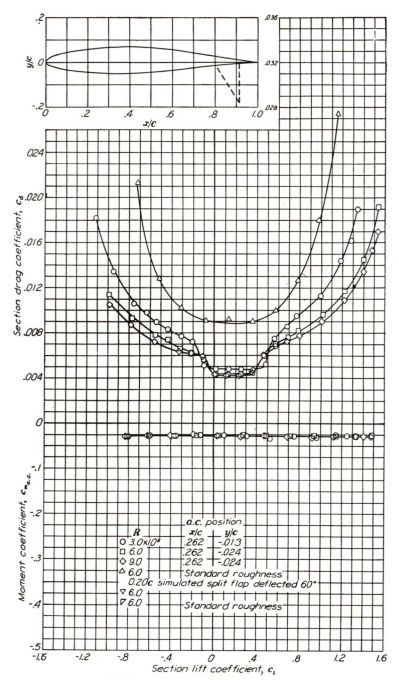

NACA 64₁-212 Wing Section (*Continued*)

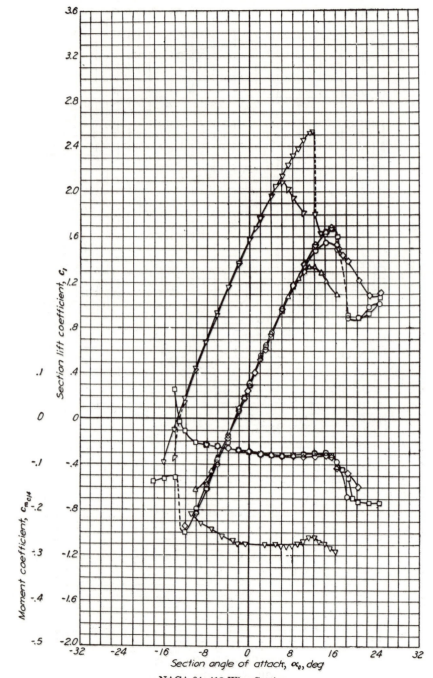

NACA 64$_1$-412 Wing Section

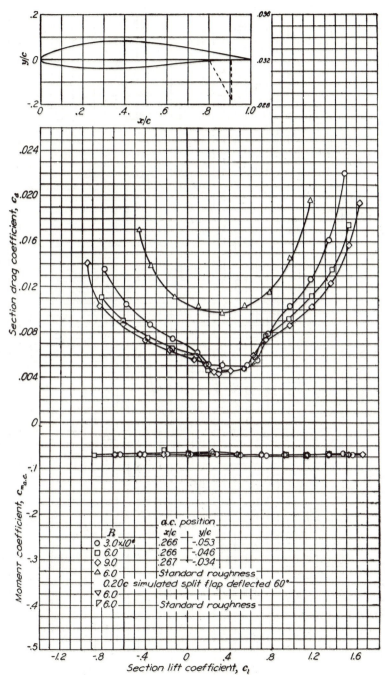

NACA 64₁-412 Wing Section (*Continued*)

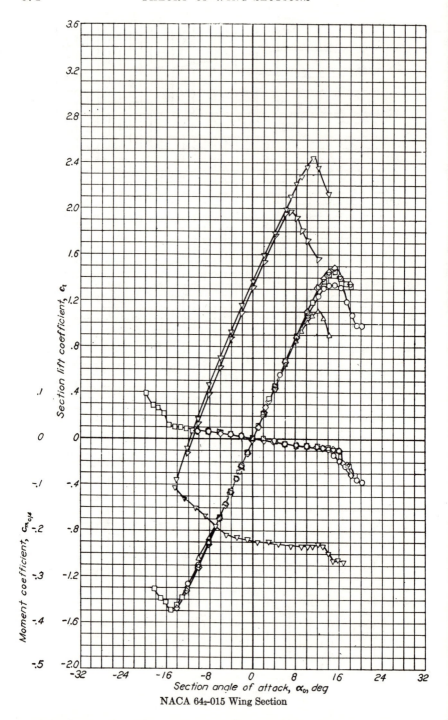

NACA 64₂-015 Wing Section

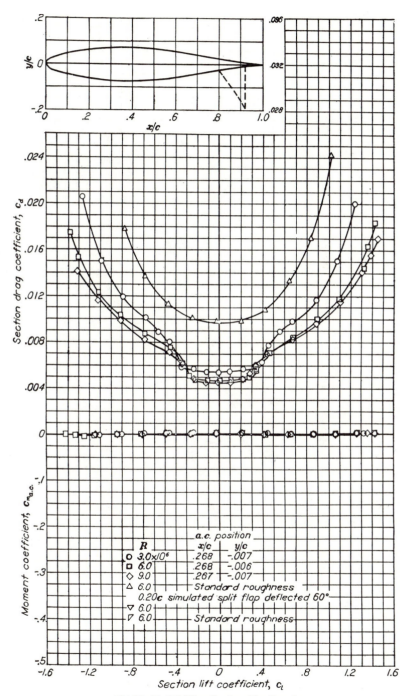

NACA 64₂-015 Wing Section (*Continued*)

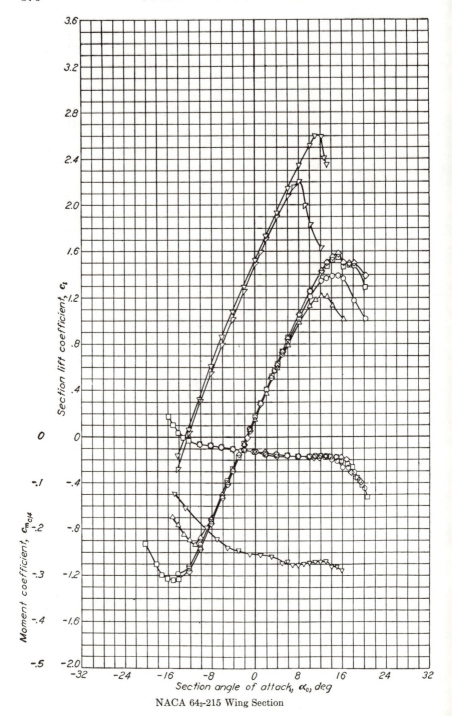

NACA 64₂-215 Wing Section

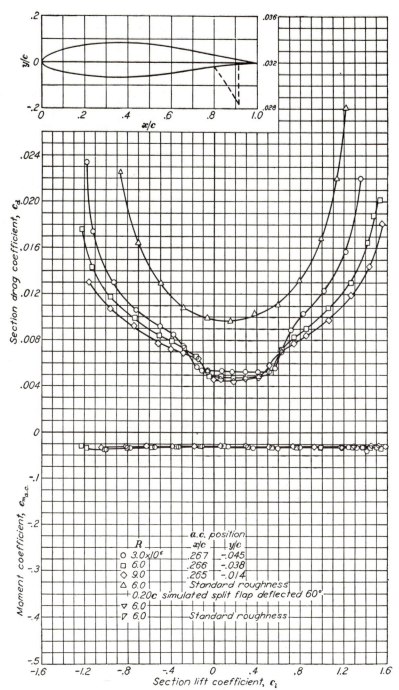

NACA 64₂-215 Wing Section (*Continued*)

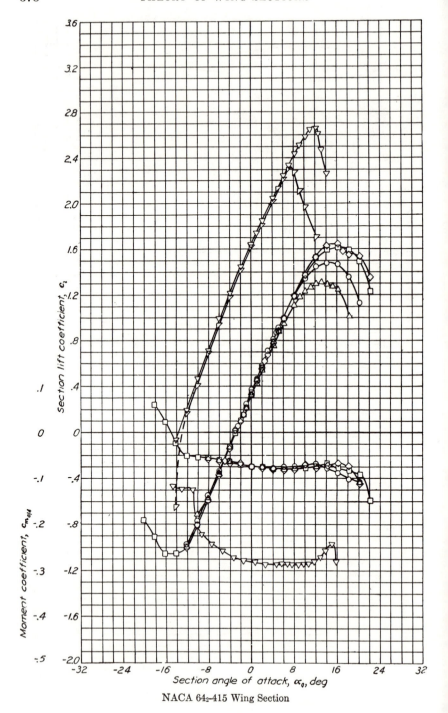

NACA 64₂-415 Wing Section

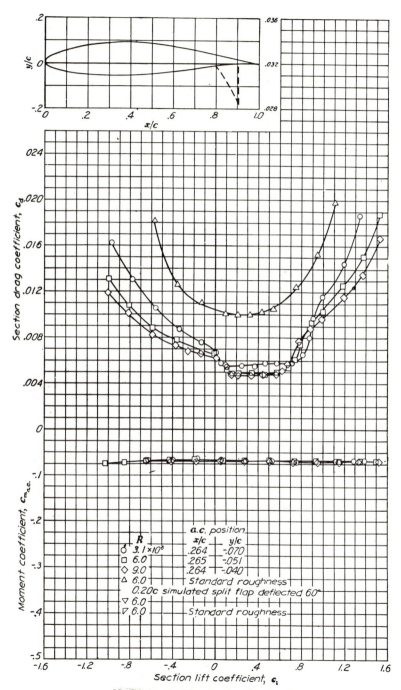

NACA 64₂-415 Wing Section (*Continued*)

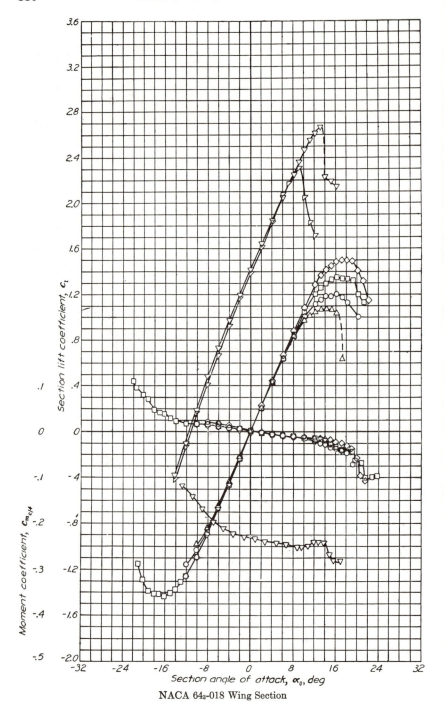

NACA 64₃-018 Wing Section

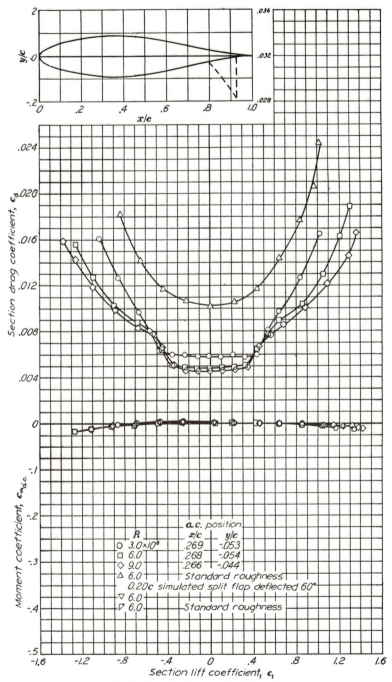

NACA 64₃-018 Wing Section (*Continued*)

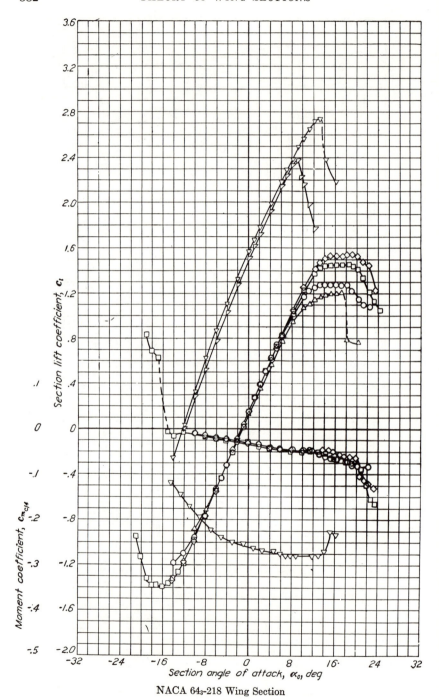

NACA 64₃-218 Wing Section

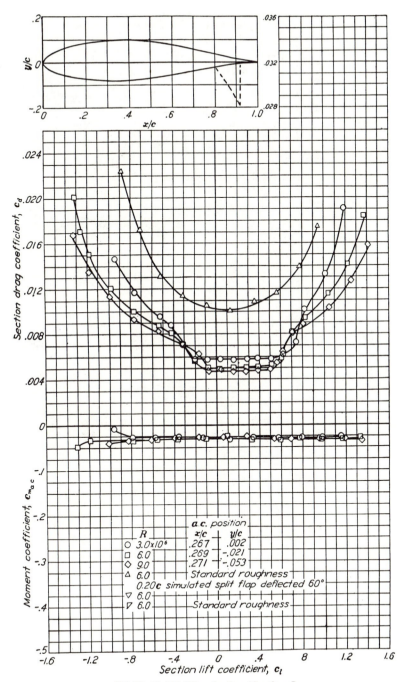

NACA 64₃-218 Wing Section (*Continued*)

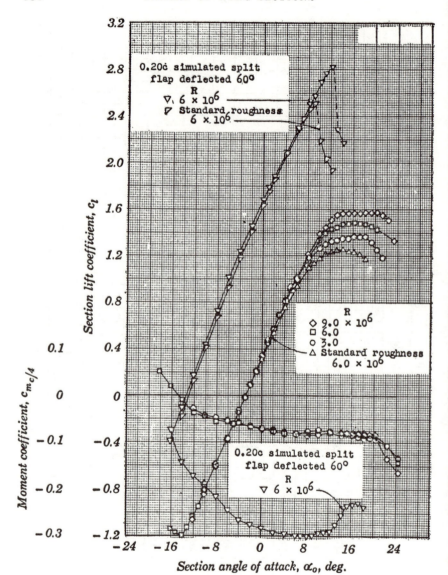

NACA 64₃-418 Wing Section

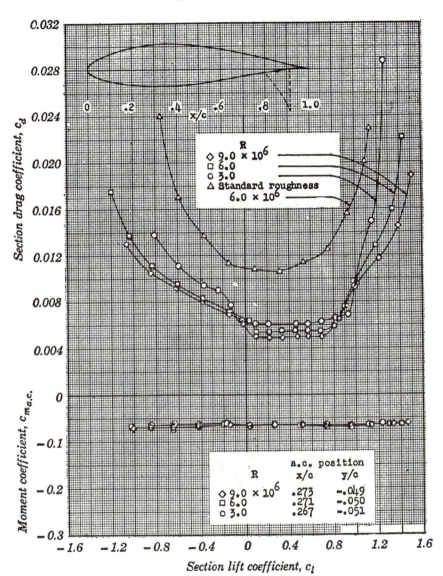

NACA 64₃-418 Wing Section (*Continued*)

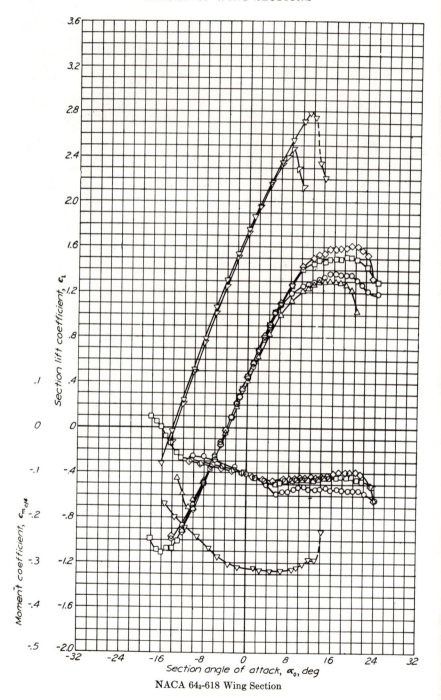

NACA 64₃-618 Wing Section

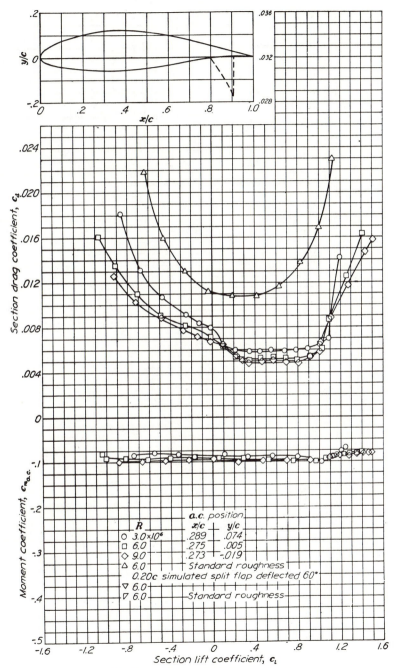

NACA 64₃-618 Wing Section (*Continued*)

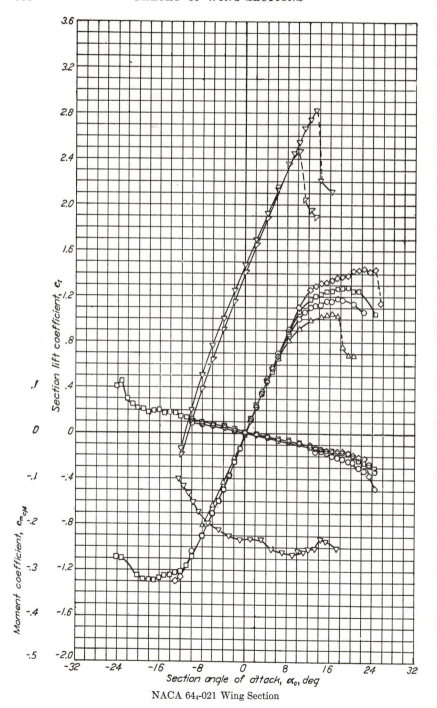

NACA 64₄-021 Wing Section

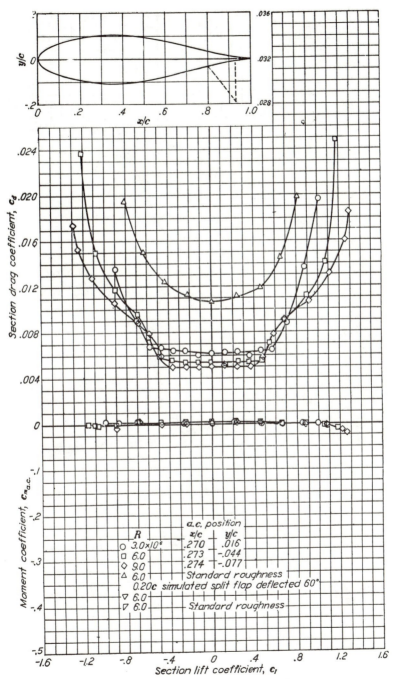

NACA 64₄-021 Wing Section (*Continued*)

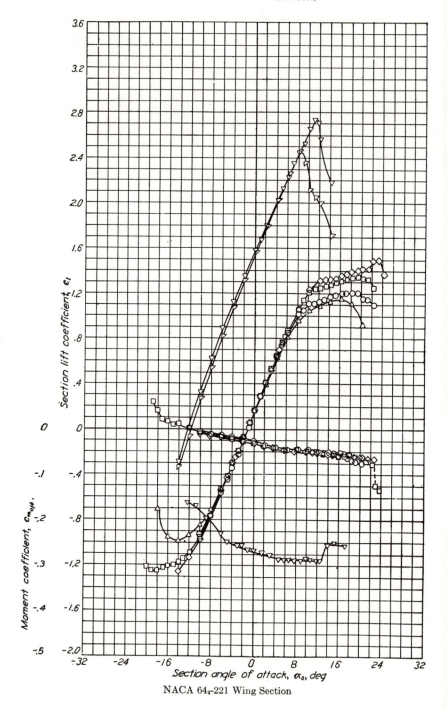

NACA 64$_4$-221 Wing Section

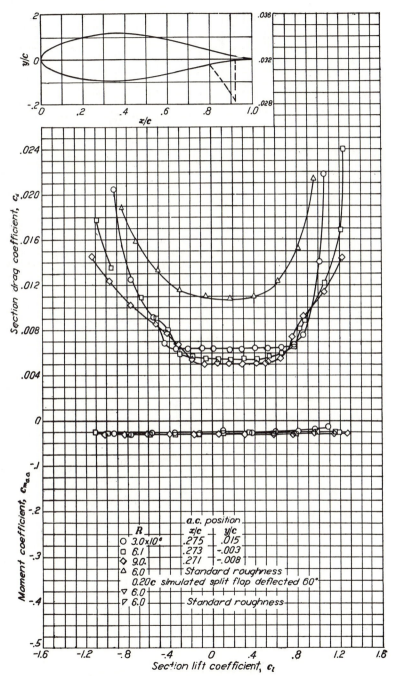

NACA 64$_4$-221 Wing Section (*Continued*)

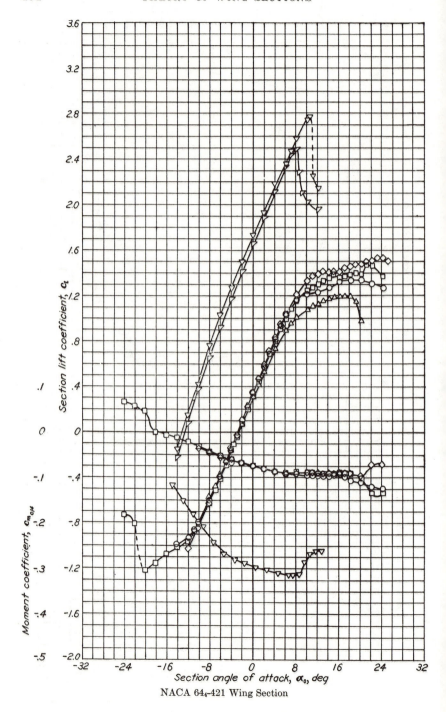

NACA 64₄-421 Wing Section

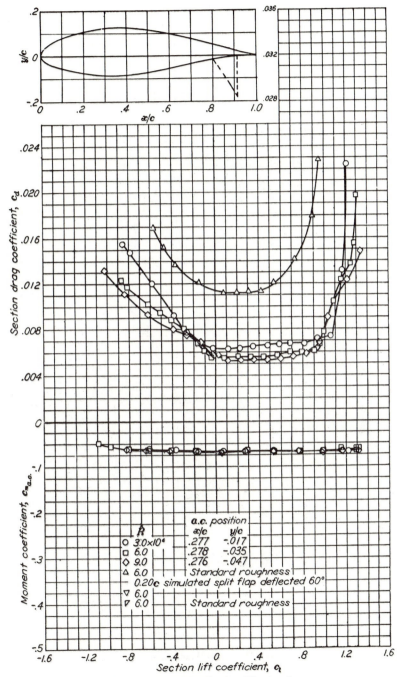

NACA 64₄-421 Wing Section (*Continued*)

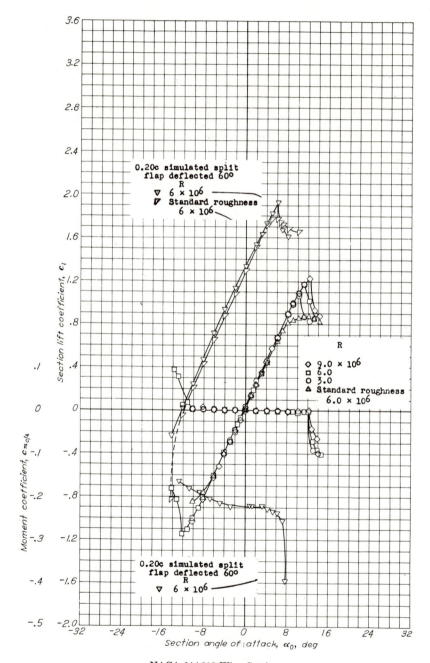

NACA 64A010 Wing Section

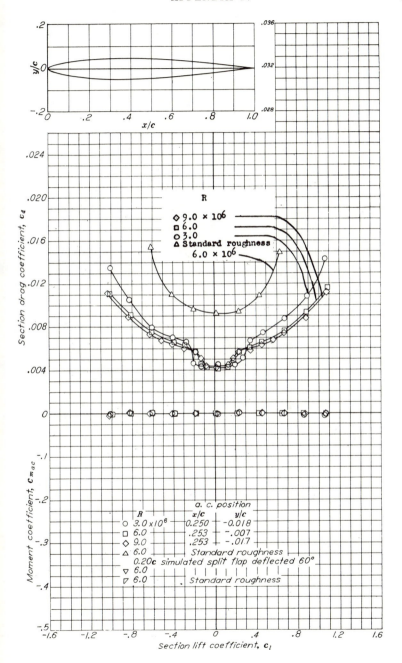

NACA 64A010 Wing Section (*Continued*)

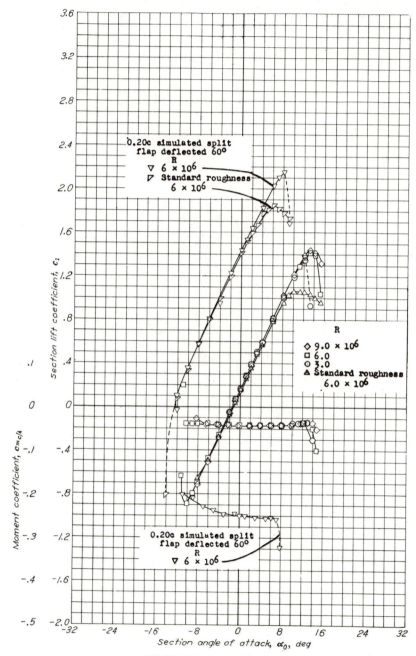

NACA 64A210 Wing Section

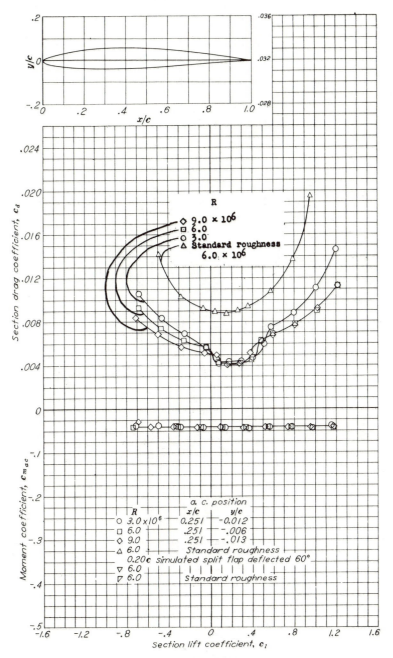

NACA 64A210 Wing Section (*Continued*)

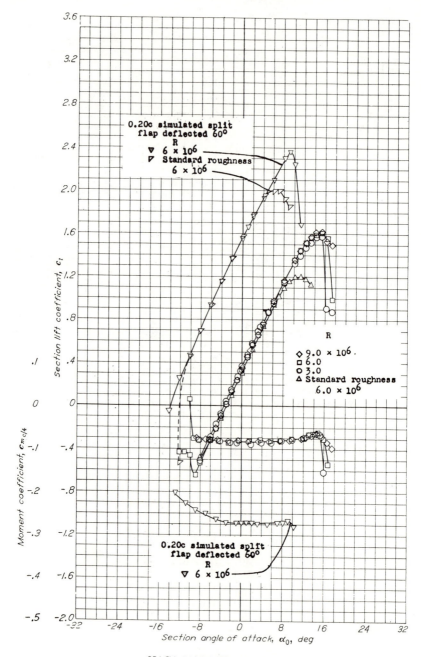

NACA 64A410 Wing Section

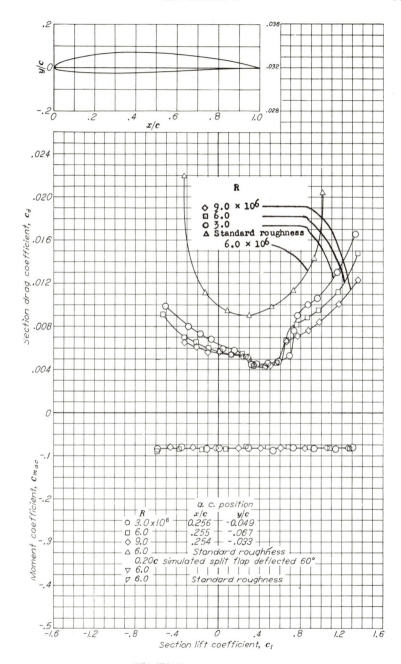

NACA 64A410 Wing Section (*Continued*)

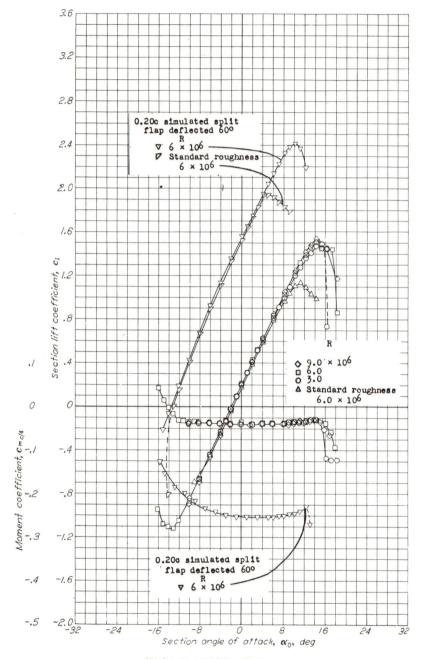

NACA 64₁A212 Wing Section

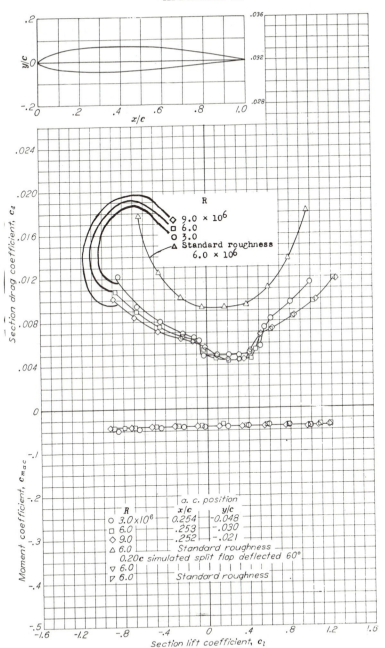

NACA 64₁A212 Wing Section (*Continued*)

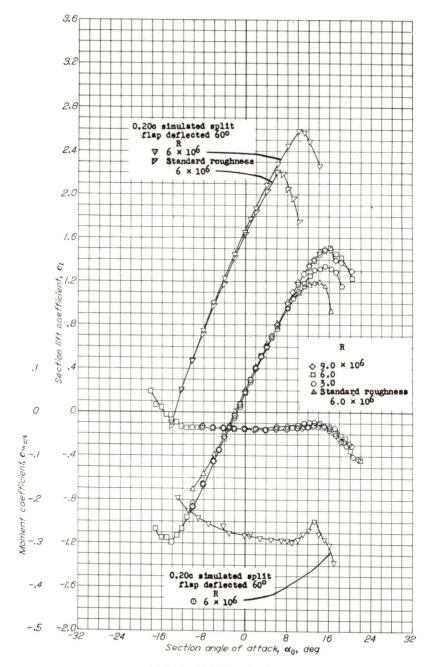

NACA 64₂A215 Wing Section

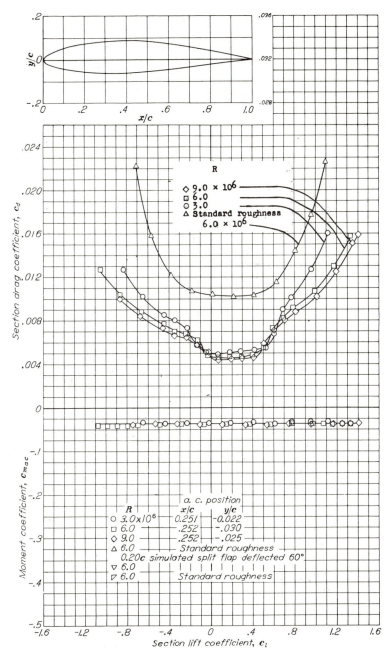

NACA 64₂A215 Wing Section (*Continued*)

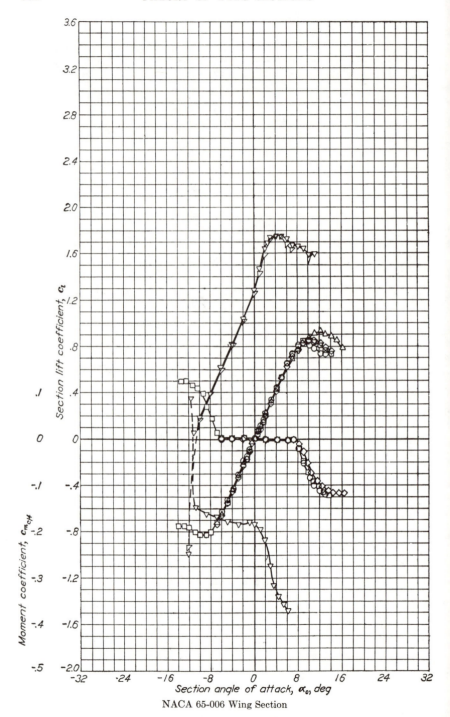

NACA 65-006 Wing Section

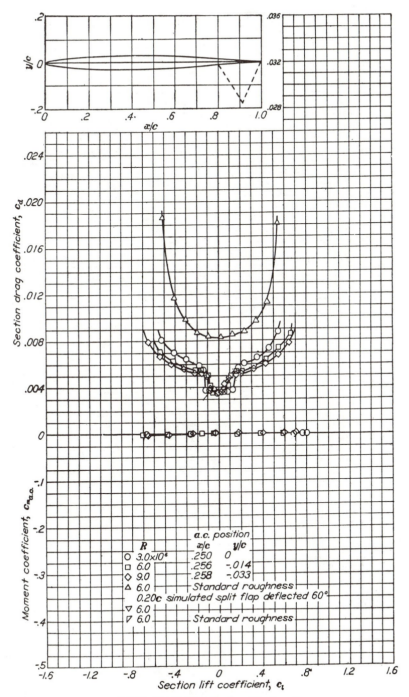

	R	x/c	y/c
O	3.0×10^6	.250	0
□	6.0	.256	-.014
◇	9.0	.258	-.033
△	6.0	Standard roughness	

0.20c simulated split flap deflected 60°

| ▽ | 6.0 | | |
| ▽ | 6.0 | Standard roughness | |

a.c. position

NACA 65-006 Wing Section (*Continued*)

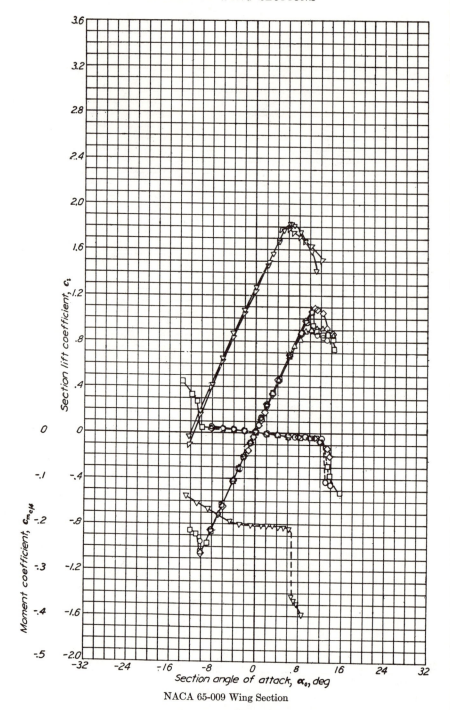

NACA 65-009 Wing Section

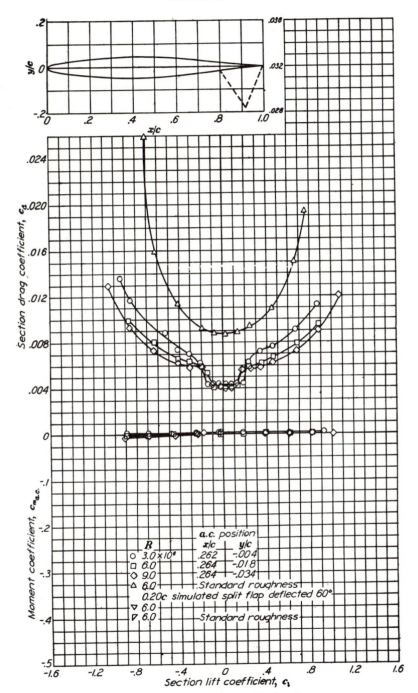

NACA 65-009 Wing Section (*Continued*)

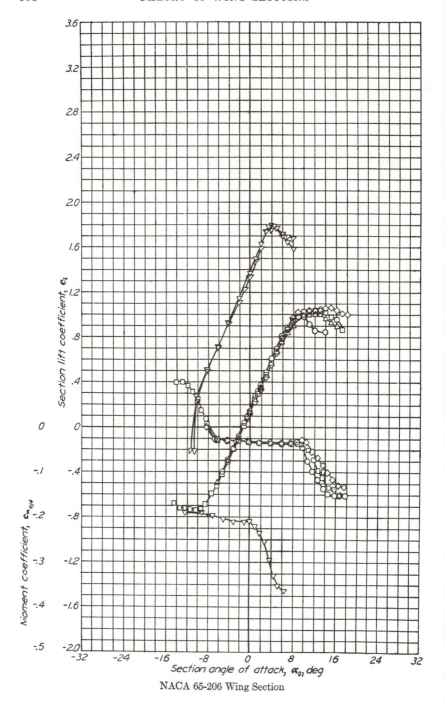

NACA 65-206 Wing Section

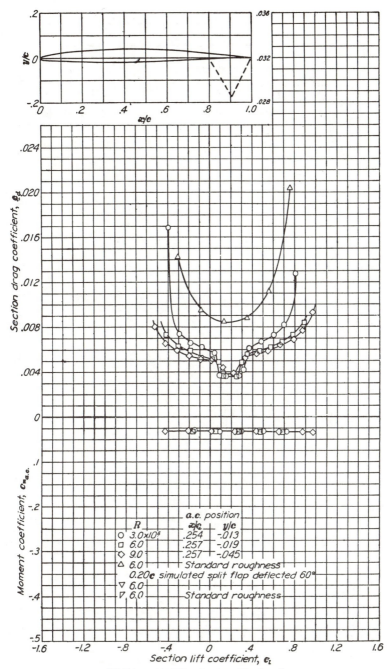

NACA 65-206 Wing Section (*Continued*)

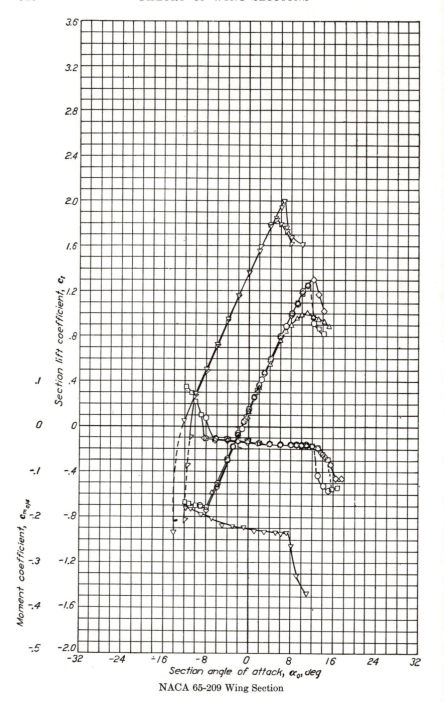

NACA 65-209 Wing Section

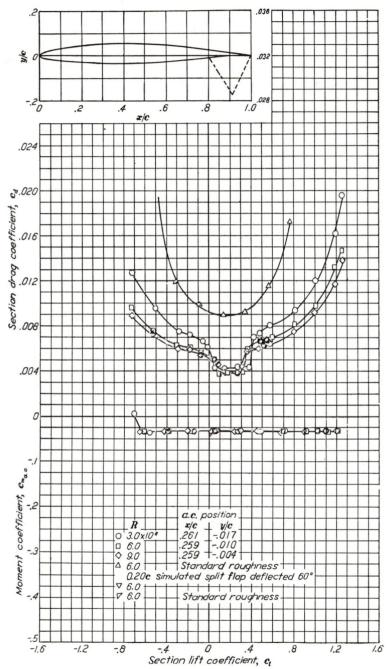

NACA 65-209 Wing Section (*Continued*)

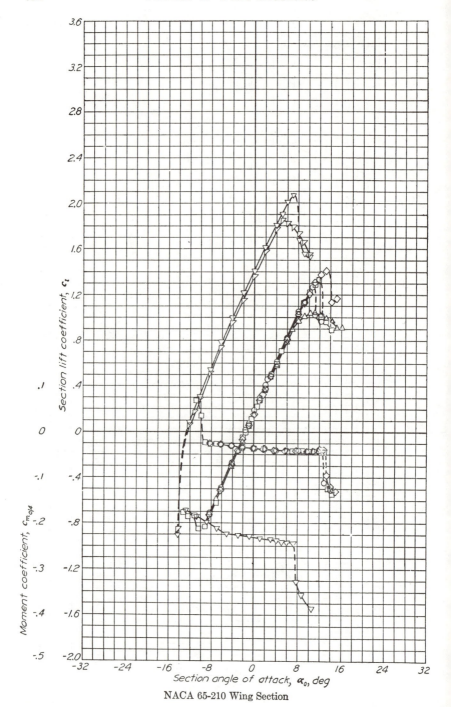

NACA 65-210 Wing Section

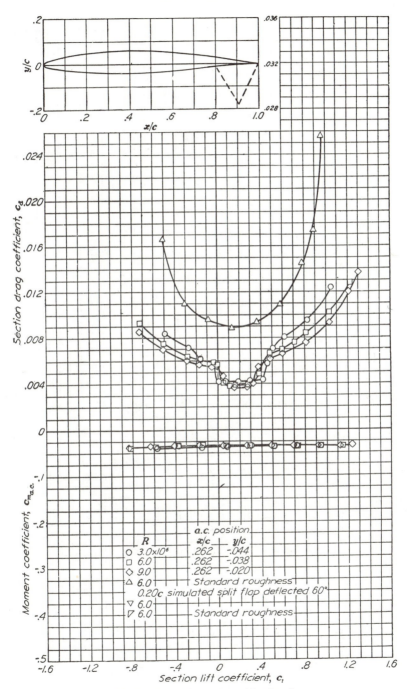

NACA 65-210 Wing Section (*Continued*)

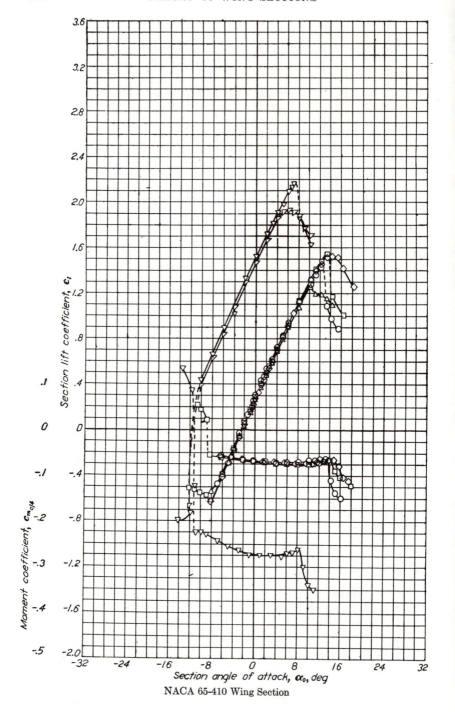

NACA 65-410 Wing Section

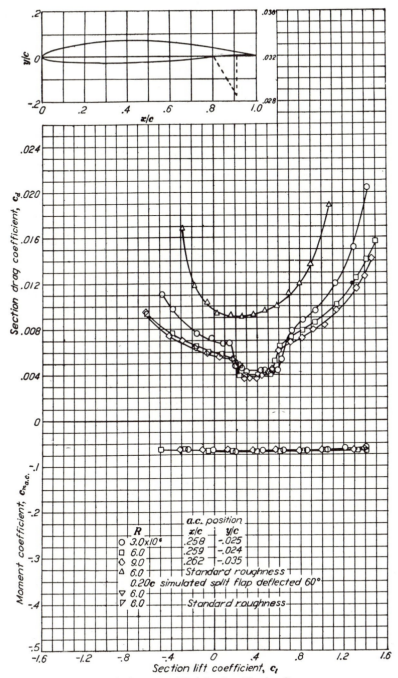

a.c. position

R	x/c	y/c
○ 3.0x10⁶	.258	-.025
□ 6.0	.259	-.024
◇ 9.0	.262	-.035
△ 6.0	Standard roughness	
0.20c simulated split flap deflected 60°		
▽ 6.0		
▽ 6.0	Standard roughness	

NACA 65-410 Wing Section (*Continued*)

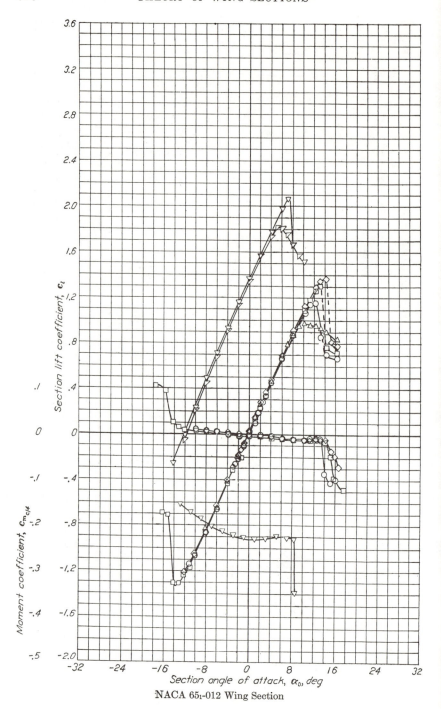

NACA 65₁-012 Wing Section

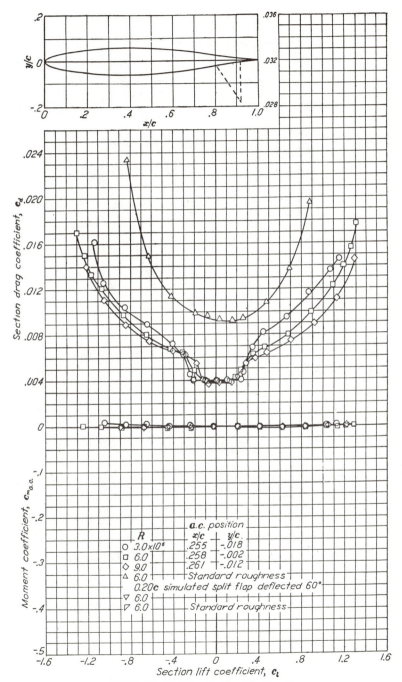

NACA 65₁-012 Wing Section (*Continued*)

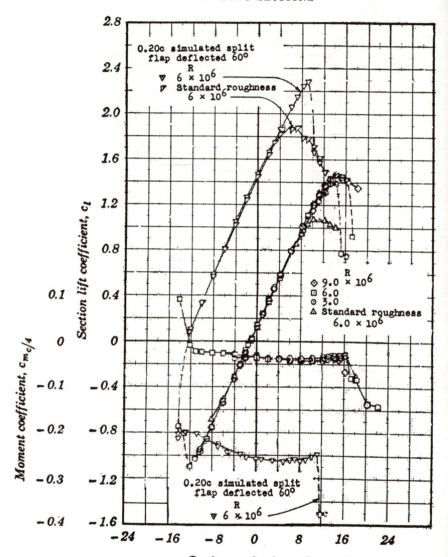

NACA 65₁-212 Wing Section

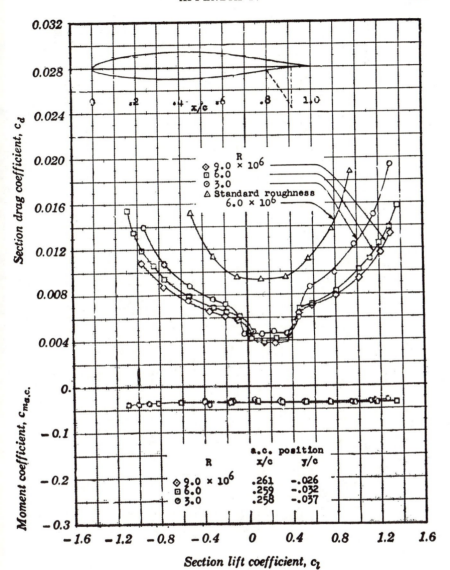

NACA 65_1-212 Wing Section (*Continued*)

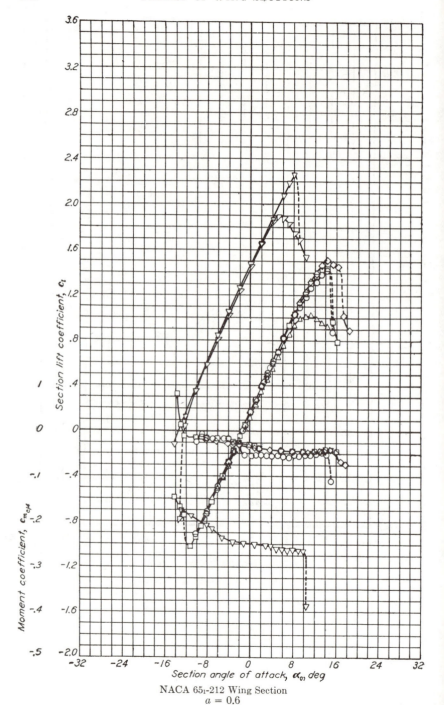

NACA 65₁-212 Wing Section
$a = 0.6$

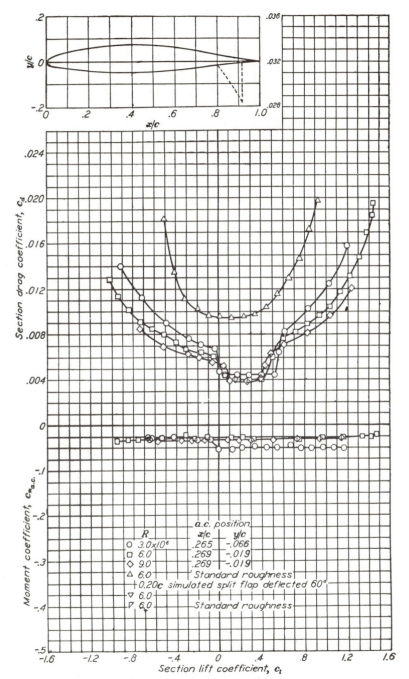

NACA 65₁-212 Wing Section
a = 0.6 (*Continued*)

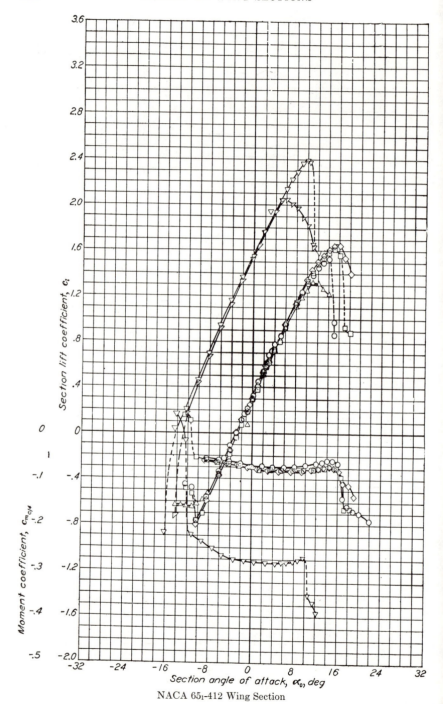

NACA 65₁-412 Wing Section

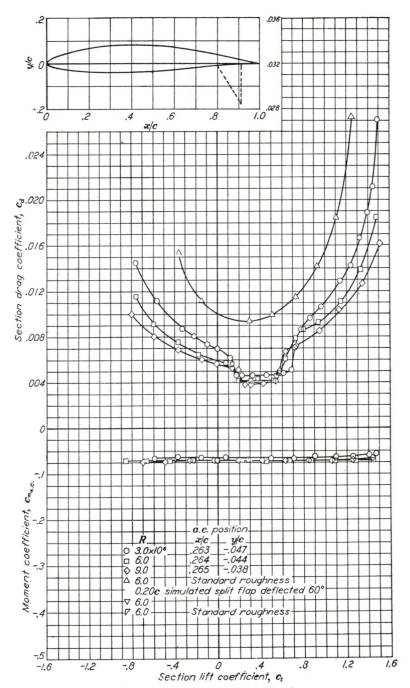

NACA 65₁-412 Wing Section (*Continued*)

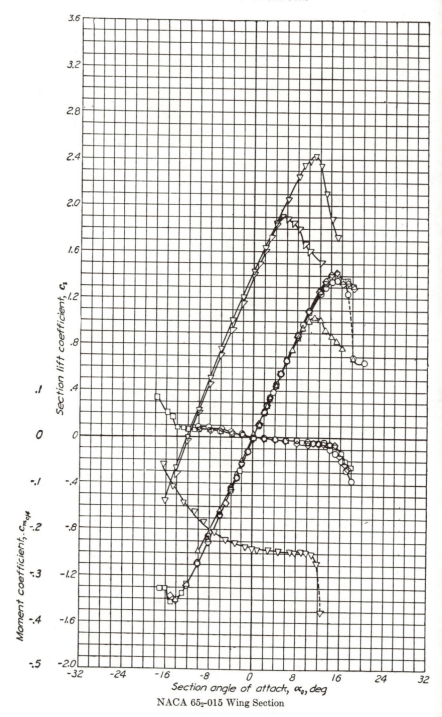

NACA 65₂-015 Wing Section

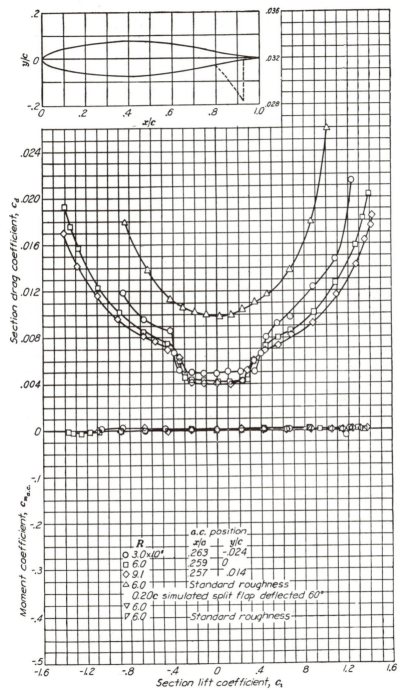

NACA 65₂-015 Wing Section (*Continued*)

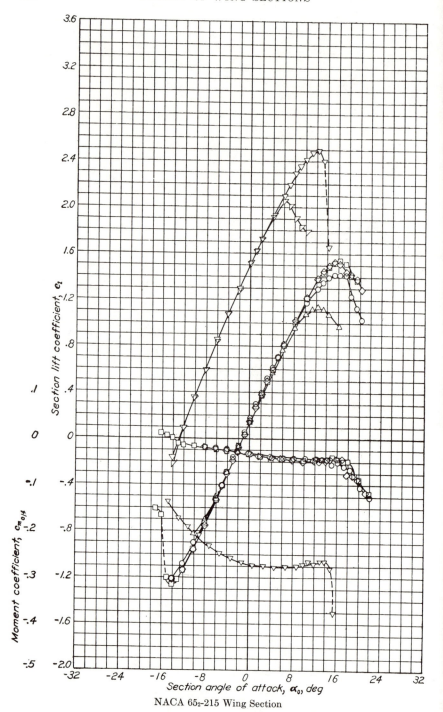

NACA 65₂-215 Wing Section

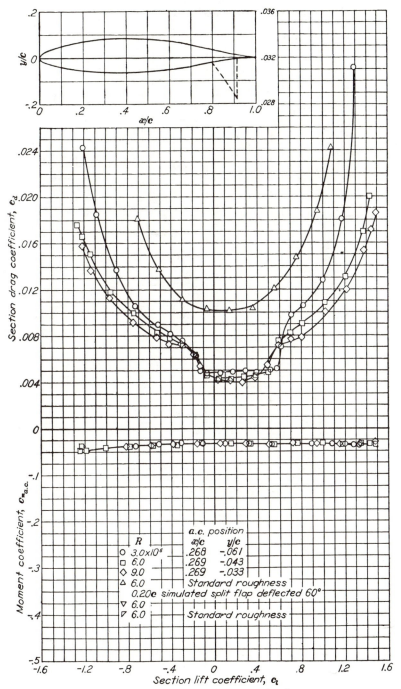

NACA 65₂-215 Wing Section (*Continued*)

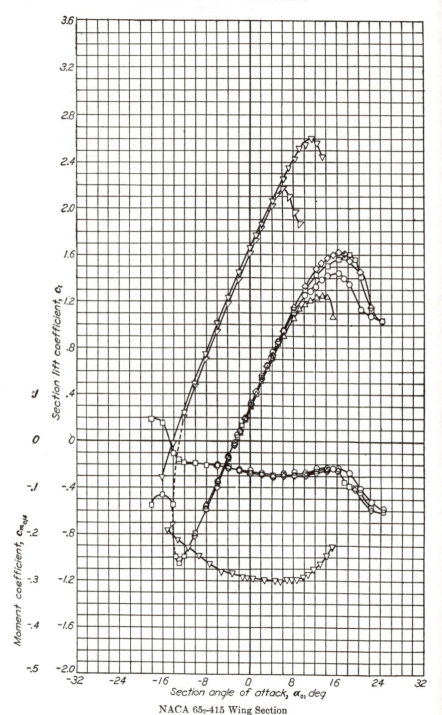

NACA 65₂-415 Wing Section

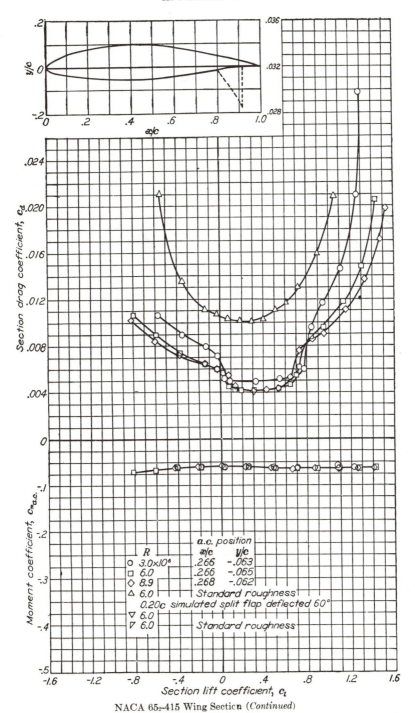

NACA 65₂-415 Wing Section (*Continued*)

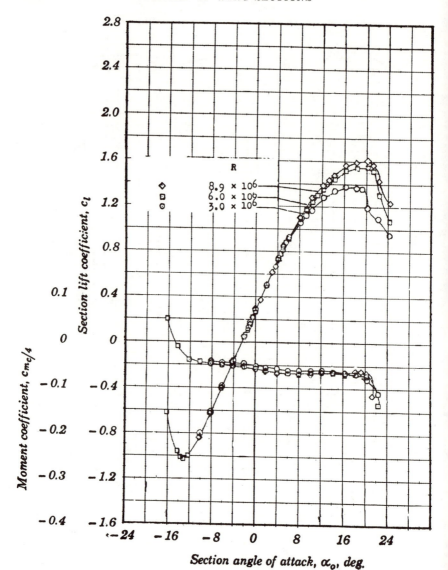

NACA 65$_2$-415 Wing Section
$a = 0.5$

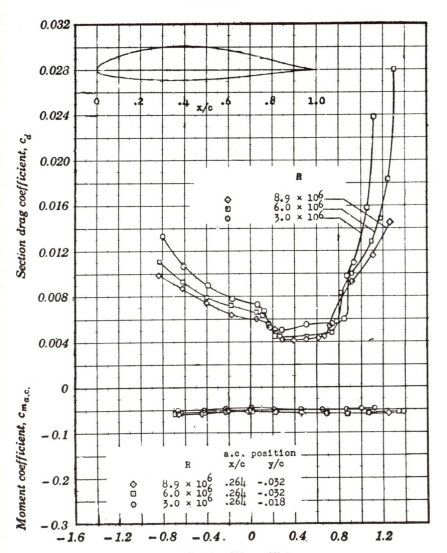

NACA 65₂-415 Wing Section
$a = 0.5$ *(Continued)*

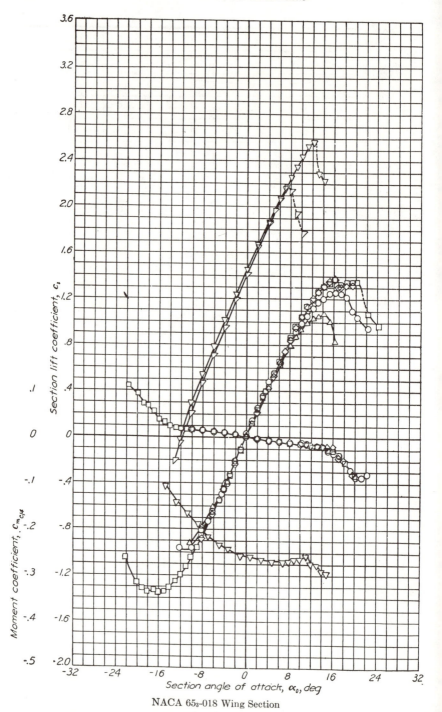

NACA 65₃-018 Wing Section

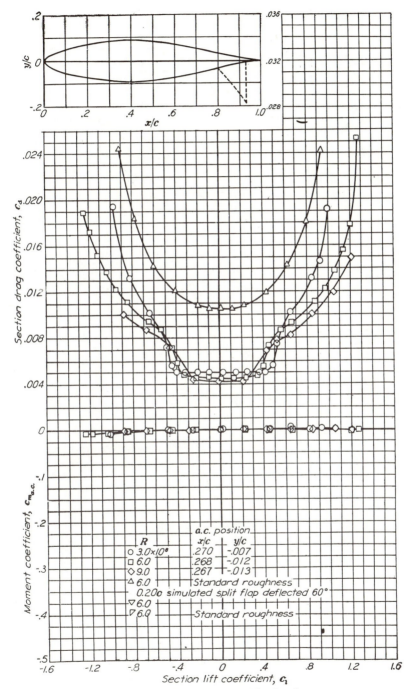

NACA 65₃-018 Wing Section (*Continued*)

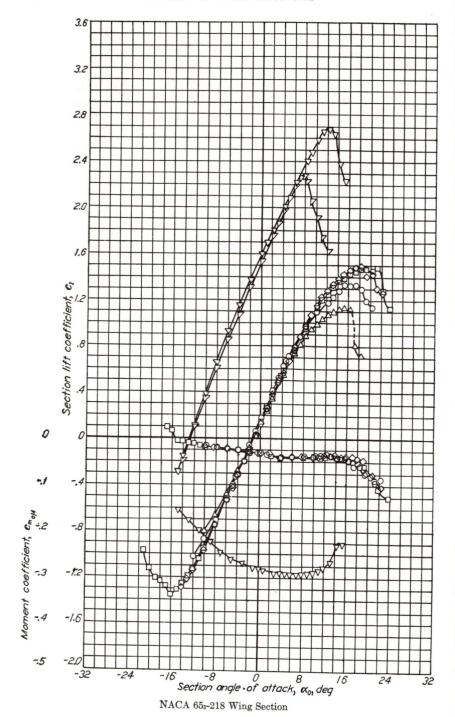

NACA 65₃-218 Wing Section

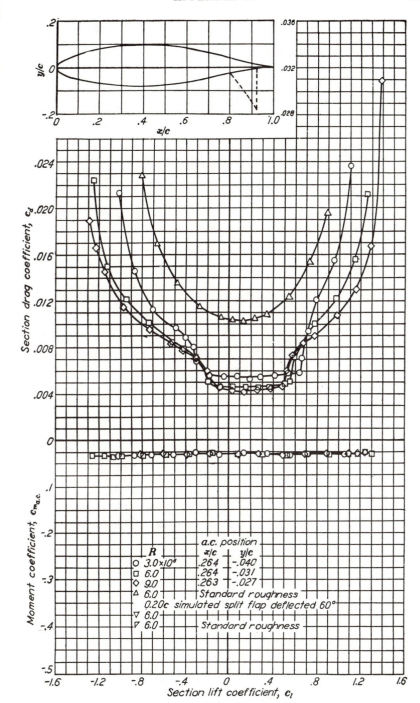

NACA 65₃-218 Wing Section (*Continued*)

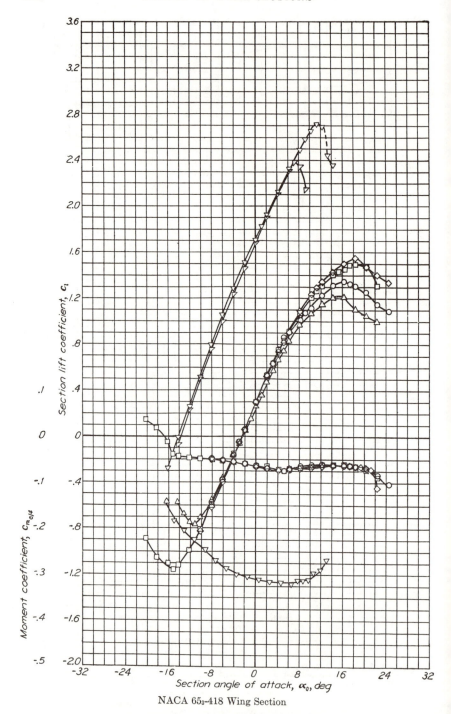

NACA 65₃-418 Wing Section

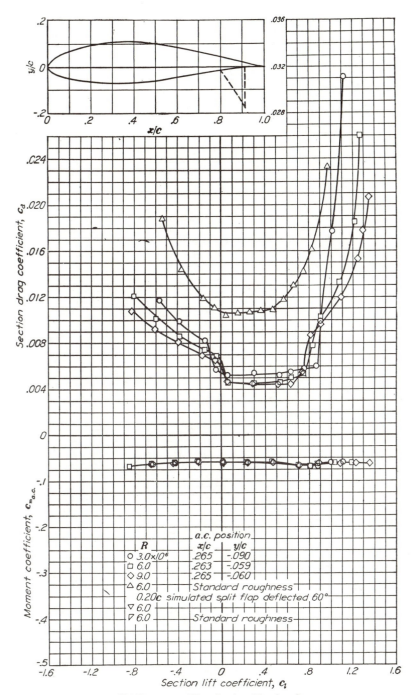

NACA 65₃-418 Wing Section (*Continued*)

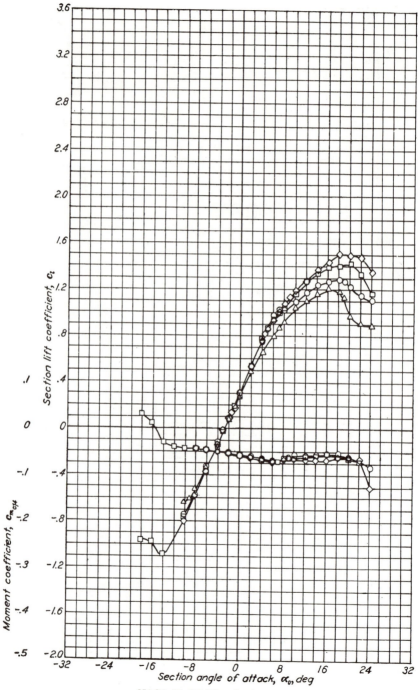

NACA 65₃-418 Wing Section
a = 0.5

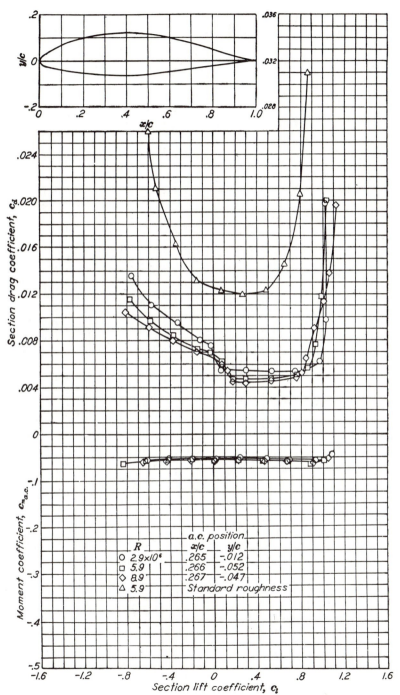

NACA 65_3-418 Wing Section
$a = 0.5$ (*Continued*)

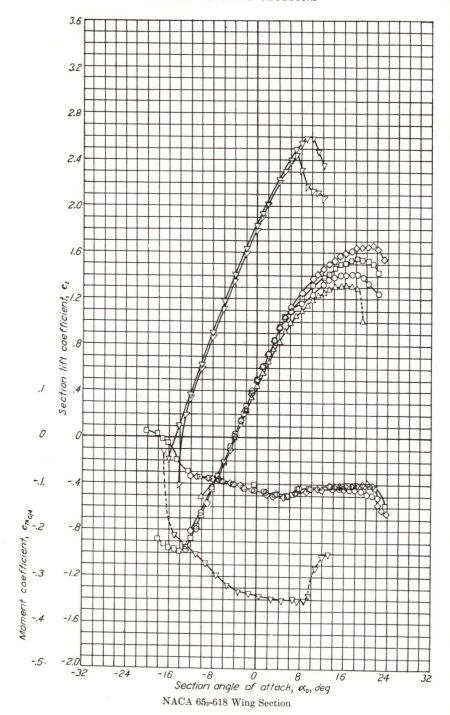

NACA 65₃-618 Wing Section

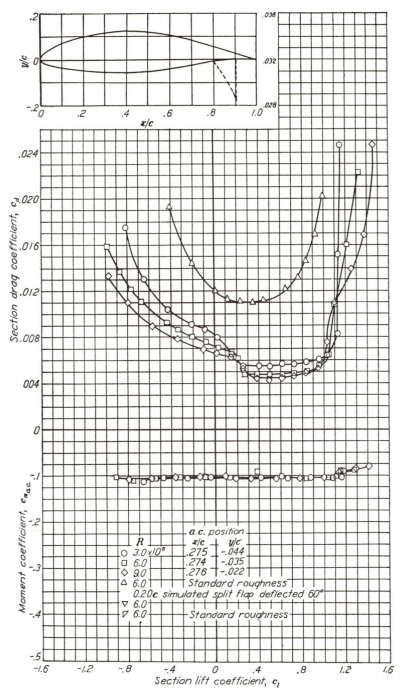

NACA 65₃-618 Wing Section (*Continued*)

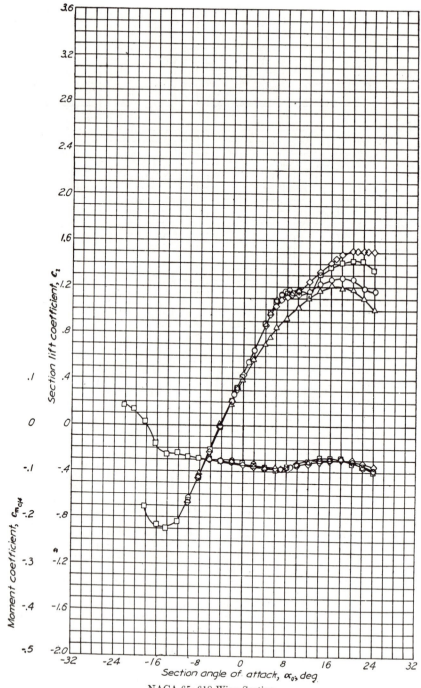

NACA 65₃-618 Wing Section
a = 0.5

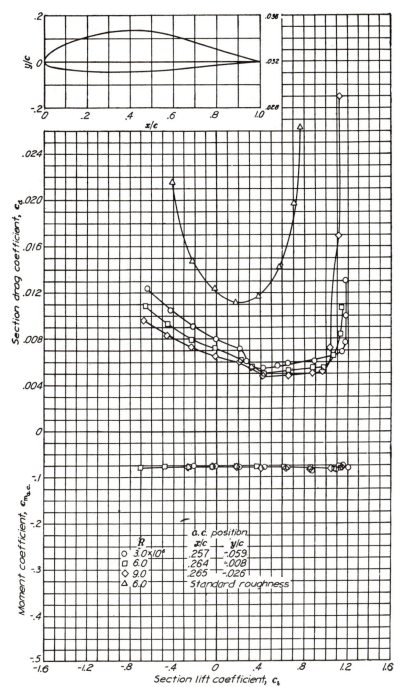

NACA 65₃-618 Wing Section
$a = 0.5$ (*Continued*)

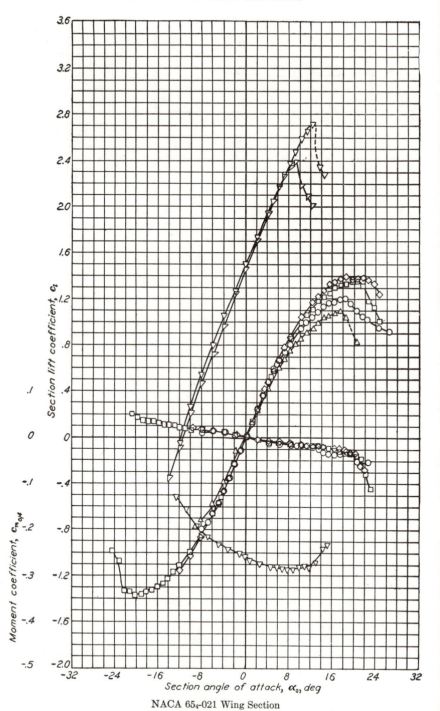

NACA 65$_4$-021 Wing Section

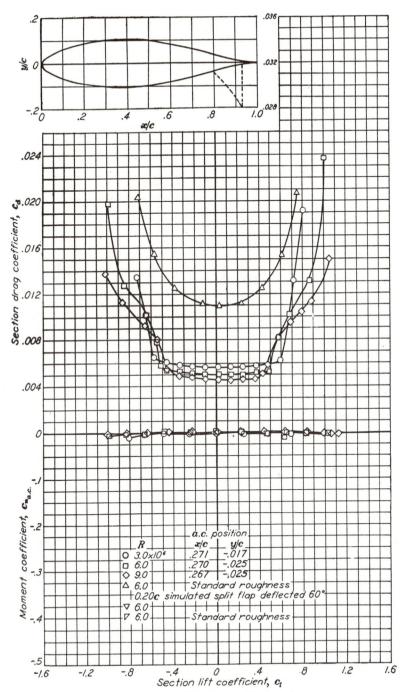

NACA 65$_4$-021 Wing Section (*Continued*)

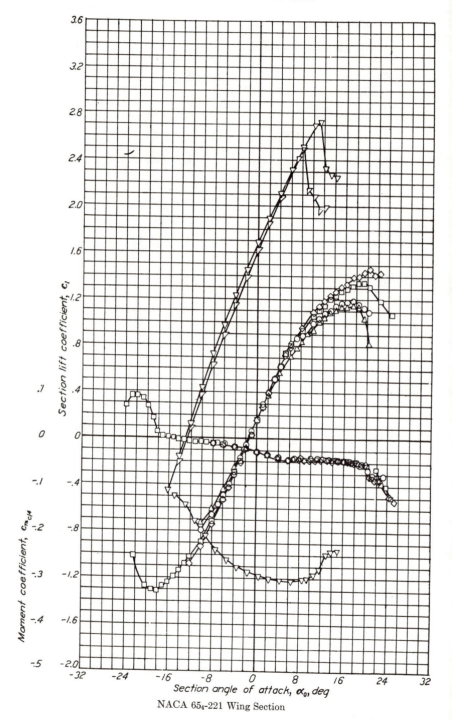

NACA 65$_4$-221 Wing Section

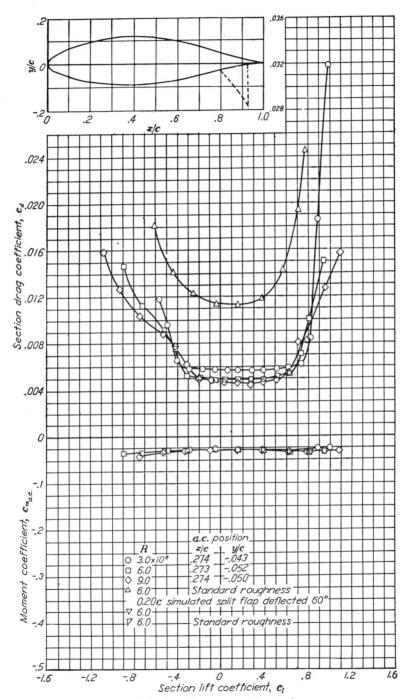

NACA 65$_4$-221 Wing Section (*Continued*)

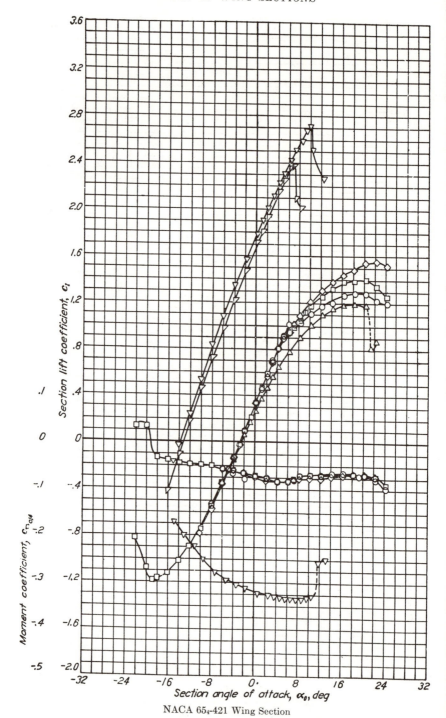

NACA 65₄-421 Wing Section

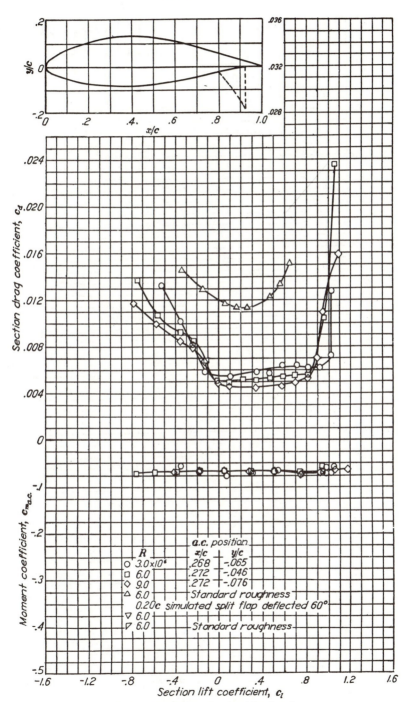

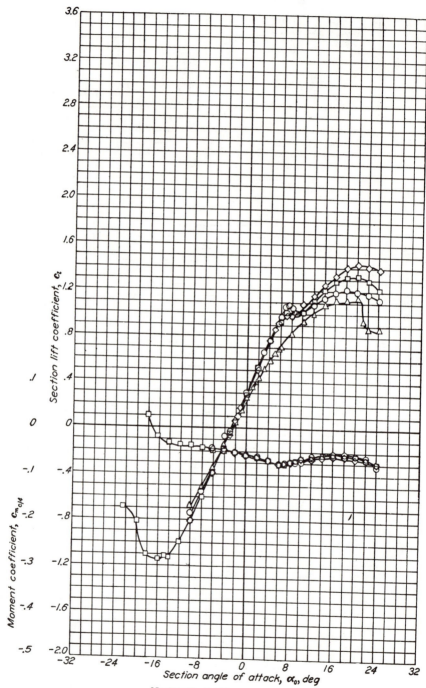

NACA 65₄-421 Wing Section
$a = 0.5$

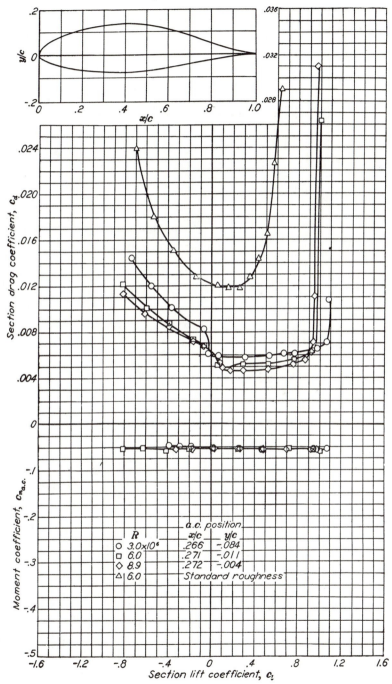

NACA 65$_4$-421 Wing Section
$a = 0.5$ (*Continued*)

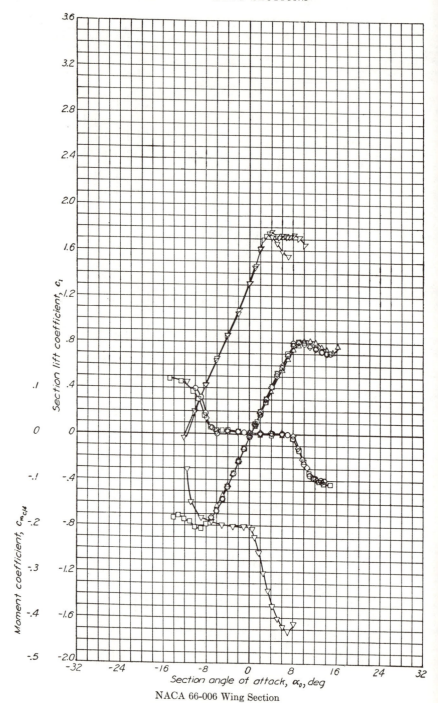

NACA 66-006 Wing Section

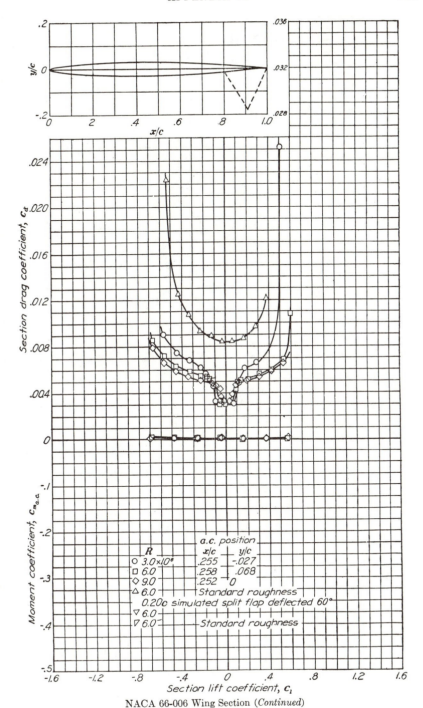

NACA 66-006 Wing Section (*Continued*)

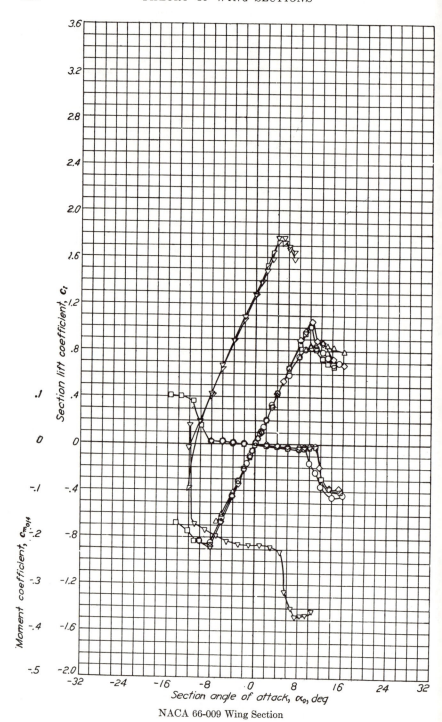

NACA 66-009 Wing Section

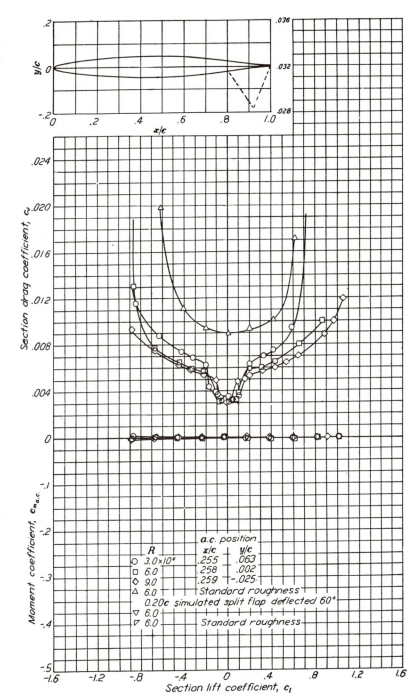

NACA 66-009 Wing Section (*Continued*)

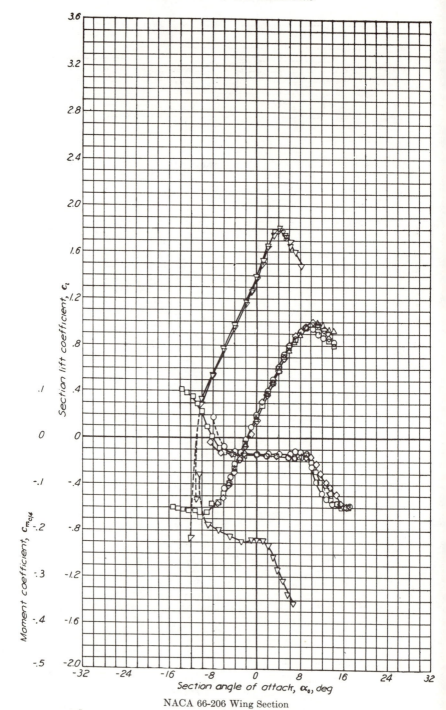

NACA 66-206 Wing Section

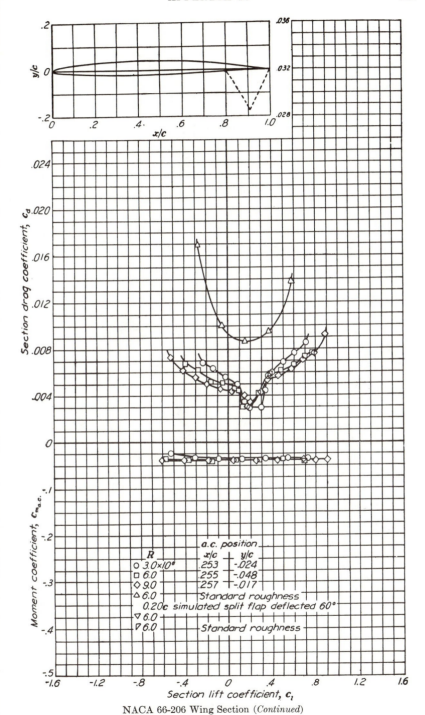

NACA 66-206 Wing Section (*Continued*)

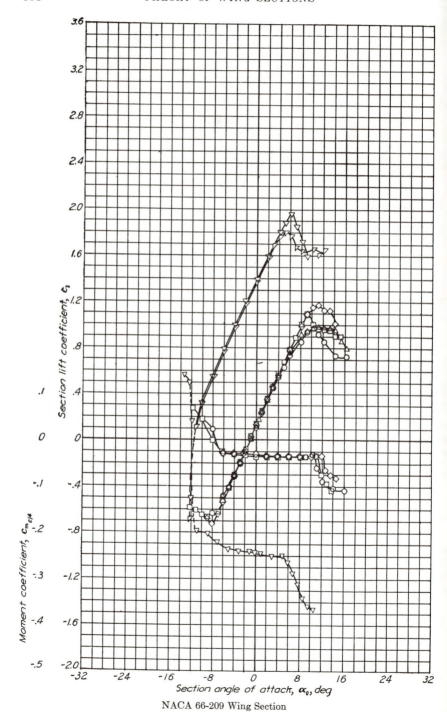

NACA 66-209 Wing Section

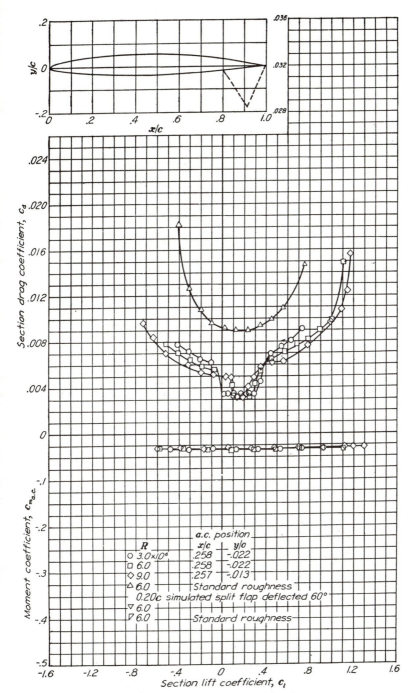

NACA 66-209 Wing Section *(Continued)*

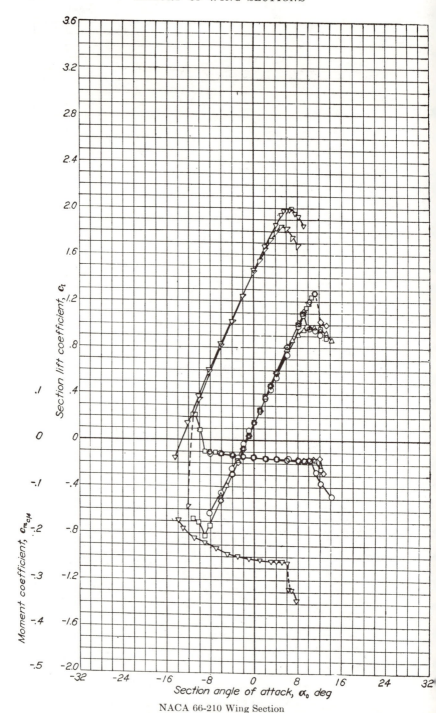

NACA 66-210 Wing Section

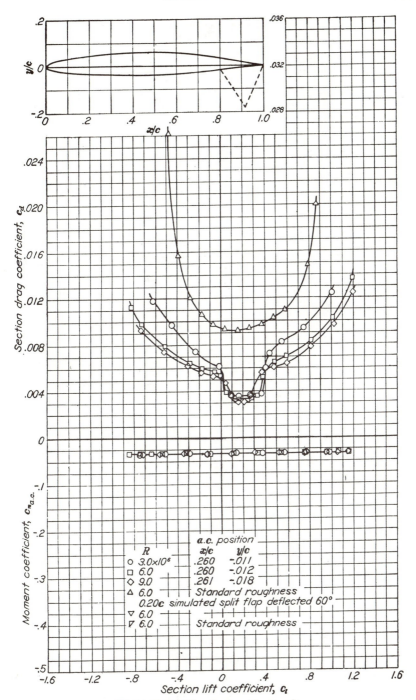

NACA 66-210 Wing Section (*Continued*)

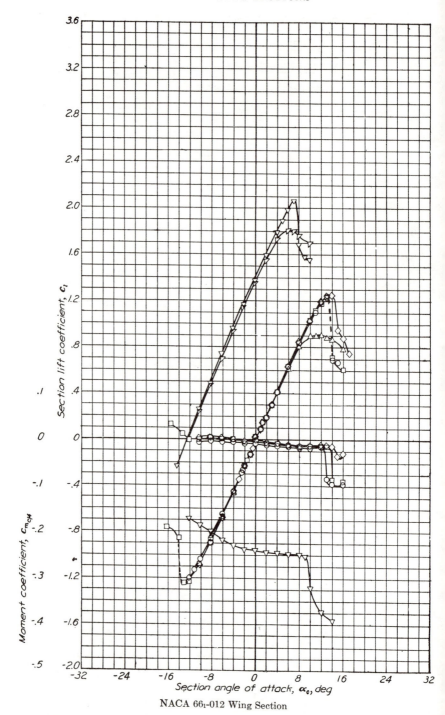

NACA 66₁-012 Wing Section

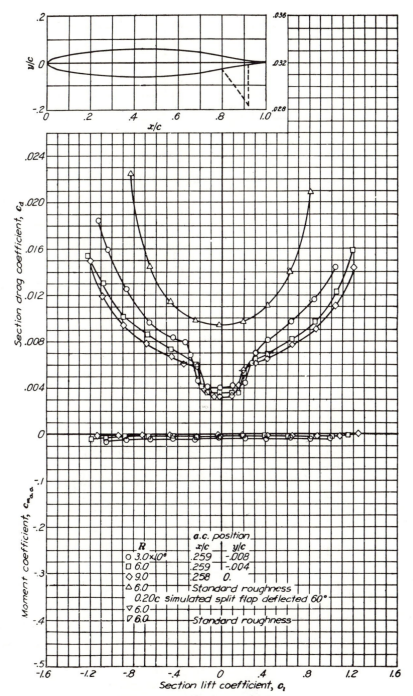

NACA 66₁-012 Wing Section (*Continued*)

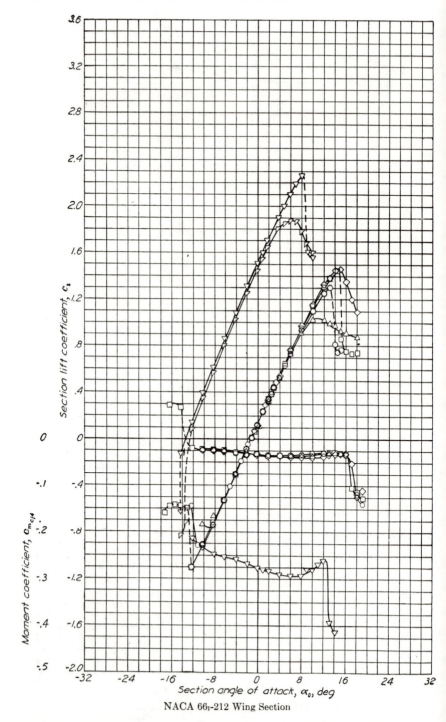

NACA 66₁-212 Wing Section

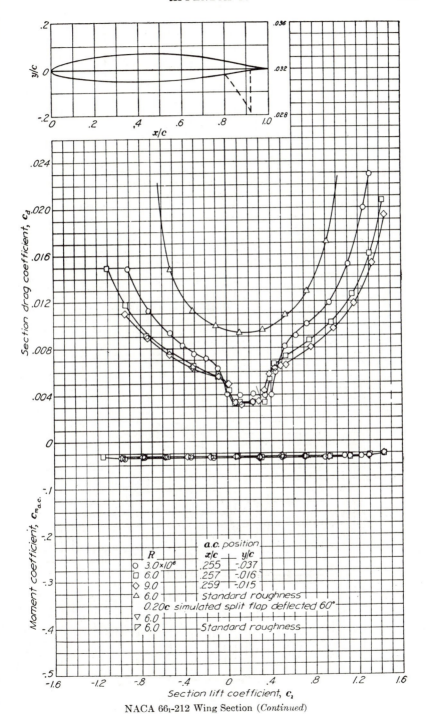

	R	*x/c*	*y/c*
○	3.0×10^6	.255	-.037
□	6.0	.257	-.016
◇	9.0	.259	-.015
△	6.0	Standard roughness	
	0.20c simulated split flap deflected 60°		
▽	6.0		
▽	6.0	Standard roughness	

a.c. position

NACA 66₁-212 Wing Section (*Continued*)

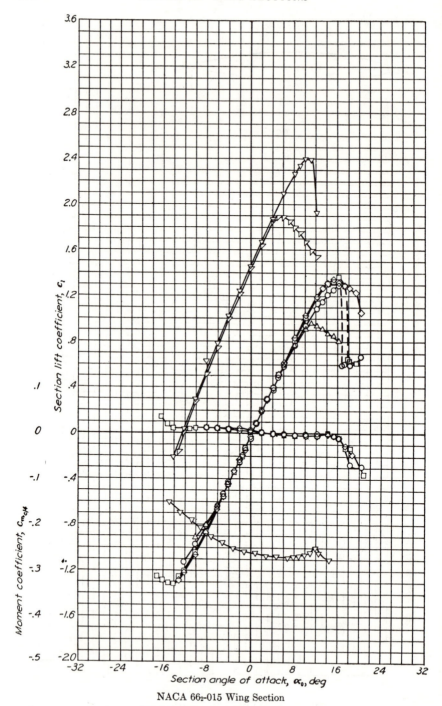

NACA 66₂-015 Wing Section

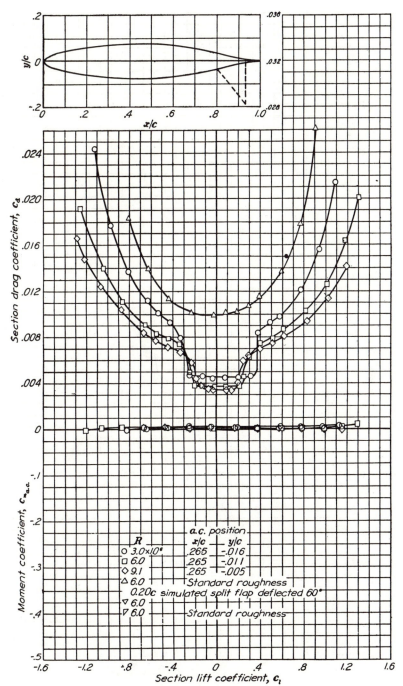

NACA 66₂-015 Wing Section (*Continued*)

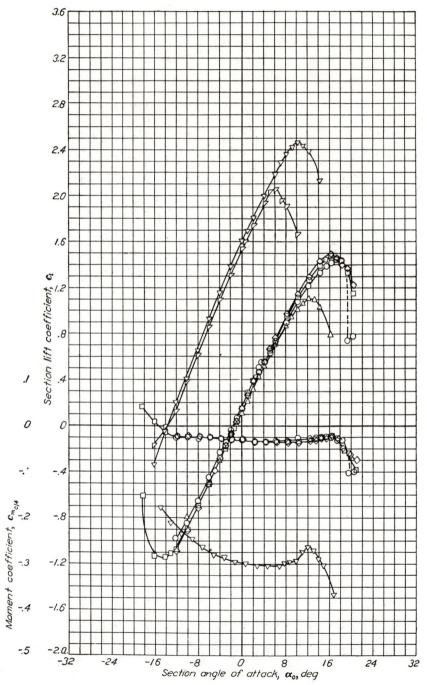

NACA 66₂-215 Wing Section

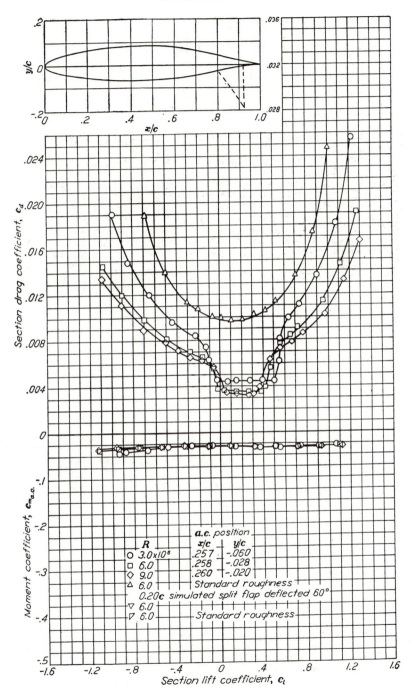

a.c. position

R	x/c	y/c
○ 3.0x10⁶	.257	-.060
□ 6.0	.258	-.028
◇ 9.0	.260	-.020
△ 6.0	Standard roughness	

0.20c simulated split flap deflected 60°

| ▽ 6.0 | | |
| ▽ 6.0 | Standard roughness | |

NACA 66₂-215 Wing Section (*Continued*)

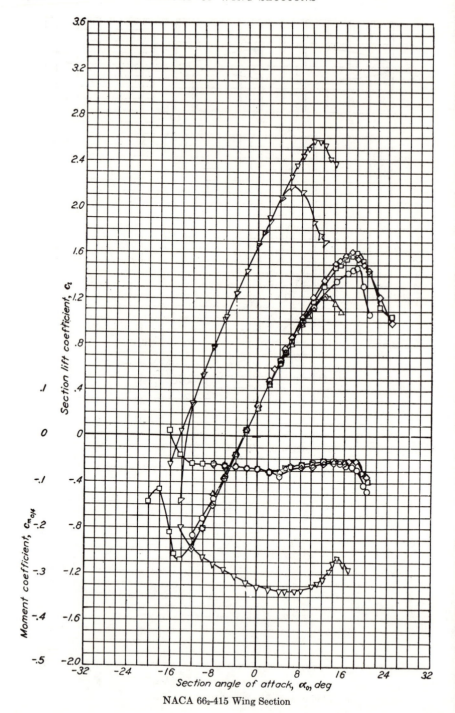

NACA 66₂–415 Wing Section

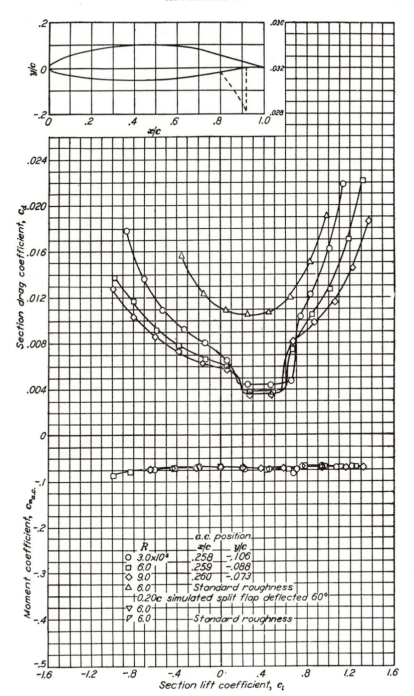

NACA 66₂-415 Wing Section (*Continued*)

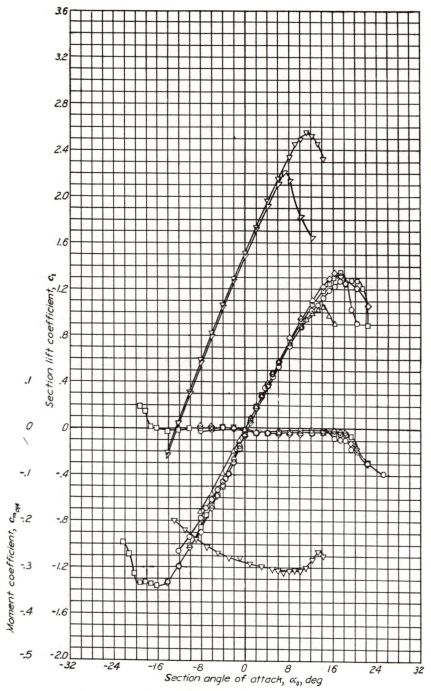

NACA 66_3-018 Wing Section

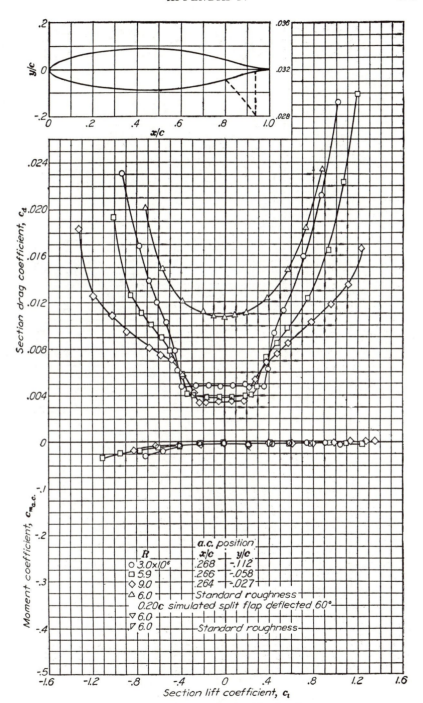

NACA 66₃-018 Wing Section (*Continued*)

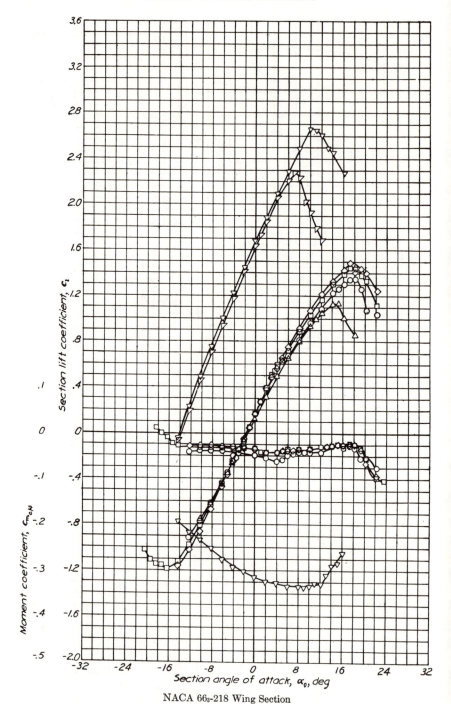

NACA 66₃-218 Wing Section

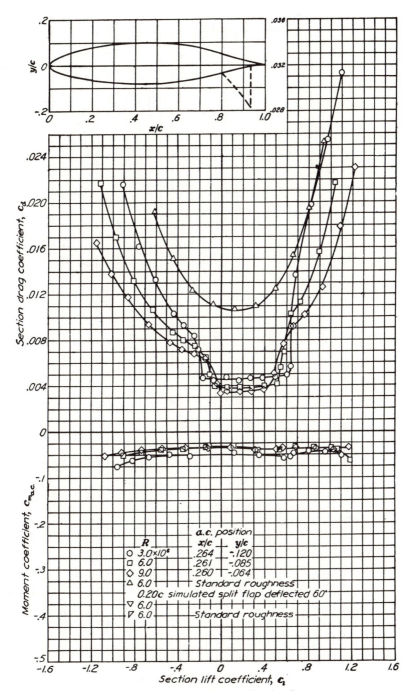

NACA 66₃-218 Wing Section (*Continued*)

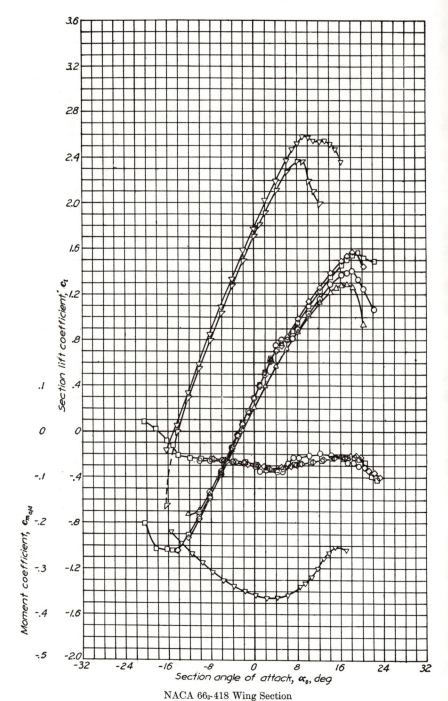

NACA 66₃-418 Wing Section

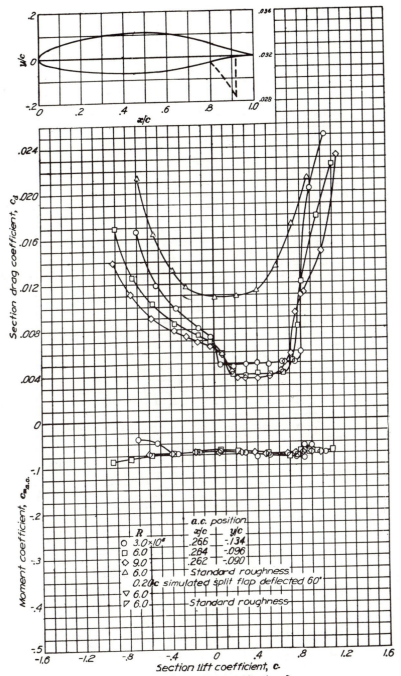

NACA 66₃-418 Wing Section (*Continued*)

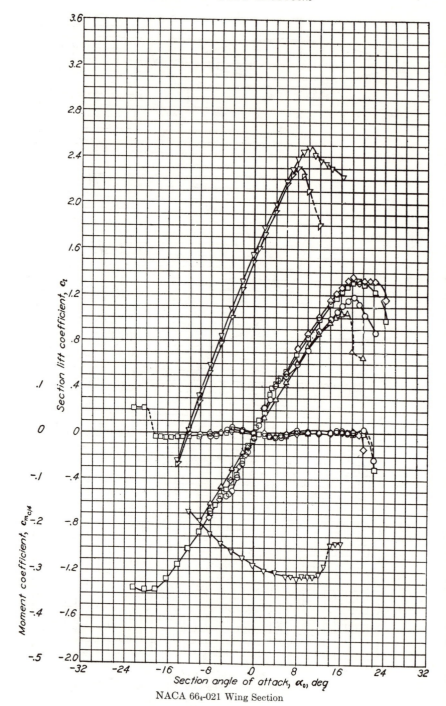

NACA 66₄-021 Wing Section

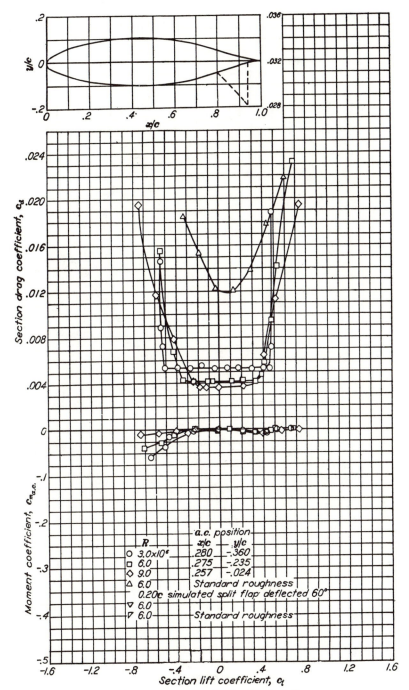

NACA 66₄-021 Wing Section (*Continued*)

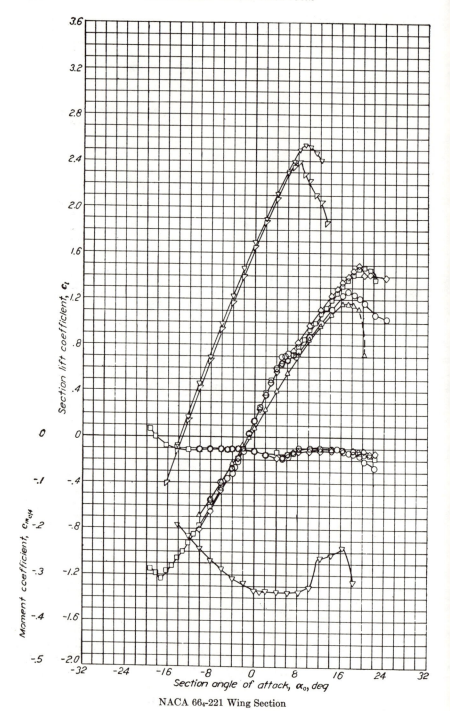

NACA 66₄-221 Wing Section

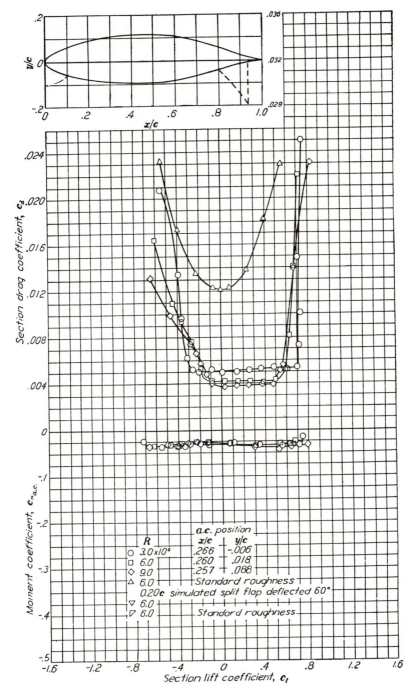

NACA 66₄-221 Wing Section (*Continued*)

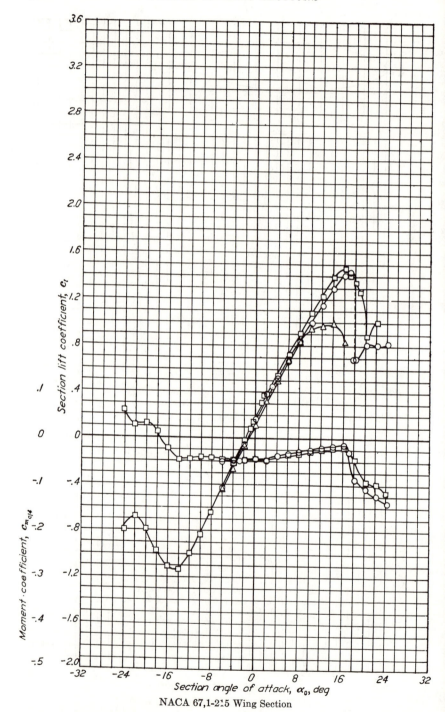

NACA 67,1-215 Wing Section

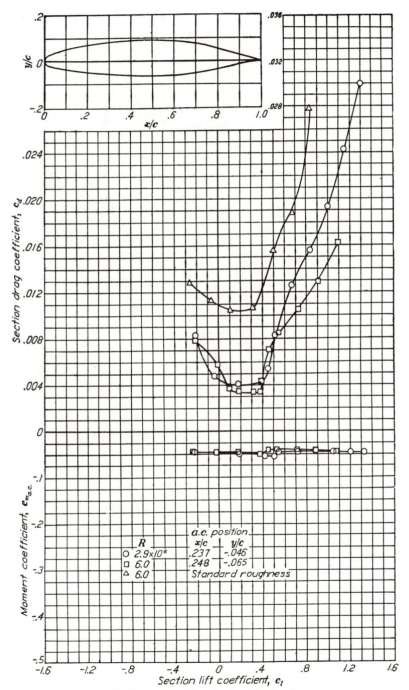

NACA 67,1-215 Wing Section (*Continued*)

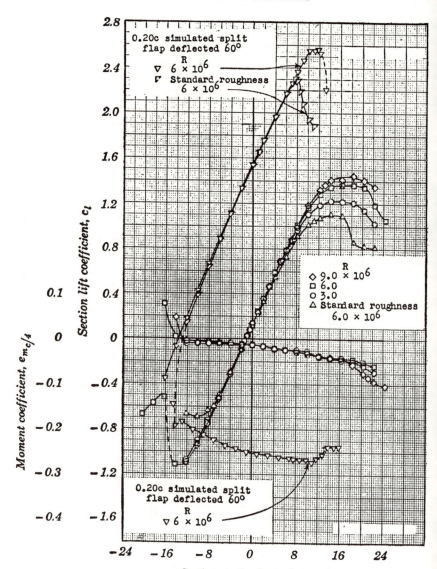

Section angle of attack, α_o, deg

NACA 747A315 Wing Section

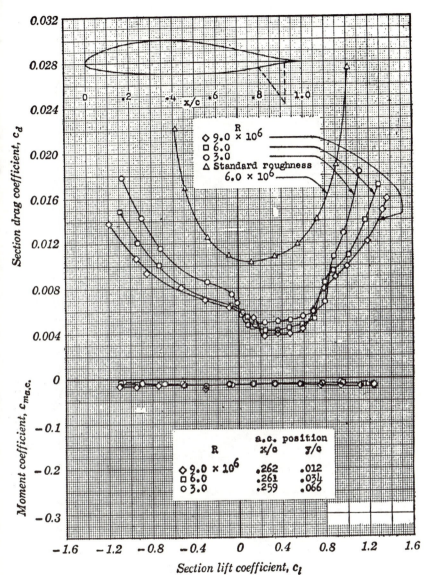

NACA 747A315 Wing Section (*Continued*)

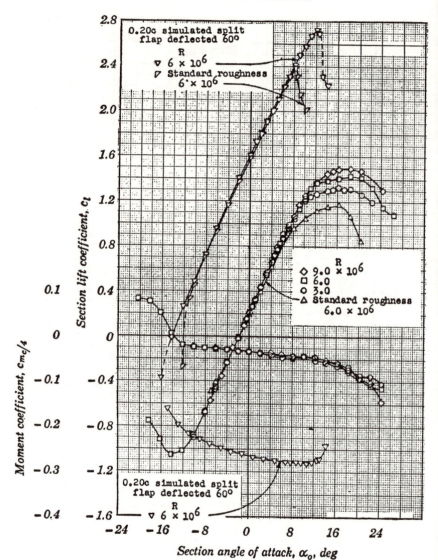

Section lift coefficient, c_l

Moment coefficient, $c_{m_{c/4}}$

0.20c simulated split
flap deflected 60°

R
▽ 6 × 10⁶
▽ Standard roughness
6 × 10⁶

R
◇ 9.0 × 10⁶
□ 6.0
○ 3.0
△ Standard roughness
6.0 × 10⁶

0.20c simulated split
flap deflected 60°
R
▽ 6 × 10⁶

Section angle of attack, α_o, deg

NACA 747A415 Wing Section

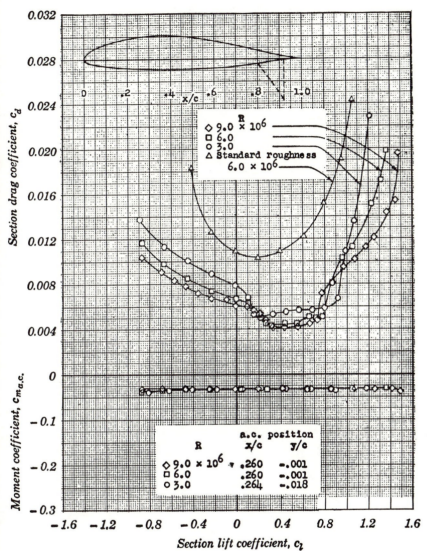

NACA 747A415 Wing Section (*Continued*)

INDEX

A

Ackeret, J., 268
Adiabatic law, 249
Aerodynamic center, experimental, 182
  theoretical, 69
  wing, 5, 19
Aerodynamic characteristics, experimental, 449
Airfoils (see Wing sections)
Allen, H. Julian, 65, 164, 194, 202, 236
Anderson, Raymond F., 9
Angle of attack, 4
  effective, 22, 25
  ideal, 70
  induced, 20
  wing, 8, 11
Angle of zero lift, experimental, 128
  theoretical, 66, 69
    approximate methods, 72
    variation with flap deflection, 192
  wing, 16
Aspect ratio, 6
  at high speeds, 296

B

Bamber, Millard J., 234
Bernoulli's equation, 35, 251
Birnbaum, Walter, 65
Blasius, H., 87, 96
Boshar, John, 20
Boundary layer, 80
  compressibility effects, 109
  concept, 81
  crossflows, 28
  laminar, 85
    Blasius solution, 87
    characteristics, 85
    separation, 83, 86, 93
    skin friction, 87
    thickness, 89, 92
    transition, 105, 157
    velocity distribution, 87

Boundary layer, momentum relation, 90
  slip, 83
  thickness, displacement, 89
    momentum, 92
  turbulent, 95
    pipe flow, 95
    separation, 83, 103
    skin friction, 97, 174
    thickness, 102
    velocity distribution, 97, 100
Boundary-layer control, 231

C

Cahill, Jones F., 206, 221
Camber, 65
  application to thickness forms, 75, 112
  design of mean lines, 73
  effect of, on aerodynamic center, 182
    on angle of zero lift, 128
    on lift curve, 132
    on maximum lift, 136
    on minimum drag, 148
    on pitching moment, 179
    on profile drag, 151, 178
  induced, 28
  shapes and theoretical characteristics, 382
Chaplygin, Sergei, 258
Chord, 113
  mean aerodynamic, 27
Circular cylinder in uniform stream, 41, 49
  with circulation, 44, 49
Circulation, 37
  lift, 45
  thin wing sections, 65
  vortex, 43
Complex number, 47
Complex variable, 47
Compressibility effects, 247
  boundary layer, 109
  drag characteristics, 281
  first order, 256

689